GEONS,
BLACK HOLES,
AND QUANTUM FOAM

GEONS,
BLACK HOLES,
AND QUANTUM FOAM

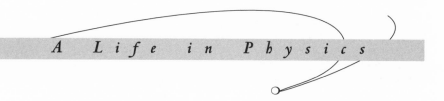

A Life in Physics

JOHN ARCHIBALD WHEELER

with

KENNETH FORD

W. W. NORTON & COMPANY

NEW YORK · LONDON

For information about permission to reproduce selections from this book, write to Permissions,
W.W. Norton & Company, Inc., 500 Fifth Avenue, New York, NY 10110.

The text of this book is composed in Electra, with the display set in Centaur.
Composition by Chelsea Dippel
Manufacturing by the Haddon Craftsmen, Inc.
Book design by Jo Anne Metsch

Library of Congress Cataloging-in-Publication Data

Wheeler, John Archibald, 1911–
    Geons, black holes, and quantum foam : a life in physics / John Archibald Wheeler with
Kenneth Ford.
        p.   cm.
    Includes index.
    ISBN 0-393-04642-7
    1. Wheeler, John Archibald, 1911– . 2. Physics—History.
3. Astronomy—History. 4. Physicists—United States—Biography.
I. Ford, Kenneth William, 1926– . II. Title.
QC16.W48A3   1998
530'.092—dc21
   [B]

-44566
  CIP

W.W. Norton & Company, Inc., 500 Fifth Avenue, New York, NY 10110
http://www.wwnorton.com

W.W. Norton & Company Ltd., 10 Coptic Street, London WC1A 1PU

1 2 3 4 5 6 7 8 9 0

To the wonderful teachers, students, and colleagues who have inspired and guided me over the years;

and

to the still unknown person(s) who will further illuminate the magic of this strange and beautiful world of ours by discovering How come the quantum? How come existence?

We will first understand how simple the universe is when we recognize how strange it is.

CONTENTS

GEONS,
BLACK HOLES,
AND QUANTUM FOAM

1

·

"HURRY UP!"

ON MONDAY, January 16, 1939, I taught my morning class at Princeton University, then took a train to New York and walked across town to the Hudson River dock where the Danish physicist Niels Bohr was scheduled to arrive on the MS *Drottningholm*. Bohr—with whom I had worked a few years earlier—was coming to give some lectures at the Institute for Advanced Study in Princeton and spend time with his friend Albert Einstein, then a professor at the Institute, and I had decided to greet him.

For a dozen years, Bohr and Einstein, probably the two most eminent physicists in the world at that time, had had a running debate on the meaning and interpretation of quantum mechanics, the subtle theory that governs motion and change in the subatomic realm. Bohr held that uncertainty and unpredictability are intrinsic features of the theory, and therefore of the world in which we live. Einstein embraced a deterministic worldview; he could not believe that God "played dice." Over the years, Einstein had proposed various thought experiments that at first appeared to expose cracks in the structure of quantum mechanics, and Bohr had been able to turn every one of them around to show more clearly than ever that his "Copenhagen interpretation" of quantum theory, with its fundamental probability, stood fast. As it turned out, however, nuclear fission, not the mysteries of the quantum, occupied most of Bohr's time during his visit. Just before embarking from Denmark, he had learned of this new phenomenon, and he had been thinking hard about it all the way across the ocean.

Albert Einstein and Niels Bohr in Brussels, one of the places
where they carried on their famous debates, 1930.
*(Photograph by Paul Ehrenfest, courtesy of AIP Emilio Segrè
Visual Archives.)*

I was not the only one who decided to welcome Bohr personally. While I
was waiting on the dock, who should turn up but the Italian physicist Enrico
Fermi and his wife, Laura, who, with their two children, had arrived in the
United States only two weeks earlier. Enrico, short, muscular, and intense,
was a man of habit and order whose mind never rested. Laura, dark and pretty, had studied engineering and science before marrying Enrico and would
later establish herself as a writer. As the story jokingly puts it, Fermi, after
receiving his Nobel Prize in Sweden in December 1938, became lost on his
way back to Italy and ended up in New York. In fact, they wanted to get away
from the Fascism of their native Italy—Laura was Jewish—and the itinerary

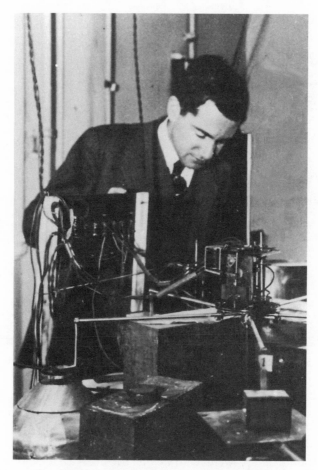

Otto Frisch in his Copenhagen laboratory, 1936.
(Courtesy of Niels Bohr Archive, Copenhagen.)

had been carefully and quietly planned to bring them to New York, where a professorship at Columbia University awaited Enrico.

Fermi came to the dock to invite Bohr to spend a day with him in New York before going to Princeton. The news of fission that Bohr had in his head would be of consuming interest to Fermi, himself a nuclear pioneer. But by chance, it would be I, not Fermi, who would become the first on these shores to hear of it.

Bohr learned of fission on January 7, just as he and his son Erik were about to board the train in Copenhagen for Gothenburg, the MS *Drottningholm*'s embarkation point. Otto Frisch, a German emigré physicist working at Bohr's University Institute for Theoretical Physics in Copenhagen, sought out Bohr to inform him of the postulate of fission that he (Frisch) and his aunt, Lise Meitner, had devised in the last week of December to explain puzzling results

Lise Meitner in animated discussion with the Italian physicist
Emilio Segrè in Copenhagen, 1937.
(*Courtesy of Niels Bohr Archive, Copenhagen.*)

found by the German chemists Otto Hahn and Fritz Strassmann in their
Berlin laboratory. When Hahn and Strassmann bombarded uranium with
neutrons (subnuclear particles with no electric charge), they found evidence
that the element barium was created. Since barium is far removed from uranium in the periodic table and has a much lighter nucleus, they could not
make sense of this result. Hahn wrote to Meitner in Sweden, describing the
puzzle, for she had been his longtime colleague in Berlin before leaving
Germany to escape persecution, and was trained in physics. When her
nephew Frisch came for a holiday visit, they took a Christmas Eve walk in
the woods—he on skis and she on foot—to ponder the Berlin results. Suddenly it became clear to them. The uranium nucleus must be breaking into
large fragments, resulting in the nuclei of other elements, including sometimes the nucleus of barium.

When Bohr heard Frisch offer this explanation, his reaction was swift and
positive. "Oh what idiots we all have been!" he said. "Oh but this is wonderful! This is just as it must be!"[1] Bohr's was a mind prepared. He knew as much

---

[1] The quotation is as recalled by Otto Frisch in his memoir, *What Little I Remember* (Cambridge, England: Cambridge University Press, 1979).

as any person alive about atomic nuclei, and could see at once that fission made sense—even though, up to that point, he and other nuclear physicists had imagined that at most only tiny fragments could break off from a nucleus.

In addition to his son Erik, Bohr brought with him a young colleague, Léon Rosenfeld. Rosenfeld was to serve as Bohr's sounding board and scribe, to help Bohr formulate his ideas, and to capture for publication whatever sparks might fly when Bohr and Einstein put their heads together. Throughout the nine-day ocean crossing, fission was probably more on Bohr's mind than the upcoming meetings with Einstein. He and Rosenfeld discussed it incessantly. (Bohr had a blackboard installed in his stateroom, to facilitate their talks.) By the time Bohr shook hands[2] with me and the Fermis on the dock, he had a pretty good idea of a direction to go in to give a theoretical account of fission. That is what would occupy us intensely for the next few months.

But at that shipside greeting, he said nothing about fission. In his characteristic way, he wanted to be sure that Meitner and Frisch got the credit they deserved before the news spread widely. Not even during his day with Fermi did Bohr breathe a word of the discovery. It must have been hard to restrain himself. Fermi had just received the Nobel Prize for his studies of neutron bombardment of nuclei, and had in fact produced fission several years earlier in his laboratory in Rome without knowing it. He had interpreted the results as evidence for the creation of elements heavier than uranium rather than the splitting of uranium. Even when the German chemist Ida Noddack suggested in 1934 that Fermi had in fact split the uranium nucleus, no one paid attention. It was, at the time, too radical a thought. (One can't help wondering whether Noddack's insight would have found a more receptive audience if it had come from a man instead of a woman.) In retrospect, the blindness of physicists and chemists to fission in the mid-1930s can be regarded as a blessing. Had scientists in Germany—and elsewhere—followed up on Ida Noddack's suggestion, it might well have been the Germans, not the Allies, who got the atomic bomb first. The history of the world could have been different.

After the greetings on the dock, Bohr and his son agreed to stay in New York with the Fermis for a day, while Rosenfeld would accompany me back to Princeton, where he could check into the Nassau Club and get settled, awaiting the Bohrs' arrival. Rosenfeld, unaware of Bohr's concern about priority for Meitner and Frisch, spilled the beans to me on the train. I was excited. Here was a whole new mode of nuclear behavior that we had overlooked.

[2] Bear hugs were not in fashion then, especially with Bohr, who addressed me always as "Wheeler." Ironically, Bohr's wife, Margrethe, had told me once that he would like it if I called him "Niels." I almost never did.

Monday, the day of Bohr's arrival, was the day of the Physics Department Journal Club. This was a regular weekly event during the academic year, an informal evening gathering at which faculty members, graduate students, or visitors described new results in physics—usually results that had just been published. I had been put in charge of the Journal Club that year, so the moment I heard about fission from Rosenfeld, I decided to rearrange the schedule. I asked Rosenfeld to give a report of twenty minutes or so on fission. He agreed. When Bohr learned the next day that we had "taken the cap off the bottle," he was upset. But in typical Bohr fashion, he was low-key and gentle. He chastised neither Rosenfeld nor me.

Rosenfeld's report caused a stir. It was immediately clear to everyone that it was more than just an interesting new bit of nuclear behavior; it held at least the possibility of a chain reaction and large-scale release of energy. In those days, physicists could not rush to their computers to spread news by e-mail across the world in seconds, and they did not make long-distance phone calls as a matter of course. Despite the excitement in that evening's Journal Club, it took some days for the news to spread to other laboratories around the country.

I. I. Rabi,[3] the noted experimental physicist from Columbia University, was in Princeton that week and heard Rosenfeld's report. Surprisingly, he did not rush the news back to his new colleague Fermi. As it turned out, Willis Lamb, a young Columbia faculty member and, like Rabi, a future Nobelist, brought Fermi the news. Lamb came down to Princeton by train on Friday morning, January 20, in part to continue working with me on some calculations we were doing together, in part to attend the afternoon theoretical seminar. He stayed for dinner and socializing with some Princeton friends, then caught a 2:00 A.M. train that put him back in New York around 4:00 A.M. "After taking the milk train and getting very little sleep," Lamb told me later, "I went over to Pupin Lab looking for [John] Dunning [the professor in charge of the Columbia cyclotron]. I didn't find him, but I did find Eugene Booth [a post-doc] and Herb Anderson [Fermi's student], so I told them about fission. Then I found Fermi and told him. This was the first he had heard of it, and he showed great interest." No doubt an understatement.

Bohr's official report on fission came on January 26, ten days after Rosenfeld's Journal Club report, at the Conference on Theoretical Physics at George Washington University. George Gamow, a Russian emigré theorist who had spent time at Bohr's institute in Denmark before coming to Ameri-

---

[3] Rabi's given name was Israel Isaac, which, through a misunderstanding when he started school, got changed to Isidor Isaac. In print, he was always "I. I.," and to his colleagues, friends, wife, and sister, he was "Rabi" or "Rab."

ca and who was then a professor at George Washington, was the principal organizer of the conference. Harry Smyth, my department chair, readily agreed to Bohr's request that I be allowed to leave Princeton for a few days to attend the conference. But I decided that my obligations to my students took precedence, so I stayed in Princeton.

By the time of the conference, Frisch in Denmark and several groups in the United States had confirmed the reality of fission by physical (rather than chemical) experiments—experiments in which the great energy of the fission fragments was detected directly. It may seem odd that only days were required to confirm the effect, when it had taken years of painstaking work to discover it. The very energy of the fission process is what made it easy. Neutron bombardment was already an art practiced at numerous laboratories. Once physicists knew what they were looking for, it was short work to find a uranium target, set up the right detector, and measure the characteristic large energy pulse of a fission event. At Columbia, Anderson did all of this in a single day, Sunday, January 29, 1939.

A few days after Bohr's report, probably on Monday morning, January 30, the physicist Luis Alvarez—another future Nobelist—was reading the *San Francisco Chronicle* while getting his hair cut in a barbershop on the campus of the University of California, Berkeley. When he came across a wire story reporting Bohr's announcement of the discovery of fission, he abruptly got up without waiting for the barber to finish and literally ran to the university's Radiation Laboratory, where he brought the news to his student Phil Abelson. By the next day, Abelson had verified the fission phenomenon. When Alvarez brought his colleague Robert Oppenheimer to the lab to see the evidence, Oppenheimer switched from doubter to believer in a few minutes. Within a quarter of an hour, according to Alvarez, Oppenheimer was visualizing the whole process in his mind and imagining chain reactions.

The field of fission physics was launched. But a solid theoretical underpinning for the phenomenon was still missing.

Like most physicists, I was interested in nuclear fission for what it revealed about basic science, not for what it might have to do with reactors or bombs. In 1939, even after we understood fission and knew that a chain reaction might be possible, even after war broke out in Europe, I was interested only in working with students, doing my research, learning more about nature. I was slow to realize that perhaps I had a duty to apply my skills to the service of my country. Two years later, in the fall of 1941, I was involved in a particularly exciting research problem with my brilliant (and fun-loving) graduate student Dick Feynman. Smyth sat me down at a lab bench one day and said, "John, you'd better finish up your work with Feynman. You'll surely be involved in war work soon." He was right. Soon after that conversation, in December 1941, the Japanese attacked Pearl Harbor and the United States entered the war. I started at once looking for ways to contribute to the war effort.

Early in 1942, I joined the exodus of physics professors and their students from the laboratories and classrooms of America's universities. Some went to the Radiation Laboratory of the Massachusetts Institute of Technology (MIT) in Cambridge, Massachusetts, to work on radar. Some went to Chicago and New York and Berkeley to work on nuclear fission. Some found war work by walking across the hall or across the campus at their home universities. Within two years there would be large concentrations of scientists on the mesas of New Mexico, among the hills of Tennessee, and in the desert of eastern Washington. After stints in Chicago and in Wilmington, Delaware, I found myself in the fall of 1944 in Richland, Washington, working on the giant reactors (atomic piles) at nearby Hanford designed to produce plutonium for atomic weapons. Many of my friends were in Los Alamos, New Mexico, and Oak Ridge, Tennessee.

By October 25, 1944—a few weeks after the first Hanford reactor was powered up—the German army in Italy had been driven well north of Rome, but hills, rain, and mud had bogged down the advance of General Mark Clark's Allied forces beyond Florence toward the Po River. My brother Joe, then thirty—three years my junior—was killed in action that day. Joe, who held a Ph.D. in history from Brown University, was a private first class in the Blue Devils Unit of Clark's army. First, we learned that he was "missing in action." Much later his death was confirmed. For eighteen months, until it was discovered in April 1946, Joe's body, disintegrating to bones, lay with that of a buddy in a foxhole on the hill where he was killed. Now Joe lies buried with 4,401 other soldiers in the 70-acre Florence American Cemetery near Florence. It's a beautiful spot, with its neat mathematical rows of white crosses vividly differentiating it from the vineyards and woods nearby. Every time I visit Joe's grave, I am reminded that he is one of many—one of many millions, I calculate, both soldiers and civilians—whose lives might have been spared if the Allies had developed the atomic bomb a year sooner.

On the day Joe was killed, the uranium separation plant at Tennessee's Clinton Engineer Works (the complex that included the new town of Oak Ridge) was partially operational and had produced grams of uranium 235 (U-235), but not yet the kilograms that would be needed for a bomb. A nuclear reactor at the same site had produced grams of plutonium 239 (Pu-239), but not the kilograms that would be delivered only after the Hanford plant became fully operational the next summer. At the Los Alamos laboratory in New Mexico, scientists and engineers had largely mastered the design of a gun-type weapon. They just had to wait for the uranium to make it work.

Only months earlier, experiments had shown that if a plutonium bomb were going to work, it would have to be an implosion-type weapon. In October 1944, a reorganized laboratory was just getting up a full head of steam to solve the implosion problems and design a weapon that could use plutonium.[4]

In the late summer or fall of 1944, I got a card from Joe, written from the front lines in Italy. Its complete message was "Hurry up!" Enough had been written in the newspapers about uranium and nuclear fission in 1939 and 1940 that anyone who cared to think about it might conclude that the Allies, and probably the Germans and Japanese too, were making efforts to develop an atomic bomb. Joe had a little extra knowledge. He knew that in 1939 Niels Bohr and I had worked together to develop the theory of fission, a theory that predicted, among other things, that the isotope U-235 (and the not-yet-discovered isotope Pu-239) would undergo fission if bombarded with slow neutrons. He knew that I had taken a leave from my job at Princeton to go to the University of Chicago to do war work, and that from there I had moved on to E. I. du Pont de Nemours & Co. in Wilmington, Delaware, and then to a remote place in the state of Washington. It didn't take too much guesswork for him to surmise what the nature of that war work might be.

Joe hoped for a miraculous means of ending a terrible war. So he told me to "hurry up."

I am convinced that the United States, with the help of its British and Canadian allies, could have had an atomic bomb sooner and ended the war sooner — perhaps a year sooner than the summer of 1945 — if scientific and political leaders had committed themselves to the task earlier. Between mid-1944 and mid-1945, more than 3 million lives were lost in battle and in bombings. Government-sanctioned murders accounted for at least 12 million more, including the intensified killing of Jews in the Holocaust. The total is so unimaginably great, the loss so horrible, that it staggers the mind. Yet one cannot escape the conclusion that an atomic bomb program started a year earlier and concluded a year sooner would have spared 15 million lives, my brother Joe's among them.

Once General Groves took over the Manhattan Project in 1942 and scientists and industry were mobilized to give their full effort to building an atomic bomb, progress was swift. But that was three years after we understood the essential ideas of nuclear fission and three years after Albert Einstein had written a letter to President Roosevelt drawing attention to its potential military importance.

---

[4] In a gun-type weapon, one piece of uranium is fired down a cylindrical barrel into another piece of uranium. In an implosion weapon, a sphere of nuclear material is squeezed symmetrically by setting off high explosives all around it.

It does little good to second-guess history. But I cannot avoid reflecting on my own role. I could have understood the gravity of the German threat sooner than I did. I could—probably—have influenced the decision makers if I had tried. For more than fifty years, I have lived with the fact of my brother's death. I cannot easily untangle all of the influences of that event on my life, but one is clear: my obligation to accept government service when called upon to render it.

But what of my 1939 work with Bohr, when we were driven more by curiosity about the atomic nucleus than by any thought of weapons? At the time, he was fifty-three and I was twenty-seven. Bohr, a Nobel Laureate, directed the Copenhagen institute that drew physicists from all parts of the world to little Denmark. I was in my first year of an untenured assistant professorship at Princeton, where I had been hired to help move that institution into the coming world of nuclear physics (before Princeton or I had any inkling that the atomic nucleus was anything more than a fascinating little chunk of matter).

In 1933, five years before going to Princeton, I had earned my Ph.D. in theoretical physics from Johns Hopkins University. Then came one-year apprenticeships with Gregory Breit at New York University and with Bohr in Copenhagen, and three years as an assistant professor at the University of North Carolina. I had married early, and had two small children when my wife, Janette, and I moved to Princeton in 1938. There we are still, after sixty years interrupted by numerous leaves of absence and by a happy decade at the University of Texas in Austin.

Bohr's conversations with Rosenfeld on the ocean crossing were typical of the way he worked. He liked to be on his feet, talking, pacing, writing on a blackboard, almost always with a junior colleague at hand. He behaved no differently on shipboard than back at his institute in Copenhagen. The conviction he had reached by the time he set foot on the dock in New York was that the liquid-droplet model of the nucleus should be able to account for the fission phenomenon. That a nucleus bears some resemblance to a drop of liquid was suggested first by George Gamow. Bohr extended the idea, using it as a way to describe the behavior of a nucleus to which extra energy has been added by a bombarding particle—an example of what he called the "compound nucleus."

Despite Bohr's initial hesitancy to speak of fission, little time was wasted getting early reports into print. In three successive issues in February 1939, the journal *Nature* carried Letters on the subject. In the first of these, appearing in the February 11 issue (having been submitted on January 16, the day of Bohr's arrival in New York), Meitner and Frisch proposed fission as the mechanism accounting for the production of barium by neutron bombardment of uranium. While Bohr was en route, Frisch had gone to his laboratory and obtained the large energy "signature" of fission. His separate report on this experiment, also submitted on January 16, appeared on February 18. Bohr himself spent his first few days in Princeton writing a short paper setting forth his general idea about fission. Wanting to be sure that he didn't accidentally upstage Meitner and Frisch, he sent his paper, dated January 20, to Frisch with the request that it be forwarded to *Nature*. It appeared on February 25. With these three papers, fission physics was under way.

Almost at once, Bohr asked me if I would like to work with him on a more detailed theory of fission. It was a subject in which Rosenfeld had less interest and less experience. Besides, he wanted to keep Rosenfeld available to write up notes on his lectures and his discussions with Einstein (which did take place, but on a scale reduced from the original expectation[5]). I was the logical choice, having worked in nuclear physics since 1934 and being well known to Bohr from my postdoctoral year with him in 1934–1935. I accepted readily, even though the fission work would pull me away from a subject I was deeply interested in at the time, action at a distance. For some time, the idea of particles acting at a distance on one another had seemed to me a simpler, more satisfying description of electromagnetism than the standard "field theory," which assigned "substance" to electric and magnetic fields existing in space.

So Bohr and I each changed course, he away (temporarily) from pursuit of the quantum, I away (temporarily) from pursuit of electromagnetism.

We worked well together. It was an exciting time. I have been told that my style of work, and even some of my mannerisms, resemble those of Niels Bohr. It is probably true. I, too, like to work in free-wheeling talk sessions with colleagues, with more questions than answers flying back and forth. I, too, try always to emphasize the positive in my junior colleagues' work, give them all credit that is due, and build their confidence. But was I drawn to Bohr (and he to me) because my approach to physics and to people was similar to his—or did I acquire my style from him? Some of both, I suspect.

The word *fission*, borrowed from cell biology, was suggested by Frisch to describe this newly discovered nuclear process, the splitting of an atomic nucleus into two large fragments. Frisch had acquired the word by asking William Arnold, an American biologist then working in Copenhagen, what cell division is called. Bohr was not enchanted with the word. "If *fission* is a

---

[5] Oswald Veblen, one of Bohr's hosts at the Institute, told me of his surprise when the gladiatorial combat he had been expecting between the two giants did not occur. No jumping up and down. No pounding of fists. Einstein sat quietly, listening to Bohr lecture, saving his discussion for their private meetings.

Bohr and Einstein, c. 1930. Because of nuclear
fission, they had little time for this kind of relaxed
comradeship in 1939.
*(Photograph by Paul Ehrenfest, courtesy of AIP Emilio Segrè
Visual Archives.)*

noun," he said to me, "what is the verb? You can't say 'a nucleus fishes.'" No
sooner had our collaboration begun than we raced up the stairs from our
offices on the second floor of Fine Hall to the mathematics-physics library
on the third floor, where we spent more than an hour looking through dic-
tionaries and reference books in search of a term more to Bohr's liking. Our
search was fruitless. After considering and rejecting several possibilities we fell
back on *fission*, which has stuck. (For a while, Bohr called a nucleus capable
of undergoing fission a "splitter." I'm glad that didn't take hold.)

To me, just as to Bohr, fission seemed immediately believable. I felt stupid not to have realized, several years before, that nuclei should be able to split. My student Katharine Way at the University of North Carolina had investigated magnetic properties of nuclei using the liquid-droplet model. Her equations had no solution when the nucleus rotated too fast. This told us that rapid spin could make a nucleus become unstable and fall apart. It would have been natural to ask ourselves whether there were other ways to make a nucleus come apart. Had we followed her lead, we might have thought of fission.

The carpeted office that Princeton provided to Bohr, Fine Hall Room 208, contained paneling and built-in bookshelves on one wall, a blackboard on another wall, and a bank of five windows looking out on trees on a third wall. At eighteen by eighteen feet, it was reasonably spacious although not elegant. Fine Hall, named after Dean Henry Burchard Fine, was principally a mathematics building, but it also housed some physicists and the wonderful Fine Hall Library for both mathematics and physics. My office, Fine Hall 214, a near-duplicate of Bohr's office, was a few doors away. That made it easy for us to collaborate in the face-to-face way we enjoyed. A work session might start with Bohr sitting or standing near the blackboard in my office. He would outline an idea based on his compound-nucleus model and sketch something on the board. Soon we would be trading a piece of chalk back and forth, drawing pictures and writing equations on the board. If my office began to seem confining, Bohr would lead us round and round the loop of hallway that circled the second floor of Fine Hall, continuing to talk as we walked. These circuits might end at Bohr's office, where more ideas would be exchanged at the blackboard until we decided to separate for private thinking and calculating. As Bohr became more animated, the chalk in his hand was likely to break as he stabbed at the board. On the left side of the board one thing remained always neatly in place, Bohr's list of things to do, a reminder of his outside obligations. When we finished a session or broke for tea, Bohr would lift the edge of the rug in his office and kick broken bits of chalk under it. He had learned that otherwise he would be scolded by the janitor.

Bombs and reactors were only in the backs of our minds as we worked together. We were trying to understand a new nuclear phenomenon, not design anything. One thing we realized right away: for a heavy nucleus such as uranium to split into large fragments, it had to undergo considerable deformation first. (We assumed that the nucleus was spherical before it absorbed a neutron. We know now that even when unexcited, the uranium nucleus, and indeed most nuclei, are prolate spheroids—little footballs. But fission requires a temporary deformation beyond that of the normal shape.)

When you cut an orange in half, the two halves fall apart. This is not true of a nucleus. Imagine a uranium nucleus hypothetically cut into two hemispheres. The powerful nuclear forces between the particles in one half and the particles in the other half will prevent the hemispheres from separating. But if a small separation is achieved in some way, to get beyond the short range of the attractive nuclear forces, the two positively charged halves will repel each other electrically and fly apart at high speed. We say that there is an energy "barrier" standing in the way of cleavage. The energy required to surmount this metaphorical barrier depends on the "route" followed by the nucleus on its way to separation, just as the altitude to which hikers must ascend in going from one place to another depends on what route they traverse between the two points. What Bohr and I showed is that the height of the energy barrier to be surmounted is lowest if the nucleus, instead of falling apart like the two halves of an orange, is deformed through a sequence of other shapes—from orange to cucumber to large peanut. This "path" is analogous to that of hikers who have found the lowest pass over a mountain range, and thus minimized their expenditure of energy to get through. Once the nucleus has acquired just enough extra energy and has deformed into just the right shape, it is perched atop the energy barrier. Then it comes "unglued" and its parts separate, blown apart by their mutual electrical repulsion.

What makes the nucleus deform in the first place? Its act of absorbing a neutron gives it extra energy. Because of this extra energy, we say the nucleus is "excited." Excitation can affect the nucleus in various ways, one of which is to set it into a deforming kind of vibrational motion—much as a raindrop, with energy added, can oscillate from sphere to egg shape and back again. If the vibration of the nucleus carries it up and over an energy barrier, it doesn't pull itself back to its original shape but instead comes apart. In the act of fission, the excited nucleus wriggles through its orange-to-cucumber-to-peanut sequence of shapes in about 1 millionth of 1 billionth of a second (10^{-15} s). It could lose its extra energy in other ways, such as by emitting a gamma ray (a high-energy quantum of electromagnetic energy), but that is less likely. Once excited with enough extra energy, a uranium nucleus is more likely to undergo fission than to do anything else.

Eugene Wigner, a physics faculty member who was to be a key figure in the Manhattan Project and would become my lifelong friend, occupied Fine Hall Room 209, next to Bohr's office. This large corner office, complete with fireplace, had been Einstein's before he moved across town in 1938 to more modest quarters in a new building of the Institute for Advanced Study. Wigner, nine years my senior, was a Hungarian expatriate who had been trained as a chemical engineer but found his true calling in mathematical physics. He was known as much for his unfailing politeness as for the precision of his thought. Graduate students at Princeton, seeing how Wigner always held the door for others, borrowed from the Biblical story of how hard it is for a rich man to get into heaven, saying that it is harder to get through a door behind

Wigner than to pass through the eye of a needle. In 1963, Wigner's achievements in physics were recognized with the Nobel Prize.

When Bohr and I got going on the energy hills and valleys and mountain passes of the deforming uranium nucleus, we naturally had to address the question: What is the chance that a nucleus, having gained extra energy by absorbing a neutron, will squirm through the sequence of shapes leading to fission rather than doing something else with that extra energy? It occurred to me that this question was not unlike a question that one might ask about a complex molecule endowed with extra energy that could disintegrate into smaller fragments. I knew that Wigner had worked on such questions with the physical chemist Michael Polanyi in Berlin. So I wanted to talk to Wigner to see if he could provide any helpful leads. (Bohr and I had no hesitation in seeking advice from our colleagues on vexing questions that arose in our work.) As it happened, Wigner was at that moment in the university infirmary suffering from jaundice, which he apparently had contracted from eating contaminated oysters. Despite his yellow complexion, Wigner greeted me warmly when I showed up at his bedside, and he steered me in the right direction to figure out the answer. I was able to go back to Bohr in a day or two with a formula for the probability of fission.

My office was separated from Wigner's and Bohr's by the central gathering point in Fine Hall, the second-floor lounge, or Tea Room, where faculty and graduate students of mathematics and physics gathered every afternoon for tea. As I later heard Robert Oppenheimer put it, "Tea is where we explain to each other what we do not understand." Bohr and I were regulars at the afternoon teas. At the other end of the looping hallway from the Tea Room is another large room, Fine Hall 202, then the "professors' room" and now a lounge called Jones Hall 202, serving East Asian Studies. Today, in that room, Einstein's words are chiseled in stone above the fireplace:

Raffiniert ist der Herr Gott   Aber Boshaft ist Er nicht
[God is subtle. But He's not malicious.]

In other words, there is hope of figuring things out.

Adjoining Fine Hall was Palmer Physical Laboratory, which housed other physics faculty offices, lecture halls, teaching laboratories, shops, storerooms, and research laboratories. (I was later to have my office there.) In the attic of Palmer Lab was a small accelerator that could accelerate deuterons (nuclei of heavy hydrogen[6]). These charged particles could be directed against a target, stimulating a nuclear reaction that released neutrons. By adjusting the

[6] The nucleus of ordinary hydrogen is a single, positively charged proton. The deuteron consists of a proton and a neutron bound together. (A still heavier form of hydrogen has for its nucleus a triton, consisting of a proton and two neutrons.)

Benjamin Franklin and Joseph Henry long flanked the entrance to
Princeton's Palmer Physical Laboratory.
(Photograph by Robert Matthews, courtesy of Princeton University.)

energy of the deuterons, the energy of the neutrons could be controlled, and
the neutrons, in turn, could be used to bombard other targets. Beginning in
January, as Bohr and I were undertaking our theoretical work, two graduate
students, Henry Barschall and Morton Kanner, with their professor, Rudolph
Ladenburg, started a series of experiments in the Palmer Lab attic to find out
how the probability of fission in uranium (or the nuclear target "cross section") varies as the energy of the bombarding neutrons is changed. Their
results were puzzling. They found, as expected, that the cross section is large
for high-energy neutrons and diminishes as the neutron energy diminishes.
But, surprisingly, at very low neutron energy, the cross section becomes large
again.

One morning that winter, George Placzek joined Bohr and Rosenfeld for
breakfast at the Nassau Club. Placzek, thirty-three, with dark, wavy hair, glasses, a large nose, intense eyes, and a quick wit, could have been an actor playing the part of a brilliant Czech scientist and master of many languages,
which is exactly what he was. Born in Moravia, he earned his Ph.D. in the

Netherlands, and, in the space of less than ten years, had worked with some of the world's most eminent physicists: Paul Ehrenfest in Leiden, Bohr in Copenhagen, Fermi in Rome, and Lev Landau in Kharkov, USSR. He had also taught at the Hebrew University in Jerusalem and had just finished a stint in Paris with the Austrian physicist Hans von Halban. No one was surprised to see him turn up in Princeton.

I had met Placzek first in Copenhagen in 1934. He was awkward in the way that some intellectuals are, but gregarious and always amusing. He was, above all, a provocative questioner. Only a few weeks before breakfasting with Bohr in Princeton, he had been in Copenhagen and had suggested to Otto Frisch how he might most easily confirm the existence of fission — which Frisch promptly did. Now, with Bohr, he saw that the results of Barschall, Kanner, and Ladenburg created a problem for interpreting fission. "What kind of crazy thing is this big cross section for both fast and slow?" Placzek said, in effect, to Bohr. "How can you reconcile it with your view of nuclear reactions?"

Walking across campus after breakfast to Fine Hall, where he would short-

George Placzek, 1946.
(Courtesy of AIP Emilio Segrè Visual Archives,
Rudolf Peierls Collection.)

ly be meeting with me, Bohr talked over Placzek's question with Rosenfeld. Suddenly he said, "Now I have it!" As soon as he reached my office, he told me his idea: The substantial cross section at low energy must be due to the rare isotope U-235, present to only three-quarters of 1 percent in normal uranium. At high energy, the abundant isotope U-238 can also undergo fission, and its cross section increases as the neutron energy increases (up to a certain point, where it levels off).[7] The behavior at low energy is influenced by the wave nature of neutrons. The lower the energy of the neutron, the greater is its wavelength, and the more it can "reach out" to interact with a target nucleus. The chance of fission occurring therefore increases as the energy decreases, provided the nucleus in question—in this case, U-235—is sufficiently excited by absorbing a low-energy neutron to undergo fission at all.

Bohr and I reviewed the picture of the fission process as we then saw it, and the new idea fitted in beautifully. There were subtle differences between one isotope and another, enough to determine whether the isotope would or would not undergo fission after absorbing a low-energy neutron. U-235 would; U-238 would not. This line of reasoning led us to consider what other nuclei might be subject to fission by low-energy neutrons. We could predict with some confidence which isotopes of other elements, known or unknown, would also undergo fission under low-energy neutron bombardment. It was my Princeton colleague Louis Turner who first saw the great potential significance of one such (still undiscovered) isotope, belonging to an element two places higher in the periodic table (94 instead of 92) and with mass 239 (instead of 235). That element, discovered in 1941 and named plutonium, indeed had the property we predicted. In one of the most remarkable industrial developments of all time, plutonium, unknown in nature, was manufactured in kilogram quantities at Hanford during World War II and served as the cores of the bomb tested at Alamogordo and the bomb dropped on Nagasaki. (The Hiroshima bomb used uranium.)

A great deal about fission, indeed about fusion and a whole array of other nuclear properties as well, can be understood in terms of certain simple properties of the neutrons and protons of which nuclei are composed. Neutrons and protons (collectively known as nucleons) attract each other with essentially the same force. They belong to a family of particles called fermions. (Enrico Fermi delineated their properties back in the 1920s.) One important characteristic of such particles is that no two identical fermions can move in

---

[7] An *element* (such as uranium) has a particular number of protons in its nucleus. *Isotopes* (such as U-235 and U-238) have the same number of protons but different numbers of neutrons in their nuclei. Isotopes differ physically but not chemically.

exactly the same way at the same time in the same space. It is as if they "don't like each other." Any two protons can coexist in a nucleus only if their motions are different in some way—and having their spins oppositely directed is enough of a difference. Likewise, any two neutrons must move differently, or spin differently, if they are to share the same space. However, there is no bar to a proton and a neutron moving harmoniously as a pair. It is as if ballet dancers dressed in red were required to dance around each other without touching, and ballet dancers dressed in blue had the same restriction, whereas a red dancer and blue one could embrace and dance across the stage together.

The general consequence of protons and neutrons being fermions is that light nuclei contain equal, or nearly equal, numbers of neutrons and protons. The nuclei of nature's two most common isotopes of nitrogen and oxygen (the principal components of air), for example, are N-14 and O-16. The first contains seven protons and seven neutrons. The second contains eight protons and eight neutrons. But heavy nuclei break this equal-number rule. The nucleus of U-238, for instance, contains 92 protons and 146 neutrons. Why not an equal number? Because protons repel each other electrically, whereas neutrons do not. For the lighter nuclei (up to sixteen protons or so), this electrical repulsion within the nucleus is not significant enough to undermine the effects of the attractive nuclear force. But as more protons are added, their mutual repulsion does undermine the tendency of protons and neutrons to cluster in equal number. As we move to heavier and heavier nuclei, the "neutron excess" gets greater and greater.

Electrical repulsion not only accounts for the neutron excess in heavy nuclei; it also explains why the periodic table of the elements comes to an end. Beyond a certain point, there are no stable nuclei at all. Nuclei heavier than U-238 have been created in the laboratory—Pu-239, for example, and other nuclei containing up to 112 protons—but they are unstable, with lifetimes much shorter than Earth's several-billion-year age.

This pattern of nuclear composition has crucial consequences for nuclear energy. In brief, the rule is this: For light nuclei, *fusion* (combining two nuclei to make a heavier nucleus) releases energy; for heavy nuclei, *fission* (breaking a nucleus into pieces) releases energy. For light nuclei, the attractive nuclear force rules. Because of it, nuclei—up to a point—become more stable as more nucleons are added. Fusing two oxygen nuclei to make a nucleus of sulfur would release energy if it were practical. It isn't, but fusing two hydrogen nuclei to form helium is. That is what happens in a thermonuclear weapon and what is the focus of intense work for future practical power generation.

For heavy nuclei, the same electrical repulsion that accounts for the neutron excess and that puts an end to the periodic table makes nuclei *less* stable as more nucleons are added. Then fission releases energy. It was clear to us and others almost immediately when fission was discovered that fission not only releases energy, it is likely to release neutrons as well. A nucleus undergoing fission has more neutrons than the resulting fission fragments need in order to be viable nuclei. There are neutrons to spare. One way or another, we assumed, some of them would be released in fission. What actually happens is that unstable, neutron-rich nuclei are formed, along with, typically, a few neutrons. Because of these extra neutrons, a chain reaction is possible.

Some heavy nuclei undergo fission upon the absorption of a slow neutron, and some do not. Why? There are two factors at work. One relates to how much charge is present relative to the total mass of the nucleus. The more charge there is for a given mass (and volume), the greater is the effect of the electrical repulsion among the protons—the more delicately, we can say, is the nucleus perched on the side of stability. With each bit more of charge or bit less of mass, the nucleus is pushed toward instability; its barrier against fission is lowered. Bohr and I found that this effect depends on a single calculated quantity, the square of the number of protons divided by the total number of nucleons. For U-236, formed when U-235 absorbs a neutron, this parameter is $\frac{92^2}{236}$, or 35.86. For U-239, formed when U-238 absorbs a neutron, it is $\frac{92^2}{239}$, or 35.41. A small difference, but enough to have huge consequences! We estimated the barrier against fission of U-235 to be some 16 percent less than the barrier against fission of U-238.

The second factor is the preference of nuclei to contain an even rather than an odd number of neutrons (or protons). This preference arises from the fact that nucleons are fermions, each with a spin that can point in either of two directions. A neutron striking U-235 is "welcomed" more strongly than one striking U-238, because in the first case, its absorption creates an even number of neutrons, whereas in the second case, its absorption creates an odd number of neutrons. The extra binding energy for U-235, Bohr and I estimated, is equal to another 16 percent of the "barrier height."

These seem like subtle effects, but think of their practical consequences. Because the rare isotope U-235 is fissionable by slow neutrons and the plentiful isotope U-238 is not, it was necessary in World War II to devise ways to separate U-235 from U-238 on a large scale, a task so difficult that the Tennessee factory capable of doing it cost more than a billion dollars (more than $13 billion in today's dollars).

For the nucleus that we now call plutonium 239 (still hypothetical in 1939, when we called it simply "number 94"), we could predict fissionability by slow neutrons with a very high level of confidence. Its charge-squared-over-mass parameter (after including the absorbed neutron) is $\frac{94^2}{240}$, or 36.82, even greater than for U-236, and therefore more favorable for fission. Moreover, the preference for an even number of neutrons also works in Pu-239's favor, just as it does for U-235. In both cases, the nucleus formed when a neutron is absorbed contains an even number of neutrons (146 and 144, respectively).

Neither Bohr nor I recognized at first how important was the implication of our work that Pu-239 is fissionable—although we had no reason at all to doubt the prediction. What Louis Turner emphasized was the importance of the *chemical* difference between number 94 and uranium. The separation of U-235 from U-238, although successful, was an almost impossible job with the technology of the 1940s. Once plutonium was created in the reactor at Hanford, on the other hand, its chemical separation from uranium and other elements became not nearly so difficult. Enormous though the Hanford operation was, its cost in World War II was only one-third the cost of the U-235 separation plant.

In the postwar period, powerful centrifuges provided the means to separate U-235 at much less cost, and Pu-239 also became cheaper to produce. As a result, half a dozen smaller nations tried, and some succeeded, in making atomic weapons with one or the other of these materials. These efforts and successes remain unacknowledged. As to the five declared nuclear powers (the United States, Great Britain, France, Russia, and China), the relative ease of producing nuclear materials has enabled them to make tens of thousands of nuclear weapons.

By the time of Bohr's departure in April 1939 to return to Denmark, he and I had finished the essentials of our work, but it took another two months to complete some details and for me to finish writing our paper, draw the figures, and prepare the tables. (Ever since my mechanical drawing course as an undergraduate at Johns Hopkins University, I have enjoyed being my own artist and draftsman.) It was Bohr's usual habit to go back and forth with his coauthors, often for an extended period, as he struggled for the precision, generality, and clarity that he always held forth as a goal. This time, uncharacteristically, he gave me permission to edit and submit the paper without sending the final version to him for review. Victor Weisskopf of MIT and Rudolph Peierls of the University of Birmingham, England, both of whom had worked with Bohr, later expressed amazement and envy when they learned of this efficient handling of the fission paper. I submitted the paper on June 28. The editor and referees of *Physical Review* were also speedy. Just over two months elapsed until the paper was reviewed, accepted, printed (taking up 25 two-column pages), and distributed. It appeared in the issue dated September 1, 1939, the day that Germany invaded Poland. (By coincidence, the same issue of *Physical Review* carried an article by Robert Oppenheimer and Hartland Snyder discussing a strange prediction of general relativity theory: that a burned-out star might sometimes collapse to a point. Gravitational collapse later captured my imagination, and I named the resulting entity a "black hole.") I understand that our paper found interested readers not only in the United States and Great Britain, but in Germany and the Soviet Union as well.[8]

That summer I attended the University of Michigan summer school in physics in Ann Arbor to lecture and to learn. I spoke on nuclear physics. Fermi was there too. His topic was cosmic rays. Werner Heisenberg, one of the principal architects of quantum mechanics—the theory that governs atomic and nuclear events—and whom I had come to know in Copenhagen, attended from Germany. At a Sunday afternoon picnic, he told me that he was going to leave early "for machine-gun practice in the Bavarian Alps."

Following our 1939 paper and two follow-up papers published in 1940, Bohr and I thought we were finished with fission. He wanted to go back to pondering the quantum. I wanted to go back to electrodynamics and pursue ideas of a world made of particles without fields. On that subject I had the help of the extraordinary graduate student Richard Feynman. He completed his Ph.D. dissertation, "The Principle of Least Action in Quantum Mechanics," in 1942—three years after starting his graduate work—but, as it turned out, our joint papers on electrodynamics were not to be completed and published until 1945 and 1949. Another idea I had been pursuing for some years, building all matter from electrons and positrons (antielectrons), also got sidetracked. Finally, in 1946, I published a paper on polyelectrons, as I called the almost endless family of entities built of electrons and positrons. (Subsequently, two of these entities were created in the laboratory. It will be exciting to see more.)

Of course, neither Bohr nor I could be finished with fission. It was too closely aligned with war. In mid-July 1939, six weeks before our paper appeared, Leo Szilard and his fellow Hungarian Wigner visited Einstein at his summer home on Long Island to enlist his aid in alerting certain heads of state to the possibilities of a chain-reacting uranium weapon. Szilard wanted Einstein to write to Queen Elizabeth of Belgium, whom Einstein knew, to ask her to prevent uranium mined in the Belgian Congo from being sold to Germany. Wigner wanted Einstein to lend his famous name to an approach to the United States government. Einstein was reluctant to approach the Belgian

---

[8] The September 1, 1939, issue of *Physical Review* has become a collector's item, fetching, I am told, $400 in the secondhand book market.

Werner Heisenberg, 1936.
(Courtesy of AIP Emilio Segrè Visual Archives.)

queen directly but drafted a letter to the Belgian ambassador to the United States (which was never delivered — in the end an exclusive approach to President Roosevelt seemed preferable). Szilard, whose contacts extended outside of physics to the worlds of business and finance, got in touch with the Russian-born economist and financier Alexander Sachs, a friend and advisor to the President. Sachs, moved by the importance of the mission, agreed to carry a letter to Roosevelt.

Two weeks after his first visit to Einstein, Szilard returned, accompanied this time by yet another Hungarian, Edward Teller. (On the two visits, Wigner and Teller served as chauffeurs as well as physicist colleagues. Szilard, who preferred hotels to houses and big cities to the countryside, had never learned to drive.) With Szilard as the leader, the three Hungarians prepared a letter to go from Einstein to Roosevelt. After further consultation by mail, the letter was completed and sent to Sachs in mid-August. Because the war in Europe intervened to put demands on the President's time, Sachs was not able to get an audience with Roosevelt until mid-October. Then in his own words, he delivered the case for the importance of fission.

Einstein opened his letter by citing the work of Fermi and Szilard that led him to expect that "the element uranium may be turned into a new and important source of energy in the immediate future." He found it conceivable "that extremely powerful bombs of a new type may thus be constructed." He

Leo Szilard, 1949.
(Photograph from Argonne National Laboratory, courtesy of AIP
Emilio Segrè Visual Archives.)

drew attention to the Belgian Congo as an important source of uranium and
pointed out that Germany had reportedly stopped the sale of uranium from
the mines in Czechoslovakia that it now controlled. The central element of
Einstein's letter was a plea for government liaison with the scientists working
on uranium. Very soon—thanks in great part to the effectiveness of Sachs as
the messenger—such a liaison was established. Yet the sense of alarm and
urgency that permeated Einstein's letter took a long time to take root in the
government.

Still working in an ivory tower, and with no connection to the political or
scientific movers and shakers, I had no part in these events. I was aware, however, of some of the discussions going on between Szilard and Wigner,

because they took place within earshot in Princeton. Alas, they spoke usually in Hungarian as they paced a hall or campus path, so I missed most of what they said, even when I overheard it. Like Teller, these Hungarian expatriates were somewhat older than I, and certainly more worldly. Their concerns about the danger of Hitler and Fascism were more keenly developed than mine. Nevertheless, I have often asked myself why I did not try to get involved in the politics of uranium earlier. My work with Bohr and my conversations with other physicists made me as aware as anyone of the possible implications of fission.

I feel a sense of sadness now that I hung back and did not at least try to find some avenue of influence to stimulate the government's commitment to fission. I cannot be excused for youth. Twenty-seven is old enough to play a role in great events. I cannot be excused for being on the sidelines. I was as central as anyone in fission physics. I simply lacked the vision to see that the vital interests of the United States were at stake.

So, for a time, I went back to my own physics and teaching. In 1940, Barschall and I wrote a paper on the scattering of neutrons by helium nuclei, which, as he had the wit to see later, gave the first clear indication of a coupling between the orbiting motion and spinning motion of nucleons within the nucleus, something that proved to be important for understanding the structure of all nuclei. In 1941, I published two papers on nuclear physics and one, with Rudolph Ladenburg, on cosmic-ray mesons (particles intermediate in mass between electrons and protons). In February of that year, Glenn Seaborg and his collaborators at the University of California, Berkeley, discovered plutonium. A year later, in early 1942, they had microgram quantities to work with and could confirm its fissionability with slow neutrons. In October 1941, two years after Sachs had presented Einstein's letter to him, Roosevelt approved a full-scale effort to develop an atomic bomb. On December 7, the Japanese attacked Pearl Harbor. For me and for nearly every other physicist in America, all hesitation about pursuing war work then vanished. We were committed. We joined up. A few weeks after Pearl Harbor, I agreed to join Arthur Compton's Metallurgical Laboratory at the University of Chicago. I moved there in late January 1942.

2
·
THE MANHATTAN PROJECT

WHEN AMERICA entered the war in December 1941, my daughter Letitia (Tita) was five, my son James (Jamie) was three, and my daughter-to-be, Alison, was on the way. Janette and I and the two young children had occupied our new home in Princeton for only two years. I was absorbed in research, including my work with Feynman on a new formulation of electrodynamics. It was not an ideal time to pull up stakes and move. But there was no question or debate about it. I was prepared to serve the war effort and to go wherever I was asked to go. Several of my Princeton colleagues, especially those with experimental skills in electronics, went off to the Massachusetts Institute of Technology to work on radar. Since I had worked on fission, I was not surprised when Arthur Compton invited me to contribute to the nuclear program. This led me soon to the University of Chicago.

In the latter part of 1941, but before the December 7 attack on Pearl Harbor, work on nuclear fission in the United States was scattered in various places, principally Columbia University in New York; the University of California, Berkeley; Princeton University; and the University of Chicago. Vannevar Bush, head of the Office of Scientific Research and Development (OSRD) in Washington, was in overall charge. (This doesn't mean that he directly supervised any of the physicists—we were employed by our various institutions. Rather, he controlled the allocation of government funds to support the work.) Bush had created a section of OSRD called S-1 to deal with

Arthur Compton, at the time of his appointment
as chancellor of Washington University,
St. Louis, 1946.
(Courtesy of Hagley Museum and Library.)

nuclear research, and he named various people, most of them at Columbia, to oversee different parts of the work.

Shortly after Pearl Harbor, Bush recognized the need for more coordination if progress were to be accelerated. He divided nuclear research into three broad areas and asked three notable scientists, each a Nobel Prize winner (and each in a different part of the country), to take charge. Ernest Lawrence at UC Berkeley was to be responsible for electromagnetic separation of uranium 235 from uranium 238 and for plutonium research. Lawrence was a natural choice for both responsibilities. In 1932, he had invented the cyclotron, close sibling to the device that would be needed for electromagnetic separation; and it was at his institution that Glenn Seaborg and his colleagues had recently discovered plutonium. Harold Urey at Columbia University was to be responsible for heavy water research and other methods of isotope separation. Again, a very logical choice; Urey, a chemist, had discovered heavy water and knew about separating isotopes. Finally, Arthur Compton at the University of Chicago was to be responsible for chain-reaction studies and—initially—weapon theory.

Compton, at forty-nine, was the oldest of the three. His beautiful experiment on scattering of X rays in 1923 had answered a question that physicists had been debating since 1905: is the quantum of electromagnetic energy that Einstein had proposed (what we now call a photon) *really* a particle? Compton's evidence made clear that the answer is yes. It is as much a particle, we now know, as an electron or positron or quark (the subnucleon particle within protons and neutrons). Compton was one of those rare scientists who combined great skill in research with comparable skill in administration and management. A midwesterner with a Mennonite background, he was admired by everyone who worked with him. Bush could not have made a better choice.

Compton's first action was to consolidate theoretical work in Chicago. In addition to me, he invited my Princeton colleague Eugene Wigner to come to his newly named "Metallurgical Laboratory," and also invited, among others, Edward Teller, then at Columbia. (Teller moved later to Los Alamos.) Throughout my life, it seems that I have been destined to work with Hungarians and often to be guided by them. Teller, like Wigner, became a lifelong friend.

Compton at first intended to let the experimental work stay at Columbia, where Fermi's group had made such a good start. Some of the Columbia people were already scouting sites in the New York area in search of a place to construct the first reactor. But by late January 1942—just about the time that I was making ready to go to Chicago—Compton concluded that all the work for which he was responsible should be consolidated. Reluctantly but without argument, Fermi agreed to relocate to Chicago. Leo Szilard and others also moved from New York, and within the first few months of 1942, the Metallurgical Laboratory (or Met Lab, as we called it) housed an effective working team, covering theoretical and practical aspects of reactor design. Right from the start, I put my effort into the design of plutonium production reactors, while Fermi took charge of the design and construction of a small, low-power first reactor that could be built in Chicago to test principles.

When I boarded the Broadway Limited for the trip to Chicago at the end of January, I was, unfortunately, alone. We had no sure housing in Chicago. Janette, seven months pregnant, wanted to remain near her parents in Baltimore. Her father, Robert Hegner, a parasitologist at the Johns Hopkins School of Hygiene and Public Health and known for his work on malaria, was seriously ill at the time. On March 7—just two days, as it turned out, before her father died—Janette and the two children moved to Baltimore. She stayed on with her mother while Letitia and Jamie moved in with my parents about three miles away, to be cared for by my mother. Our new daughter Alison arrived on March 31 during a late winter blizzard. My father, who had been scheduled to drive Janette to the hospital, called her to say he couldn't get his car out. So, accompanied by her mother, Janette rode the only streetcar to the hospital—and got there on time. When I arrived for a brief visit a few days later, all the excitement was past.

Then began three and a half years of a very peripatetic life for the Wheeler family. By ringing doorbells in the University of Chicago vicinity, I found a house to rent for the summer of 1942. Janette and the three children arrived in June. It was a hard summer for us. Jamie suffered from rheumatic fever, from which he took months to recover, and Janette was in and out of the hospital because of complications following Alison's birth. In September we moved into a so-called Pullman apartment, but stayed there only half a year. In March 1943, we all moved back to the East Coast, to Wilmington, Delaware, so that I could more effectively fill the role I had been assigned by Compton—principal liaison scientist to Du Pont. By the summer of 1944, it was necessary to gather the family and head for the other end of the country. We were to spend the last year of the war living in Richland, Washington, while I worked on the plutonium production reactors in neighboring Hanford.

The administrative structure at the Met Lab was loose and informal. I may have been assigned officially to Wigner's group, but I did not "report" to him. Compton probably sensed that I liked to work outside any confining structure. He let me function, in effect, as a group of one. Of course, I talked to many of my fellow scientists regularly, but there were no lines of authority—other than to Compton—and no management hierarchy of which I was aware. I did have the help of Mrs. Ardis Munk, a faculty wife, for carrying out calculations. She used a turn-the-crank mechanical calculator, the state-of-the-art computer of the time.

I was never much for small talk, and to some of my fellow scientists in Chicago, I may have seemed a little distant. I usually declined invitations to join groups going out to lunch at nearby restaurants, so that, for an hour, I could lay aside the war work and do the research that I called my "Princeton work" while I ate a sandwich at my desk.

Even before Pearl Harbor and before moving to Chicago, Wigner had been caught up in Szilard's enthusiasm for making a nuclear chain reaction. These two Hungarians could hardly have been more different, yet they worked well together and saw eye to eye on the importance of pushing toward a fission weapon. Szilard, unmarried and often unemployed, bounced around from one lab to another, a gadfly whom everyone recognized as both brilliant and difficult. He seemed to like to live out of a suitcase. Before he moved to Chicago, around the same time as I, he lived at the King's Crown Hotel just south of Columbia University in New York. He slept at odd times and reportedly took hour-long baths. According to one story, he sometimes left the toilet to be flushed by the maid, because he could not be bothered to do it himself. I admired Szilard for his flashes of brilliance and his enthusiasm, but I confess that I was among those who had trouble exercising patience with his eccentricities.

Szilard and Wigner met often, sometimes in New York, sometimes in Princeton. While Szilard was rushing back and forth between the worlds of science and business, promoting bomb work and trying to round up tons of purified graphite for the first reactor, Wigner set about calculating. He had an extraordinary intellect for both mathematics and physics, and was an organized, methodical worker. He was then thirty-nine.

Wigner's calculations addressed the multiplication factor. Given one fission event produced by a neutron, how many neutrons (or what fraction of a neutron) would, on average, eventually be available to cause other fission events? The answer rested not only on the question of how many neutrons were emitted during and immediately following the fission event. That number was known to be substantially greater than 1. Fermi and others were determining it experimentally. The critical question was what happened to these neutrons after they were emitted. Some could escape from the reactor, some could be absorbed by various nuclei—especially U-238—without causing fission, and some could cause further fission. Reactor design aimed to minimize losses and absorption in order to maximize further fission. Slowing the neutrons down quickly was a key to getting a multiplication factor greater than 1, because a slow neutron is most effective in causing fission of U-235. The reactor therefore had to contain material to decelerate the neutrons. Fermi, in his straightforward way, called this material a "slower downer." That term got under my skin so much that I was forced to think up a different term. I chose "moderator" and it has stuck.

Very early in the history of fission, probably around the time that Bohr and I were finishing our paper on the theory of fission in the summer of 1939, several scientists in America and in France independently realized that a reactor needed a lattice design. Its moderator and its uranium fuel should be in chunks rather than mixed together in tiny pieces like the ingredients of bread dough. No one was quite sure, however, exactly how big the chunks should be or how they should be arrayed. Wigner tried to find out.

But his calculations could go only so far without experimental data. In addition to knowing how many neutrons are emitted, on average, in a fission event, he needed to know more about the details of the slowing-down process as the neutrons make their way through the fuel and the moderator, and also had to know the chance that a neutron would be absorbed in the moderator and thereby be lost to the chain-reaction process. To answer these questions, Wigner got Princeton's cyclotron group busy. Milton White, head of the group, was happy to authorize the needed experiments. (Although this was before the United States entered the war, the importance of uranium research was clear.) One of the real live wires in White's group was a young man named Edward Creutz. He later served as a group leader at Los Alamos, and after the war became a professor at the Carnegie Institute of Technology in Pittsburgh. Ed Creutz was a pleasure to work with. He was ready to sweep floors if that's what it took to get a job done. On the basis of his skill and can-do attitude, I later recommended him to Frederic De Hoffmann for a senior post at General Atomics (a division of General Dynamics Corporation), where he successfully designed and built a simple, marketable nuclear reactor.

As a result of experiments such as those carried out by White and Creutz, those of us thinking about reactor design concluded that carbon in the form of graphite would be a suitable moderator, if it could be secured in pure enough form. We knew that "heavy water" (D_2O) — in which deuterium replaces ordinary hydrogen to make D_2O instead of ordinary "light" water (H_2O) — would be even better, but it would be costlier to make. Indeed, only a few liters of it were available at that time in the United States, and it cost more than $2,500 per liter. Fortunately for us, our German counterparts reached a different conclusion. They thought that neutron absorption in graphite made it an unsuitable moderator, and they therefore turned their attention to acquiring large quantities of heavy water.

General Leslie Groves, having been briefed on the potential importance of heavy water to the Germans, made an immediate move after assuming command of the Manhattan Project. He requested Allied action to destroy a heavy-water plant at Vemork, Norway. The British, who already had Vemork on their list of targets for commando raids, moved up the date of their planned attack. After a disaster caused by bad weather that cost them thirty-four lives on their first try in November 1942, they tried again in March 1943, this time with just six Norwegians who parachuted into Norway and successfully joined up with four of their countrymen who had parachuted in the previous October and somehow survived. This small raiding party was dramatically successful. With no loss of life, they largely destroyed the plant, setting back its heavy-water production by about a year.

Our French colleagues were no less aware of the significance of heavy water. In 1940, Lieutenant Allier of the French Deuxième Bureau had arranged to acquire 185 liters of heavy water in Norway, most of the world's stock at that time. He double-booked himself and his twenty-six cans of heavy water on flights to Amsterdam and Scotland. The flight to Amsterdam was forced down in Hamburg by the German Luftwaffe and searched. He and the heavy water were on the flight to Scotland. By land and sea, he shepherded the precious cargo to Paris, where, however, it could not safely remain for long. The physicist Hans von Halban loaded his wife and one-year-old daughter into the front of their car, stowed a gram of Marie Curie's radium in the back, placed the heavy water, as a shield, in between, and drove to the south of France. Almost at once, he and his French team were instructed to transfer their work to England. On the crossing from Bordeaux, the cans of heavy water were strapped to wooden pallets to assure their flotation in the event the ship was sunk. Finally, in 1943, the French scientists and their 185 liters of heavy water, still the largest quantity in Allied hands, moved to Montreal, Canada. By then, a heavy-water plant in Trail, British Columbia, was gearing up for large-scale production. In 1944, it produced some 6 tons of heavy water priced at one-hundredth the 1941 price.

The German misjudgment about graphite was one of the things that slowed their progress toward a bomb. At the end of the war in Europe, two and a half years after Fermi's pile went critical, the Germans did not yet have an operating reactor, much less any clear idea of how to build a bomb.

I have to admit to a misjudgment of my own. In the months before Pearl Harbor, we had a radio in the Fine Hall Tea Room so that we could listen to the war news and discuss the war's progress. I reached the conclusion that a German-dominated Europe might be the best way to assure long-term peace in Europe. I was a member of the German Physical Society and had many German physicist friends. I admired the strength and efficiency of the German state. I cannot claim to have been a naive young man out of touch with the real world, for I had been avidly interested in history and foreign affairs since my student days. I had learned from German Jewish scientists whom I met in Copenhagen in 1934–1935 what a threat they considered Hitler's Germany to be. I had heard Einstein, Teller, Wigner, and other emigrés speak of the Nazi intolerance that led them to settle in America. One of these emigrés, my Princeton colleague Valentine Bargmann, reproached me for even reading German propaganda material that came to me because of my membership in the German Physical Society.

It is hard now, more than fifty years later, to recapture my frame of mind at that time. I discounted the fears of my friends, believing that no civilized people could translate poisonous rhetoric into inhuman action. My parents considered my German sympathy disreputable. This sympathy did not vanish instantly when America entered the war. It evaporated slowly as I learned more. Even when I was doing everything I could to help defeat Germany, I clung to the belief that people are fundamentally decent everywhere, that German atrocities were as unthinkable as American atrocities. By the end of the war I knew better. But not until I visited Auschwitz in 1947 was the full horror of German barbarism brought home to me.

Despite my realization, finally, of the depths to which humans could sink under despotic rule, I never abandoned my sense of comradeship with fellow scientists in all countries, including Germany. I first met Werner Heisenberg, later to be the leader of Germany's wartime uranium project, in Copenhagen in early 1935. I was twenty-three, trying to sharpen my skills as

a theoretical physicist. He, at thirty-three, was already famous for having discovered quantum mechanics almost a decade earlier. His name is attached
now to the Heisenberg uncertainty principle, a principle that captures the
essence of unpredictability for individual events in the small-scale world. At
that time, Heisenberg and I walked and talked outdoors on several pleasant
spring days. He was not an easy person to get to know. He was a bit withdrawn,
even wistful, I thought, rather like a child who had been let out to play with
the neighboring children but knew that he soon had to return home.

After Heisenberg abruptly left the University of Michigan summer school
in 1939 for "machine-gun practice in the Bavarian Alps," a dozen years passed
before I saw him again. One day in the summer of 1951, my wife and I were
sitting in a modest restaurant near the waterfront in Copenhagen when we
saw a lone man enter and be seated by the hostess at a table by himself.
"That's Heisenberg," I said to Janette. "Let's invite him to join us." She agreed,
of course. He seemed happy to accept. Over dinner we spoke of some of our
recent experiences, but not a word about uranium or reactors or bombs. He
told us how, just as Germany was about to fall to the advancing Allied forces,
he left his research laboratory without authorization, by bicycle, pedaling
toward Munich to be with his wife and children as the war drew to its chaotic conclusion. On the way he encountered a sentry whose orders were to
shoot anyone deserting his duty post. With the help of an offered cigarette,
Heisenberg talked his way through. "One cigarette for one life," he said. Later
I invited Heisenberg to Princeton when some of my colleagues there still considered him a pariah. Heisenberg died in 1976 at the age of seventy-four,
with fewer friends than he deserved. I remain as uncertain as anyone about
his motivations during the war, when, according to some observers, he may
have deliberately stalled the German atomic-bomb effort. The more credi-
ble view now is that he and his German colleagues did not press for massive
resources to go forward with a bomb project because they imagined the time
scale for the project to be greater than the time scale of the war, and had no
inkling that the Allies were building a bomb.[1]

In the early days of fission, it was generally assumed that making a bomb
would require the use of the rare isotope U-235, even though we all knew that
separating this isotope from the chemically identical isotope U-238 would be
extraordinarily difficult. At a meeting in Wigner's large corner office on
March 16, 1939, Bohr, Wigner, Szilard, and I were discussing the prospects of

---

[1] See David Cassidy, *Uncertainty: The Life and Science of Werner Heisenberg* (New York:
W. H. Freeman, 1992), and Jeremy Bernstein, *Hitler's Uranium Club* (Woodbury, N.Y.:
AIP Press, 1996).

making a bomb. It had been at most a few weeks since Bohr, in response to George Placzek's breakfast question, had realized that U-235 is the isotope responsible for fission by slow neutrons. Our thinking about the possibility of a bomb therefore centered on getting enough separated U-235. Bohr, recognizing the enormity of that task, said, "Yes, it would be possible to make a bomb, but it would take the entire efforts of a nation to do it."

Following my work with Bohr that year, a few people had the vision to see that a plutonium-based bomb might be a way to circumvent the enormous problem of separating U-235 from U-238. Louis Turner in the United States, as I mentioned earlier, and Karl Friedrich von Weizsäcker in Germany foresaw this possibility—before plutonium had been named, much less seen! Bohr and I, although we had predicted the fissionability of plutonium, did not immediately propose it for the core of a nuclear weapon. But by late 1941, the idea of using plutonium was taken seriously. (Yet neither Compton nor anyone else was so courageous—or so rash—as to suggest that plans to separate uranium should be abandoned and all hope pinned on plutonium.)

When I went to Chicago in January 1942, it was to work on the design of a reactor that could make kilograms—eventually tons—of plutonium. This was less than a year after Seaborg's team had discovered plutonium. By the time that microgram quantities of plutonium were available for study, the gargantuan construction project for the Hanford reactors was already under way.

The question of a moderator having been settled in favor of graphite, there remained the question of how to cool the reactor. This was one of the problems I tackled in my first months in Chicago. It was not relevant to Fermi's low-power pile, which required no more cooling than would be provided by the air in the squash court where it was built. That first reactor was to prove the feasibility of a chain reaction, not to generate power or make new isotopes. The reactors planned for plutonium production, on the other hand, would require that some gas or liquid be pumped continuously through them to carry away heat.

I was among those (a majority, I suspect) who believed that helium would be the best coolant. It does not absorb neutrons at all, so it would do nothing to interfere with the chain reaction. I did some experiments (among the few in my lifetime as a theorist!) in which I pumped not helium but air through channels in graphite to see if the fast-flowing gas would erode the graphite walls of the channel. Erosion was negligible, a result that supported the use of helium.

But helium had one notable drawback. It was a relatively poor heat-transfer medium. Water was much better for heat transfer (and, of course, it was cheaper). Wigner favored water. The number of neutrons it absorbs, he argued, would not be enough to prevent a chain reaction, especially if sufficiently pure uranium could be obtained for fuel. In fact, there was continual improvement in the chemical quality of the uranium available to us for study, lending support to Wigner's position. Probably in the late summer of 1942, we made the decision to design a water-cooled reactor.

Before that decision was made, however, preliminary planning for a plutonium production reactor was under way. Compton had chosen Stone and Webster, a large engineering and construction company, to begin design work. The principal engineer of Stone and Webster had a lot of trouble getting along with the scientists. Some of the friction was related to the question of the coolant. He had been told to plan on helium, and he was not interested in discussing alternatives. His relations with me were fine, however, so I acquired a reputation as someone who could get along with engineers. Stone and Webster did not stay with the project very long. In November, at the request of General Groves, who had taken over as head of the Manhattan Project on September 17, Du Pont agreed to design, build, and operate the plutonium production plant. Probably because of my smooth relations with the Stone and Webster engineer, Compton asked me to provide scientific liaison with Du Pont. This became my principal assignment for the balance of the war.

On Thanksgiving Day 1942, while Fermi and his team were making ready the pile under the West Stands of the university stadium, I sat at a table with Du Pont colleagues in Wilmington, analyzing possible sites for the plutonium production reactors. The places under consideration were in Florida and South Carolina and the state of Washington. All had plentiful water and were sufficiently isolated to provide safety in the event of an accident. Another requirement was a reliable electrical supply for the pumps that would be circulating cooling water through the reactors. I had asked a friend to find statistics on thunderstorms in these areas, because thunderstorms would pose the risk of knocking out the electricity. Based on those data, we ruled out Florida. I can't remember what tipped the scales in favor of Washington over South Carolina. As it turned out, Savannah River, South Carolina was chosen not many years later for another major reactor, this one to produce both plutonium and tritium.[2] (Tritium, the heaviest isotope of hydrogen, is useful in thermonuclear reactions.)

The story of Fermi's pile going critical on December 2 has been told often. It is a story imbued with drama. There is the methodical Fermi breaking for

---

[2] The Savannah River reactor provided a fringe benefit for pure physics. In 1956, Frederick Reines and Clyde Cowan detected the neutrino there. The neutrino, an elementary particle with no electric charge and little if any mass (a particle with no mass at all is possible!), had been hypothesized in 1931 but went undetected for twenty-five years because it interacts so weakly with matter. The flux of neutrinos from the Savannah River reactor was so great that Reines and Cowan were able to capture and identify a few.

lunch even on the verge of criticality. There is the bottle of Chianti provided by Wigner and signed by all who were there. There is the phone call from Compton in Chicago to James Conant in Washington reporting that "the Italian navigator has just landed in the new world." Compton went on to assure Conant that the natives were friendly. Strangely enough, to me it seemed a very routine affair at the time, just a necessary stepping stone toward making large production reactors. I had no doubt that Fermi's experiment would work as planned. I remained in Wilmington that day, working with Du Pont colleagues on the reactors that lay ahead.

As I got more and more involved with Du Pont in late 1942 and early 1943, I found myself commuting almost weekly from Chicago to Wilmington, Delaware. Finally, in March 1943, my Du Pont friends said, "Don't you think you should just move to Wilmington?" So I did. Again I had to go ahead of the family and find a house. Janette and the children packed up and followed by train. Thanks to Du Pont influence, they got sleeping accommodations on the train. Alison, not quite a year old, was having her second train ride on the way to her fourth house.

Wilmington was the best base of operations for the next year, but eventually construction and operation of the reactors in Hanford would dictate that we move to the other end of the country. Even during the time we were in Wilmington, I found myself going frequently to Chicago (about as often as I had gone to Wilmington when I lived in Chicago!), and about once a month to the state of Washington. It's a good thing I liked train travel. I can still remember taking the train up from Wilmington to North Philadelphia, crossing to the southbound track, boarding the 7:19 P.M. Broadway Limited for Chicago, changing there the next day to a Union Pacific train, getting off at Pendleton, Oregon, a day and a half later at 5:35 A.M., and being met for the seventy-mile drive to Hanford.

Richland, where we settled, was a community of houses, stores, and schools that had been erected in a matter of months by the Army Corps of Engineers. The sidewalks were tar that had been squeezed out like toothpaste. Asparagus sprouted through cracks in the sidewalks from the farm that had been there before the Corps' bulldozers moved in. Our two-story house was adequate for my family. Eight-year-old Letitia and six-year-old Jamie went to the Sacajawea School, named after the Indian guide who had served the Lewis and Clark expedition. Two-year-old Alison played at home. The older children can still remember the blowing sand in Richland that sometimes cut their legs. Janette and I look back fondly on our year in Richland. We were, in a sense, roughing it, but it was a place full of young, enthusiastic people. Spirits were high. Complaints were almost unknown. The bathtub in our house

Pennsylvania RR
THE BROADWAY LIMITED

WEST—Read down   EAST—Read up

Train No. 28 Daily		STATIONS		Train No. 29 Daily
		Eastern Standard Time		
6.00	Lv	New York	Ar	9.30
7.19	Lv	No. Philadelphia	Ar	8.03
1.55	Lv	Pittsburgh	Ar	1.20
		Central Standard Time		
9.00	Ar	Chicago	Lv	**4.30**

P.M. *times shown in boldface*

Union Pacific
The Streamliner   CITY OF PORTLAND

WEST—Read down   EAST—Read up

Streamliner No.1		Between CHICAGO and PORTLAND No Extra Fare				Streamliner No.2
		C. & N. W.				
6.00	Lv	Chicago (C.T.)	Ill.		Ar	**12.15**
2.00	Ar	Omaha	Neb.		Lv	4.15
		UNION PACIFIC				
2.10	Lv	Omaha	Neb.		Ar	4.05
5.24	Lv	North Platte (M.T.)	Neb.		Ar	**10.49**
9.10	Lv	Cheyenne	Wyo.		Lv	**7.30**
7.40	Lv	Pocatello	Ida.		Lv	8.52
12.57	Lv	Huntington (P.T.)	Ore.		Ar	1.40
5.35	Lv	Pendleton	Ore.		Lv	**8.56**

P.M. *times shown in boldface*

Portions of the railroad timetables for the Broadway Limited and
City of Portland in the mid-1940s.
(Courtesy of Michael W. Flick and Caroline Eisenhood.)

When this picture of the once-tiny village of Richland, Washington, was taken in
1944, it was home to 17,000 people.
(Photograph by Robley L. Johnson, courtesy of U.S. Department of Energy.)

was made out of concrete, three or four inches thick. Toward the end of our
stay, a shipment of more conventional steel bathtubs arrived. But then the
Columbia River flooded. Everything available, including the crated bathtubs,
had to be tossed into the dikes to hold back the river.

We scientists had our offices in the so-called 300 Area about 10 miles from
Richland, and had to drive frequently to the plant that was under construction beyond Hanford, some 25 miles farther. Every now and again, our car
would kill a rabbit on the trip. I remember Fermi on one trip calculating how
many rabbits there must be per square mile based on the speed of the car and
the number of rabbits killed. It gave him great pleasure to figure out approximate answers based on rough data and reasonable assumptions. Questions like
How many piano tuners are there in Chicago? and How many blades of grass
are there on a football field? have come to be known as Fermi questions.

One of the key Du Pont engineers, who accompanied me sometimes on
train trips between Wilmington and Hanford, was Charles Cooper, a former

Enrico Fermi in Los Alamos, 1945.
(Courtesy of AIP Emilio Segrè Visual Archives.)

Maine guide, tall, with a booming voice. His great contribution was learning how to encase the small cylinders of uranium fuel (fuel "slugs") in cans that would resist corrosion and would prevent the escape of radioactive materials into the cooling water. Cooper liked Fermi questions too. He told me that he had once taught a course on estimating, for which the single final-exam question was "How far can a wild goose fly?" After the war, Charlie Cooper and his wife, along with Princeton astrophysicist Martin Schwarzschild and his wife, Barbara, were our guests in Vermont during one Christmas vacation. An ice storm had knocked out the electric power. We managed to cook and keep warm with a wood-burning stove and fireplace. It was Cooper who advised us how to make oversized paraffin candles and who rigged a rope to the electric water pump in such a way that we could use muscle power to get water.

The Du Ponter in charge of the plant during its period of construction was Walter Simon, whose mild manner sheltered a zest for getting things

General Leslie Groves, 1946.
(Courtesy of Hagley Museum and Library.)

done. On one visit to Hanford, Simon told me, General Groves was barking in his usual blunt style: "Simon, why isn't that wall finished yet?" or "Simon, what's holding up the installation of the chemical processing equipment?" To every such outburst, Simon responded quietly and politely, which only seemed to irritate the general. At the end of the day, Groves said, "Simon, are you a man or a worm? Can't you speak up?" Simon, as restrained and polite as ever, answered, "General, we have people in Wilmington to answer questions like that."

The Du Pont people, long accustomed to dealing with high explosives, gave proper attention to safety against explosions in designing the plant. But now they had to deal with a new and unfamiliar hazard, radioactivity—the spontaneous emission of energetic radiation by atoms. From me and other physicists at the site they got a crash course in nuclear physics, learning about alpha, beta, and gamma rays, half-lives,[3] and the effects of radiation on living tissue. This knowledge they had to blend with their own well-developed safety consciousness and with their own calculus of risk versus cost. The reactor

---

[3] Radioactive isotopes may emit helium nuclei (alpha rays), electrons (beta rays), or photons of electromagnetic energy (gamma rays)—or some combination of these radiations. Each radioactive isotope has a characteristic half-life, the time it takes for half of its atoms to decay. Half-lives range from tiny fractions of a second to billions of years.

design included provisions for automatic shutdown in case cooling water stopped flowing or if various other out-of-the-ordinary conditions arose. Chemical processing of fuel slugs to remove plutonium from them was entirely automated from remote-control sites. Everyone working near the reactors or the chemical plants was regularly monitored for radiation exposure. During World War II, there was no known accident or injury associated with radioactivity at Hanford. (In the rush to get the job done quickly as well as safely at Hanford, however, not enough attention was given to long-term storage of radioactive waste. Although the stored waste caused no injuries, it did cause headaches for the next generation, who had to clean it up.)

Safety procedures that are now standard at all nuclear installations are modeled after those developed by Du Pont at Hanford. It was a Du Pont staff member, Roger Williams, who invented the term *health physics*. To be sure, his term was chosen for its positive imagery, to avoid reference to "radiation hazard" or "radiation protection."[4] But it is a good term and now defines a professional field.

Another safety concern was whether enough plutonium might accumulate in one place during the automatic separation process to produce a critical mass. If this were to happen, a self-sustaining reaction would get under way, causing, at the very least, boiling and spillage of radioactive material and contamination of a sizable area within a building. The straightforward way to reduce this risk would have been to build another chemical separation plant like the two already being planned, thereby reducing the amount of plutonium in each processing stream by a third. But the buildings were so large (and accordingly nicknamed the Queen Marys after the Cunard ocean liner) and so expensive (some $100 million each) that choosing to add another was no easy decision.

One day, returning from Wilmington, I was handed a telegram when the Great Northern's westbound Empire Builder stopped in Fargo, North Dakota. I was instructed to go at once to Los Alamos to see what I could learn from people there about the hazard of accumulating a quantity of plutonium in one place. I got off the train, had lunch in the station (overhearing the pleasant sonority of Norwegian being spoken), boarded the next eastbound train, rode back to Chicago, changed to the Santa Fe Chief, and rode to Lamy, New Mexico, the stop that served Santa Fe and Los Alamos. Dick Feynman and others filled me in on their latest calculations of the critical mass of plutonium. When I got back to Hanford, I went over the numbers and

---

[4] In more recent times, the medical profession has removed the dreaded word *nuclear* from *nuclear magnetic resonance* (NMR). In diagnostic applications, it is now *magnetic resonance imaging* (MRI).

THE EMPIRE BUILDER

WEST—Read Down   EAST—Read Up

Train No. 1 Daily		Central Time		Train No. 2 Daily
		BURLINGTON ROUTE		
11.15	Lv	Chicago	Ar	8.40
8.30	Ar	St. Paul	Lv	**10.55**
		GREAT NORTHERN RY.		
8.50	Lv	St. Paul	Ar	**10.30**
3.20	Lv	Fargo	Ar	**3.55**
		Pacific Time		
9.30	Ar	Spokane	Lv	7.30
		SPOKANE, PORTLAND, & SEATTLE RY.		
9.45	Lv	Spokane	Ar	6.50
1.35	Ar	Pasco	Lv	1.50

P.M. times shown in boldface

Santa Fe RR
THE CHIEF

WEST—Read down   EAST—Read up

Train No. 19 Daily		Stations		Train No. 20 Daily
		Central Standard Time		
12.01	Lv	Chicago	Ar	**1.00**
10.10	Lv	Kansas City	Lv	3.35
		Mountain Standard Time		
3.17	Lv	Dodge City	Lv	**9.20**
1.20	Ar	Lamy	Lv	10.14

P.M. times shown in boldface

Portions of the railroad timetables for the Empire Builder and
Chief in the mid-1940s.
(Courtesy of Minnesota Historical Society and Caroline Eisenhood.)

the risk factors with the new plant manager, Bill Mackey. We decided that we could not totally eliminate the chance of accumulating a critical mass of plutonium, but that enough precautions and safeguards could be built into the processing to reduce the risk to an acceptably small value. I recall Bill saying, "The Du Pont Company pays me to make decisions based on incomplete evidence. I am deciding that we won't build another Queen Mary." As it turned out, the plutonium in the processing plant never approached criticality.

Crawford Greenewalt was Du Pont's chief engineer at Hanford. We started working together when I was still in Chicago, and continued a close collaboration at Hanford. He was a charming, bright, and curious MIT graduate who loved to soak up physics—while I loved to absorb more engineering. Greenewalt actually lived the fairy tale of marrying the boss's daughter and eventually becoming the boss himself. Soon after joining Du Pont, he courted and married Margaretta du Pont, daughter of the company president,

Crawford Greenewalt, c. 1945.
(Courtesy of Hagley Museum and Library.)

Irénée du Pont. As he told it one night at a party I attended, his prospective father-in-law presented him with a mathematical problem to solve one evening at the dinner table. He had the good fortune to be able to solve it on the spot, suitably impressing the older man. Greenewalt then turned to me with the problem and asked me if I could solve it. I took a look at it and said, "I'm already married."

Going from Fermi's modest pile in Chicago to the huge and powerful Hanford reactors was too gigantic a change of scale to be accomplished in a single step. A reactor of intermediate size, also capable of producing some plutonium, was designed and built at Clinton, Tennessee. The Clinton reactor went critical in November 1943, less than a year after Fermi's achievement in Chicago.[5] By April 1944, it had produced a few grams of plutonium, vitally needed for exploring the properties of this new element. The first Hanford reactor went critical in September of that year, and by January had produced a kilogram of plutonium. Seven months later, a plutonium weapon, "Fat Man," was detonated in the New Mexico desert.

Fermi supervised the loading of uranium fuel slugs into the first Hanford reactor (the B reactor), and on September 15, 1944, brought it to dry criticality—a self-sustaining condition of such low power that cooling water was not required. Over the next ten days, the team added more fuel slugs and conducted more experiments. The observed performance was in near-perfect accord with what Fermi and I and other members of the physics team had calculated.

The first run at significant power was scheduled for Tuesday night, September 26. I left work at my normal time, went home, spent some time with the family, and went to bed, expecting to hear of the successful run the next morning. Since I was scheduled for some late-night baby sitting of the reactor later in the week, I decided against spending this night in the Hanford control room, drama or no. I figured that my role would be only that of an observer. When I reached my office in the 300 Area the next morning, I learned that the reactor was not exactly following the script. It had started flawlessly and reached a record power of 9 million watts. Then, inexplicably, it started losing reactivity—meaning that the control rods had to be pulled farther and farther out to maintain power. It was as if the engine of your car got sick as you were driving along a level road, and you had to push farther and

[5] Early in 1943, Fermi's pile (CP-1, for Chicago Pile No. 1) was disassembled, then reassembled at a site outside of Chicago that was to grow later into Argonne National Laboratory. CP-2, as the reassembled reactor was called, provided a test bed for numerous important experiments in 1943. It was followed by a more powerful (but still, by Hanford standards, very low energy) reactor, CP-3, in 1944.

farther down on the accelerator pedal to maintain speed; eventually, the pedal would be all the way to the floor and the car would start to slow down. Soon the control rods were as far out as they would come, and the reactor nevertheless lost power. That happened Wednesday afternoon.

Fermi's first reactor, in Chicago, had produced scarcely 1 watt of power. The Clinton reactor in Tennessee produced 1 million watts. Each of the three Hanford reactors was designed to produce 250 million watts. The steps were large. (For all of these reactors, the power appeared in the form of heat, which was dissipated, or "thrown away." Later reactors were designed to convert a portion of that heat into electricity—as much as the laws of thermodynamics permit, which is about 40 percent.)

Shortly after midnight on that first night, while I was sleeping peacefully, the Hanford reactor went critical. (Now it was Wednesday, September 27.) The plan was to pause at the 9-million-watt plateau, then increase power slowly over the next couple of days until the design power of 250 million watts was reached. As it turned out, that power was reached more than four months later, on February 4, 1945. In the predawn hours of September 27, however, only relief and satisfaction were in order. Fermi and his vivacious young colleague Leona Marshall (who officially was my assistant), the operators, the engineers, and the Du Pont managers could look at each other and smile. Two and a half years of thinking, calculating, designing, and building had come to successful fruition. Many people had had a hand in it. Not just the physicists but also the engineers, the technicians, the chemists, the managers, the military leadership, and the contractors.

By the time I got to work that morning, it was becoming clear that the motion of the control rods was, as Du Pont's Dale Babcock put it, "more out than in." (Pushing the rods all the way in shuts down the reactor because of neutron absorption in the rods. Pulling the rods out allows the reactor power to build up. Maintaining constant power normally requires small in-and-out adjustments of the rods.) By midafternoon, the control rods had to be pulled nearly all the way out to maintain the 9 million watts. The Du Pont managers, most likely at the suggestion of Fermi, decided to reduce the power to see if the problem would go away. At around 4:00 P.M., the operators reduced the power more than twentyfold to 400,000 watts. Maintaining even that low power required moving the control rods still farther out. By 6:30 P.M., the control rods were all the way out, but the reactor nevertheless "died"—shut itself down.

Obviously something was absorbing too many neutrons. Fermi and Leona Marshall at first suspected that water leaking from one or more cooling tubes into the reactor was the culprit. I recognized this as a possibility, but I was more inclined to suspect that the reactor was poisoning itself by producing

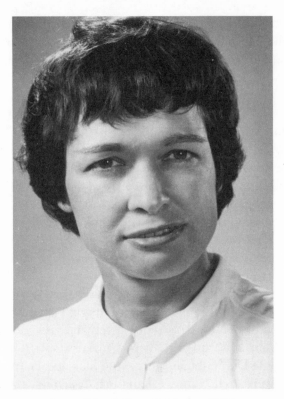

Leona Marshall, probably late 1940s
or early 1950s.
(Courtesy of AIP Emilio Segrè Visual Archives.)

an isotope with an unusually large cross section for absorbing neutrons. This possibility had been on my mind for a long time. As early as February 20, 1942, only a few weeks after my arrival at the Metallurgical Laboratory in Chicago, I had written a report drawing attention to this possible problem. In designing the Hanford reactors, we had allowed for such self-poisoning within certain limits. In an April 1942 report, I had suggested that self-poisoning would diminish reactor reactivity by no more than 1 percent provided no fission fragment had a cross section for absorbing neutrons greater than about 100,000 barns.

Let me explain the "barn," a unit of measurement introduced by Fermi. It is the area of a square with sides 1 millionth of 1 millionth of a centimeter. It is somewhat larger than the cross-sectional area of a typical nucleus. In the United States we say of a poor baseball pitcher that he can't hit the side of a barn. I don't know whether Fermi had picked up some American baseball lingo or whether the term came to him from some Italian saying. In nuclear physics, the barn is such a "large" area that no particle should miss it. If a particle is fired at a nucleus, the chance that the particle will interact with the nucleus can be expressed in terms of an effective cross-sectional area, which is a combined area of the projectile particle and the target nucleus. For a high-speed particle, this effective area is normally less than 1 barn. Such an encounter is like a marble hitting a baseball. The projectile is smaller than the target. But if the projectile is a neutron that has been slowed to thermal energy (perhaps by many collisions in a moderator), the collision is more like a giant beach ball hitting a baseball. The neutron has been "inflated" to an effective size greater than the size of the nucleus.

This inflation has to do with the wave nature of the neutron. The lower the energy, the greater the wavelength of the neutron. At thermal energy, its wavelength can be much greater than the dimension of a nucleus. The cross section for interaction is then dictated by the effective size of this neutron wave, not by the much smaller actual size of the nucleus. When I first studied possible self-poisoning of the reactor by fission fragments, I knew that some cross sections for neutron absorption were as large as 100,000 barns or more. I did not know that the reactor would produce, in significant quantity, an isotope that absorbed slow neutrons with a cross section of 3 million barns!

During the afternoon and evening of that Wednesday, September 27, hot helium was circulated through the graphite moderator to flush out any water that might have leaked into it. None was found. While this was going on, the reactor was "spontaneously" recovering from whatever had ailed it. Its neutron multiplication kept increasing. At 1:00 A.M. on Thursday, September 28, the control rods were partially removed and the reactor was put back into operation at the low power of 200,000 watts. At this power level, reactivity kept increasing, meaning that the control rods had to be moved steadily inward to maintain the fixed power. Once more, on Thursday afternoon, the control rods were pulled out until the power rose to 9 million watts. And once more, the reactivity started to drop. It was clear, again, that it would not be possible to maintain this power.

By this time, I was convinced that the problem lay with an isotope being created in the fission process, one having a large neutron cross section. To account for the spontaneous regeneration of the reactor when it was shut down, this isotope would have to be radioactive (as indeed most fission fragments are), such that it vanished gradually from the system when no more of it was being made. I looked at the curves of reactivity that were by then available from the nearly two days of operation. At once I reached two conclusions. First, the way in which the reactor regenerated itself when shut down told me that the effective half-life for the poison was about 11 hours. Second, the way in which the reactivity started to fall after high-power operation was achieved told me that the poison was itself the product of another, shorter-lived radioactive parent. Thus there was a chain of at least two radioactive decays. I then walked out into the hall to look at a chart posted on the wall opposite my office door, a chart displaying the measured half-lives of many isotopes expected to be produced in fission. I looked for one with a half-life somewhat less than, but not much less than, 11 hours. (Because of the radioactive chain, the time needed for the reactor, once shut down, to regain its reactivity is longer than the actual half-life of the poison.) Immediately, xenon 135, with a half-life of 9.2 hours, jumped out at me as the likely poison. It is the "daughter" (radioactive product) of iodine 135. There was no other candidate that readily accounted for the reactor's behavior.

On Friday, September 29, Greenewalt—a meticulous diarist—wrote in his diary that I had estimated the cross section of xenon 135 for neutron absorption to be 3 million barns. This turned out to be correct, but when I summarized my conclusions the next week in a report dated October 4, I gave 7 million barns for the cross section, based on an assumption that 2.6 percent of all fission events produce xenon 135. As it turned out, I was right the first time when I had assumed that xenon 135 is produced in 6 percent of all fission events. With the larger amount of material, a cross section of "only" 3 million barns is needed to explain the poisoning effect. This enormous cross section is larger than we imagined we would have to cope with when the reactor was designed. (Despite this huge cross section, xenon 135 went unnoticed at Clinton, where the reactor power was much lower and the poisoning effect much less.)

Curing the problem, once it was understood, was relatively simple. Back in Chicago, Greenewalt's chief deputy, George Graves, had urged that the design allow for the addition of 500 more uranium fuel rods in case they proved to be needed. Graves was guided in part by my advice, but perhaps even more by his own industrial experience. He had pioneered nylon production and ammonia synthesis, two of Du Pont's largest projects prior to the plutonium reactors, and knew that a surprise or two was likely. After conversations with Graves and me, Greenewalt backed us up.

The fuel was added in increments over the following weeks, and, at the same time, the control system was improved. Better control was needed because the reactor had two different modes of operation, with quite different reactivities. When it was running at high power, overcoming the xenon poison, it needed all the extra fuel. When it was starting up, free of the xenon poison, it had "too much" fuel and so needed very capable control rods to avoid getting out of hand. It was like a racehorse needing restraint as it eagerly waited to leap from the starting gate, but then needing goading to keep going after it had rounded the first curve and was tiring.

In some published accounts, the flexibility of the reactor design, allowing

Hanford, c. 1945. The multistory building on the right houses the B reactor (the one that caused moments of anxiety when it was started up in September 1944). On the left is a plant built to treat the cooling water from the Columbia River should it prove necessary (it didn't) and to hold extra cooling water for possible emergency use (not needed). The river is in the background.
(Courtesy of U.S. Department of Energy.)

for added reactivity, is attributed to blind luck—a tendency of Du Pont to overdesign as a matter of principle or tradition. Not so. We were aware from the beginning that our calculated allowance for self-poisoning could prove to be insufficient and that other calculations, as well, carried margins of uncertainty that made it prudent—indeed, absolutely necessary—to plan for possible surprises and be ready to cope with them.

A reactor is simple (at least conceptually) until it starts operating. Its fuel consists of slugs of uranium metal or a simple uranium compound, encased in cans within the graphite moderator. Water-filled channels penetrate the reactor core for cooling. Control rods, made of cadmium or other material that absorbs neutrons, are positioned so that they can be pushed into the reactor to slow or stop the reaction rate and can be pulled part way out of the reactor to increase the reaction rate and the power level. What changes dramatically once the reactor starts operating is the composition of the fuel slugs. Part of what was uranium becomes a complex mixture of many isotopes of many elements.

The largest constituent, by far, in a reactor fuel rod is the isotope U-238. It makes up more than 99 percent of the fuel if natural uranium is used. When a U-238 nucleus absorbs a neutron, it becomes a nucleus of U-239 (92 protons and 147 neutrons). As discussed in Chapter 1, the U-239 nucleus does not undergo fission (unless the neutron that strikes it is of high energy). It is, however, unstable (radioactive). With a half-life of 24 minutes, it undergoes what is called beta decay: it emits an electron and an antineutrino,[6] changing one of its neutrons to a proton. The result is a nucleus of neptunium 239 (93 protons and 146 neutrons). This isotope is, in turn, unstable, with a half-life of 2.4 days. It also undergoes beta decay, creating a nucleus of Pu-239 (94 protons and 145 neutrons). This is the prized isotope whose production is the whole reason for the reactor's operation. Pu-239 is itself unstable (radioactive), but its half-life of more than 24,000 years ensures that most of it will remain available for extraction and use. So the U-238 content of the reactor gets converted gradually into a mixture of uranium, neptunium, and plutonium (plus smaller quantities of other elements that haven't been mentioned here).

What about the rarer isotope U-235? Even though it makes up only 1 part in 140 of the fuel,[7] it is responsible for the chain reaction and the energy generation. The nucleus U-236 (92 protons and 144 neutrons), formed when a nucleus of U-235 absorbs a neutron, can do many things. It can lose energy by emitting a gamma ray. It can emit an alpha particle (which consists of two protons and two neutrons), leaving behind a nucleus of thorium 232 (90 protons and 142 neutrons). It can undergo beta decay, transforming itself into a nucleus of neptunium (Np-236). Or—and this is most likely—it can undergo fission, breaking into two large fragments and, at the same time, emitting one or more neutrons. Among the many possibilities is this one:

$$ {}^{236}_{92}U \rightarrow {}^{98}_{40}Zr + {}^{135}_{52}Te + 3n $$

The subscripts to the left of the chemical symbols give the number of protons in the nucleus. The superscripts give the total number of protons and neutrons in the nucleus. Zr is the element zirconium, Te is the element tellurium, and n stands for neutrons. As discussed in Chapter 1, it is these emitted neutrons that make possible the chain reaction.

The reaction shown above is only one of many possible modes of fission

---

[6] The antineutrino, although it has little if any mass and no charge, carries away some energy and has other measurable properties. Antineutrinos are emitted profusely from reactors. The "neutrinos" mentioned in footnote 2 on page 45 are actually antineutrinos.

[7] In some later reactors, the fuel is enriched to contain more U-235 than is present naturally in uranium.

(for most of which, like this one, the fragments are unequal). Yet even this one alone spawns numerous other isotopes. The zirconium 98 nucleus undergoes beta decay to niobium 98 (41 protons and 57 neutrons), which, in turn, decays to molybdenum 98 (40 protons and 58 neutrons). These processes have half-lives of 31 seconds and 51 minutes, respectively. The final molybdenum isotope is stable. The other fission product in the example above, tellurium 135, triggers a succession of four beta decays—to iodine 135, then to xenon 135, then to cesium 135, and finally to the stable isotope barium 135 (56 protons and 79 neutrons). From just *one* of a myriad of possible modes of fission of U-236 have come isotopes of eight different elements. Clearly fission stirs the elemental pot. Within hours or days after start-up, much of the pure uranium in the reactor's fuel rods has been transmuted into a mixture of more than 200 isotopes of 38 different elements.

Note the appearance of Xe-135 as one of the products of the fission event discussed above. Its tellurium grandparent has a half-life of 19 seconds. Its iodine parent has a half-life of 6.6 hours. Xe-135 itself, as mentioned above, has a half-life of 9.2 hours, which is what gave it away as the culprit in the self-poisoning of the reactor. Its offspring, Cs-135, has a 2-million-year half-life. It seems to be pure chance that Xe-135 has such a large cross section, 3 million barns, for absorbing neutrons. (Of course, there are reasons, but theory is not refined enough to permit calculating the cross section.) Because of its 9.2-hour half-life, Xe-135 (fortunately) does not keep building up within the reactor. After a day or so, an equilibrium sets in, with as much Xe-135 decaying as is being created. So its poisoning effect is self-limiting, and the extra reactivity provided by the additional fuel rods solves the problem indefinitely.

It was in late October 1944, a month after the first Hanford reactor went critical, that I got the news that my brother Joe was missing in action. The news hit me hard. I naturally feared the worst. Like many another family before us, we went through the torture of hope that he had been captured, not killed. I tried to comfort his wife, Sara, and my parents with this hope. All efforts to find out more were fruitless, and eventually we had to assume the worst. Here we were, so close to creating a nuclear weapon to end the war. I couldn't stop thinking then, and haven't stopped thinking since, of the fact that it could have happened sooner, that the war could have been over in October 1944. I took some comfort in the knowledge that Joe had seen the magnificent beauty of Rome and Florence before he got into the lethal fighting in the rain and mud of the Apennines, and I redoubled my efforts on the Manhattan Project.

As of that date, I, like all of the other Allied scientists working on the atomic bomb, still assumed that we were in a life-and-death race with the Germans. Only a few weeks later, in November 1944, thanks to the scientific sleuthing of the Dutch-American physicist Sam Goudsmit and his team, American officials learned that the German bomb program was stalled—indeed, that the Germans had scarcely any program. This information was not shared with me and most other Manhattan Project scientists until many months later—in part, apparently, because our military and political leaders didn't fully trust Goudsmit's information, in part because these same leaders did not want to erode our total commitment to the bomb. Philip Morrison, now a professor emeritus at MIT, tells me that he learned of Goudsmit's conclusions soon after they were forwarded because in addition to being a scientist at Los Alamos, he (Morrison) held a second job as liaison to military officers reporting to General Groves, head of the Manhattan Project. Robert Oppenheimer, the lab director, was informed by Groves and talked it over with Morrison, but the information went no further at that time.

The Japanese, likewise, had made little progress—but, as with the Germans, not for lack of scientific talent. The problem in both countries was lack of will. To the Allies, the bomb project had the highest priority. In Germany and Japan, it had no priority. Being in the grip of awe for German science, we were alarmed mostly by the prospect of a German bomb. We should have saved some of that alarm for the prospect of a Japanese bomb.

We experienced one Japanese "attack" at Hanford. Between late 1944 and July 1945, the Japanese launched thousands of hydrogen-filled paper balloons from their homeland with the intent that the balloons, after drifting across the Pacific on westerly winds, would drop their modest payloads (two small incendiary bombs and a 15-kilogram high-explosive bomb suspended from each balloon) to start fires and damage people and property in the western part of North America. The plan actually worked. The balloons started a number of forest fires in Oregon, Washington, and British Columbia. In eastern Oregon, a bomb that had landed before it exploded killed a woman and five children who were on an outing in the woods (the only mainland casualties of the war). During most of the time that the balloons were arriving, newspapers withheld information about them at our government's request. This avoided panic in America and discouraged the Japanese, who saw no evidence that the balloons were working. Some of us couldn't help learning about the fire balloons, however, and it was enough to make us nervous. We knew that an air raid was exceedingly unlikely, yet there was something about seeing an actual balloon borne on winds from Japan that made us apprehensive. Could the Japanese, after all, find a way to bomb our plant?

On March 10, 1945, one of the fire balloons, by pure chance, did shut down the Hanford plant. Ropes dangling from the balloon became entangled in the electrical line feeding power to the water pumps of a reactor. The reactor had to be shut down at once. Power was restored a few hours later. This was the only American plant shut down by enemy action during the war. It was, ironically, the plant producing the plutonium that would devastate Nagasaki.

Not even during the most hectic moments of war work—or later, during work on the hydrogen bomb—did I stop thinking about problems in basic physics. Fermi's presence in Richland was a stimulus. We talked about the future of cosmic-ray and elementary-particle physics. It was around this time that he developed a theory to explain why protons in the cosmic rays strike the Earth with so much energy. He imagined the protons milling around in clouds of ions (electrically charge atoms) in the galaxy, randomly gaining and losing energy, sometimes, by chance, gaining enormous energy. I felt that the future of elementary-particle physics, at least in the near term, lay with cosmic rays, because some of the incoming particles possessed energies far greater than any that could be produced in any accelerator that then existed or could be imagined to exist in the foreseeable future. The cosmic-ray particles were not numerous, but they were energetic and they were cheap. I was interested in a course of postwar particle research that would get us going without a big investment in time or money.

Dick Feynman, my colleague-at-a-distance, was another stimulus to my thinking about pure physics during the war. Still in his mid-twenties, he was a key contributor to atomic-bomb theory at Los Alamos and was already known for his exuberant personality and his practical jokes. In brief periods during my occasional visits to Los Alamos, we found time to discuss our work on electrodynamics, and we were able to complete a long paper on our action-at-a-distance formulation, which was published in *Reviews of Modern Physics* in 1945. Almost all of the other papers I wrote during the war were internal classified reports, which have never seen the light of day.

I think of my lifetime in physics as divided into three periods. In the first period, extending from the beginning of my career until the early 1950s, I was in the grip of the idea that Everything Is Particles. I was looking for ways to build all basic entities—neutrons, protons, mesons, and so on—out of the lightest, most fundamental particles, electrons and photons. In response to a contest of the New York Academy of Sciences while I was at Hanford, I submitted a paper on this subject. It won a prize and appeared later, in 1946, in *Proceedings of the New York Academy* under the title "Polyelectrons." This same vision of a world of simple particles dominated my work with Feynman. We were able to formulate electrodynamics in terms of particles acting at a distance on one another without the need for intermediate electric or magnetic fields. I will convey the essence of that work and have a bit more to say about Feynman in later chapters.

I call my second period Everything Is Fields. From the time I fell in love with general relativity and gravitation in 1952 until late in my career, I pursued the vision of a world made of fields, one in which the apparent particles are really manifestations of electric and magnetic fields, gravitational fields, and spacetime itself. This theme will be prominent in Chapters 10–14.

Now I am in the grip of a new vision, that Everything Is Information. The more I have pondered the mystery of the quantum and our strange ability to comprehend this world in which we live, the more I see possible fundamental roles for logic and information as the bedrock of physical theory. I am eighty-six as of this writing, but I continue to search. I will touch on this search in the final two chapters.

3
·
GROWING UP

IN THE summer of 1910, Joseph Lewis Wheeler, a librarian at the Washington, D.C., Public Library, made a habit of leaving work at the end of the day through a certain door. His fellow librarian Mabel Archibald left through a different door. Having prudently avoided any occasion for "talk" by their coworkers, the young lovers met in Rock Creek Park to stroll, hold hands, read poetry, and talk of marriage.

When they were married that fall, Joseph was twenty-six and Mabel was twenty-seven. I was born promptly, nine months later, the first of their four children.

My mother was tall—as tall as my father—and pretty, with naturally curly blond hair and blue eyes. She became plump as she grew older but never lost her beauty. She had a sunny disposition, always making those around her smile. Yet she was shy, never enjoying the give and take of social gatherings. As was common in those days, she gave up her professional career when she married, becoming a homemaker. She really preferred to stay at home, caring for the children and doing household chores. My father, dominant in the home and in the marriage, encouraged this trait in my mother. He did not suggest that she get a driver's license, and she never did.

Although Mabel (or Archie, as she was known to her husband and all her friends) was an intelligent woman and a voracious reader who helped my father with his book-buying decisions, she remained sensitive all her life about having no college degree. She made sure that her children were encouraged

My parents and I in 1911.

in their academic pursuits. As the firstborn son, with an inclination toward mathematics and science, I got a disproportionate share of my mother's attention. My brothers and sister felt this imbalance. I didn't feel smothered, but I was aware of the expectations that she held for me. My mathematical bent may have come from my mother, who could read a column of figures upside down in the grocery store and get the sum in her head before the grocer had arrived at the total using his pencil.

My father was about 5'8", slender, blond, and blue-eyed. He was a no-nonsense achiever and a man of principle, committed to serving the public. But, above all, he was a salesman. He devoted his life to selling libraries, and he was dramatically successful at it. He was good with the press, with affluent donors, and with legislators. When he headed the public library in Youngstown, Ohio, he managed to get a branch library built on a traffic island in the middle of the city, without the help of city funds. Later, as head of the Enoch Pratt Free Library in Baltimore, he arranged with the Archer Laundry to print advertisements for the library on the cardboard inserts in shirts, and he begged and borrowed storefront windows in downtown Baltimore for book displays. When the Pratt Library moved into the once-commercial Rouse Building, it became the first library in the nation to have large display windows of its own.

In 1918, my father wrote (he says "prepared") a book, *Your Job Back Home*, in one week for the American Library Association, and wrote more than half

My parents, Joseph and Mabel (Archie) Wheeler, c. 1935.

a dozen other books later, beginning with *The Library and the Community* in 1924, and ending with *Reconsideration of Strategic Location for Public Library Buildings* in 1967. Although his books brought him some favorable notice, they apparently didn't bring in much royalty income. My parents were always frugal, and we children were given to know that there was no money to spare.

My father liked aphorisms. One of his favorites was, "There isn't anything that can't be done better." Another that he often quoted was Theodore Roosevelt's "Do what you can, with what you have, where you are." These sayings have stuck with me and have served as touchstones for me as they did for my father. I never heard him quote the German saying, "Die Probleme existieren um überwinden zu werden" [Problems exist to be overcome], but I believe he subscribed to that too. He worked all the time, and he liked to be in charge—both in the library and at home. At the same time he was a dreamer. He was in love with his Mountain Brook Farm in west-central Vermont, not far from New York State, which he acquired when I was a boy, and he imagined that he would someday start a maple syrup business there. When he and my mother did finally retire to Vermont, it was not to the farm, which by then had been sold, but to the nearby village of Benson. In retirement, my father was so occupied with writing and speaking and consulting that he had no time for maple syrup. His papers are now housed in the Joseph Wheeler Collection at the Library School of Florida State University, Tallahassee.

My parents had a happy marriage, and I grew up in a loving home. My mother died in 1960 at the age of seventy-eight, soon after she and my father celebrated their fiftieth wedding anniversary. Losing Archie was a blow from which Joseph never fully recovered, although he never lost his zest for work. He lived ten more years, dying in 1970 at the age of eighty-six. As he was about to leave Benson for what he sensed was a final trip to the hospital in Rutland, Vermont, he called me in California and said, "I'm pushing off now. You won't be able to reach me by phone." Nor could I reach him in person before he was gone. But my children Letitia and Jamie responded to my phone calls to them and hurried to Vermont in time to say affectionate good-byes to their grandfather.

Both of my parents had ancestors in America going back many generations. It is not easy to say what influence this had on my development, my career, or my nature. It does make me a little unusual among physicists of the 1930s. Many of my American physicist colleagues at that time were newly arrived from Europe: Eugene Wigner, Edward Teller, Albert Einstein, Hans Bethe, Enrico Fermi, Leo Szilard, and George Gamow, to name some of them. Gregory Breit, my first postdoctoral mentor, arrived in the United States from Russia as a child. Robert Oppenheimer was the son of immigrant parents. Yet I was not unique in having deep roots in American soil. Ernest Lawrence, the inventor of the cyclotron, Arthur Compton, who headed the Metallurgical Laboratory in World War II, and Edward Condon, an early contributor to quantum mechanics and later director of the National Bureau of Standards, shared this kind of American heritage with me.

My father was of Puritan stock and grew up in a strict home. His father (my grandfather Wheeler) left a steady job as assistant to the vice president of the Boston & Maine Railroad to become a Swedenborgian minister, and spent the latter part of his life ministering to communities in southeastern New England, from Providence, Rhode Island, to Bridgewater, Massachusetts. His parents, in turn (my great-grandparents Ezekiel and Mehitable Wheeler), although not in the ministry, were much concerned with religion, and spent evenings in their New Hampshire home arguing theological points, each trying to save the other's soul. Further back, the Wheeler clan in this country had arrived with other Pilgrims from England via Holland, not long after the *Mayflower* landed. By 1640, a year after the town of Concord, Massachusetts, was founded, thirty-five Wheeler families lived there. Wheeler was the most common family name in the town.

Adventure is to be found on my mother's side. Her Scottish ancestors, after emigrating to Ireland in search of a better life, found themselves besieged in Londonderry in 1689 as the deposed King James II, with his Catholic followers and French troops, was trying to gain control of this Protestant town. Surviving that siege, they dreamed of a better life in the New World. With other survivors, they migrated to America and founded Londonderry, New Hampshire. After the English ousted the French from Acadia in 1755, these Archibald forebears saw another new opportunity. My great-great-great-great-grandfather Samuel Archibald, his three brothers, and their wives and children—forty-two in all—sailed in 1762 to settle in the new township of Truro, Nova Scotia. But the Archibald migration was not ended. One of Samuel's grandsons, my great-grandfather John Christie Archibald, with his wife, Jane O'Brien, succumbed to the lure of the West. Many of their descendants made their lives in Kansas, Colorado, New Mexico, and Texas. Even I eventually became a Texan.

Tension over slavery was building in the United States in the 1850s. States in the East and South had made their decisions. Some had slaves, some did not. But for some of the territories in the Midwest and West, the issue was still open. Votes would decide the issue. The idea of squatter sovereignty, later called "popular sovereignty," was written into law in the Kansas-Nebraska Act of 1854, which established the territories of Kansas and Nebraska and authorized general votes of the populations of those territories to determine which of them would be free and which would allow slaves. Some wealthy people in Massachusetts, committed to abolition, decided to subsidize emigration to Kansas by like-minded people, so that the vote there could go against slavery. John Christie and Jane Archibald, who, by this time, were living near Worcester, Massachusetts, where he practiced his trade as carpenter and house builder, heard of the opportunity and decided to act on it. They joined the Free State Party and headed for the Wakarusa River near Lawrence, Kansas, where they received a homestead allocation of 160 acres. Several of John Christie's siblings followed.

My grandfather Frederick William Archibald, whom I learned to love sixty years later, was a nine-year-old boy when he arrived in Kansas. As the story has been passed down, he narrowly escaped execution some years later at the hands of a proslavery raiding party, the Quantrill gang, from nearby Missouri. On August 21, 1863, Quantrill's men burned 200 houses and killed 150 men and boys in and around Lawrence. When one of them was about to dispatch seventeen-year-old Fred—who, by good fortune, was the only male at home at the time—his mother (my great-grandmother Jane) threw herself between the boy and the gunman, crying "Don't shoot him. He's just a boy." He was spared. Later, in the Civil War, when he was old enough to fight, he was lucky again, with an equally close call. In the "Battle of the Little Blue" in Missouri, a cannonball bounced off the ground, hit him in the head, and knocked him unconscious for three days. A local housewife came across him, found that he had a pulse, and took him in, nursing him back to health. He had a dent in his skull for the rest of his life. When he died in 1926, he was eighty years old.

Following the Civil War, Frederick enrolled in the newly opened University of Kansas, and there met another freshman, Sarah (Sallie) Reid from Ohio. Her family had been part of the Underground Railroad, helping escaped slaves reach Canada. (Still in our family is a silver spoon presented to Sallie Reid's parents by an escaping slave.) Romance blossomed. Five years later, in 1871, when Frederick was employed as a clerk in the Post Office Department in Washington, D.C., they were married, and their first two children, Robert and Jennie—my mother's older brother and sister—were born there. Some time in the late 1870s, Frederick was enticed by his older brother Albert, a lawyer and small-time rancher in Trinidad, Colorado, to move west. Possessed of the Archibald pioneer spirit, he made the move—at first alone, then followed by Sallie, the two young children, and his widowed mother, Jane O'Brien Archibald. As a teenager—a year before the 1863 incident with the Quantrill gang—Frederick had been able to see some of the Southwest and may have formed a vision of living there someday. He accompanied his uncle Putnam O'Brien, publisher of the unionist, antislavery *Santa Fe Republican*, on a trip to Santa Fe to deliver a printing press. On the way back, he passed through southeastern Colorado, probably visiting Albert, who had already settled in Trinidad.

In Trinidad, Sallie bore three more children—my mother in 1882, followed by Ralph and John Christie at two-year intervals. My grandfather Frederick and his brother Albert had a falling out, Albert went broke, and the family tried to make a go of it on a small ranch outside of Trinidad. They held out until my mother was sixteen, and then returned to Washington, where my grandfather went back into government service until his retirement.

Despite the family's financial troubles, all four of my mother's siblings went to college. Her older brother, Robert, studied law at the University of Colorado, and her sister, Jennie, received a degree in Botany there. Her younger brothers, Ralph and John Christie, both earned Engineer of Mines degrees from Lehigh University and pursued careers in mining. Somehow my mother got skipped. Perhaps, by the time she was ready to go, she declined in order to make it more likely that her younger brothers could attend college. As it turned out, John Christie had to work two years in coal mines after high school to earn enough for college. My mother's only advanced schooling was at the Library School in Albany, New York, where she met my father. In Colorado, her home had been full of books, including a set of the *Encyclopaedia Britannica* donated by her Uncle Albert.

I always wondered if my mother missed Colorado. She never spoke to me of such a yearning. She seemed to love New England as much as my father did. Yet the wide-open spaces and Hispanic culture of her girlhood must somehow have touched her soul. She liked to use Spanish words of affection for us children, and to make the sweet buns called *buñuelos* for us.

I still feel closely linked to my parents and their parents and grandparents. My father, in his later years, gathered and summarized much of the history of his and Mabel's forebears in neatly typed, bound books. I continue to look out for more information with which to embellish his record. I have always loved history, and I find the histories of the Wheelers and the Archibalds as fascinating as the histories of great leaders and great conflicts elsewhere in the world. I take pride in my American heritage.

I am, as far as I know, the lone scientist in the Wheeler-Archibald clan, but there are librarians galore.

I was born in Jacksonville, Florida, on July 9, 1911.

Why Jacksonville? The director of the Washington Library, George Bowerman, saw my father's promise and recommended him for the post of head librarian of Jacksonville's Public Library. The young couple moved to Jacksonville in April 1911, six months after their marriage, and that is where I was born. But my father was not happy there. If there was something about the job that he didn't like, he never spoke of it later. As reasons for dissatisfaction with Florida, he cited only the climate and the fact that my mother had an operation for exophthalmic goiter. Although she recovered fully, she and my father must have wondered whether her malady had something to do with the place. My father looked for another job, and in September 1912, less than eighteen months after moving to Jacksonville, he left for a better-paying job in California, as assistant director of the Los Angeles Public Library.

My father went ahead, to learn about the area and find a place to live. He found a bungalow on Casa Verdugo, near the foot of Verdugo Mountain in North Glendale. It was near an electric trolley line, which made his commuting easy. My mother and I followed by train. My earliest, dim childhood memories are of the bungalow porch in Glendale, with its white painted railing.

While my mother and I were on our way to join him, my father contracted scarlet fever. He recovered and lived a long, active life. Yet the disease left him with a weakened heart and fragile health. For me this was not all bad. The times that my father spent resting or recuperating at our farm in Vermont proved to be high points of my young life.

My brother Joe was born on August 10, 1914, not long after my third birthday. Early the following year, with Joe still an infant and me not yet four, my father lost his job—apparently because he had more ideas than tact. As he showed later when he had his own library to direct, he was a revolutionary,

My early childhood home, Glendale, California, c. 1914.

always thinking up ways to make libraries serve their users better. In Los Angeles, his ideas didn't win the applause of his library director, who came up to my father one day and said, "Your position is being terminated." Fortunately, my father found other work quickly. In April 1915 the American Library Association hired him to prepare and manage its exhibit on library services and methods at the San Francisco World's Fair, scheduled for that summer. It was temporary, but it was work. He pitched in and so pleased the A.L.A. leaders that he was given the same assignment for the Philadelphia World's Fair eleven years later, in 1926.

My father, always upbeat, liked to talk later about the excitement of the World's Fair, its wonderful exhibits, and the thrill he got from meeting former President Theodore Roosevelt. He told, too, about the storm at sea that he experienced on the voyage from Los Angeles to San Francisco and about having his billfold lifted by a pickpocket in San Francisco. He was alone for those months in San Francisco. My mother, with me and the baby, stayed briefly in Glendale to clear out the house and store our belongings. Then, in May, she took us back to live with her parents in Washington, D.C., where we stayed until the family was reunited in September in Youngstown, Ohio. I still have a dim memory of the train trip across the country with my mother and baby Joe. All my life I've had a love affair with trains. It must have started early.

While still in San Francisco, my father secured a position as head librarian of the Youngstown Public Library. I never learned how he got the job, but it proved to be just right for him. He was a person who needed to be in charge, with scope to effect change. He loved the Youngstown job and the community appreciated him. For eleven years, from 1915 to 1926, Youngstown was our base of operations, and most of my childhood memories are centered there. Another brother, Robert, joined the family in April 1917, when I was five, and my sister, Mary, arrived about a year and a half later, in November 1918, when I was seven.

I call Youngstown a "base of operations" because there were two important family excursions away from Youngstown before we moved, finally, to Baltimore, Maryland, in 1926 (when I was fifteen). The first of these was in 1917–1918, when my father was given a ten-month leave of absence to serve the U.S. war effort. Working for the Library War Service, he was put in charge of cantonment libraries in the United States and given responsibility for book selection for all overseas Armed Service libraries. My mother, with six-month-old Rob, three-year-old Joe, and me, six, moved in again with her parents at 1105 Park Road in Washington. It was not a large house, but somehow we all fitted. My father called that his home too, although he was away more than he was with us.

Youngstown had no kindergarten, so my first schooling was as a first-grader in Washington. I remember a teacher who insisted that I write with my right hand, even though I may have had more inclination toward left-handedness. When I was still a toddler, my grandmother Sallie Archibald, on a visit to California, had taught me how to hammer, and I did it with my left hand. (I learned facing her, and was trying to be her mirror image.) Today, I do woodworking with my left hand and write with my right—although,when things get busy during a lecture, I am likely to write on the blackboard with whichever hand is "handier." I love to illustrate ideas with sketches and diagrams. I usually draw with my right hand, but I am just about as good with the left.

I also remember from the first grade in wartime Washington the requirement to recite the Pledge of Allegiance. My parents didn't like the idea; they thought it came too close to religion in the school. They were liberal Protestants, with no strong church affiliation at that time. It was I who later led them to Unitarianism, when I chose to attend the Unitarian Sunday School in Youngstown. But as a first-grader, I was not one to carry my parents' convictions into the classroom. I recited the Pledge.

The home of my grandparents Fred and Sallie Archibald was warm and loving. With four and sometimes five of us Wheelers in the house, it must have been crowded, but being a six-year-old, I was unaware of it. My grandfather, by then retired, gave me his special attention. He introduced me to mathematics—even to a little algebra, if I remember correctly. We took long walks through Rock Creek Park and in the city, sometimes to visit his comrades from the Civil War. I still remember the occasion—if not the content—of a wartime speech by President Woodrow Wilson on the Mall to which my grandfather took me. Another time, with me in tow, he marched into a naval gun factory in Washington that was off limits to the public. He got past a guard by telling him a story of his Civil War service. Once inside, I got to see how an outer cylindrical casing of a large gun barrel was heated before it was fitted over an inner casing in order that the outer one might shrink to make a tight fit as it cooled.

Sunday dinners at my grandparents' home were special occasions, spiced by political argument. My great-uncle John W. Reid, my grandmother's brother, was a frequent guest at these dinners. He and my grandfather loved to debate issues of the war then in progress. Uncle John was highly critical of the Japanese, even though they were America's allies at the time. He was convinced that the Japanese were only capable of copying, not innovating. He liked to tell the apocryphal story of an American car with a bent tailpipe copied by the Japanese, resulting in their new cars all being manufactured with bent tail pipes. My grandfather was an accomplished debater. After convincing everyone of his position over a Sunday dinner, he reversed himself and argued the other side. I was old enough to appreciate the give and take. I came to love both of my Archibald grandparents, especially my grandfather, who spent so much time with me. Unfortunately, I never got to know my Wheeler grandparents, George and Mary Jane, as well.

In June 1918, with my mother again pregnant, we returned to Youngstown, where we would spend most of the next eight years. We lived at 157 Dennick Avenue in a hilly northern section of Youngstown. It was a modest home in a pleasant neighborhood, with a yard big enough for chickens—which we brought back from Vermont—and for children's building projects. As the oldest child, I had my own bedroom.

As we grew up, we four children divided into two pairs more often than not. Joe and I, the older two, played together, and I felt closest to Joe. Rob and Mary, the younger two, spent a lot of time together. All of us were expected, when we were old enough, to gather eggs, mow the lawn, and help with dishwashing and other household chores. We often paired up for these jobs as we did for our play. Later, in high school and college, it turned out that Joe and I were the more studious, while Rob and Mary, although they were good students too, led more active social lives, sometimes to the detriment of their studies.

Our childhood hobbies proved to be signposts pointing to our futures. My hobby was machinery, and I became a scientist. Joe loved books, and he became a historian. Rob collected rocks, and he became a geologist. Mary loved to arrange and catalog dolls and flowers, and she became a librarian.

By the time we were of college age, the family had moved to Baltimore, and all four of us lived at home while going to college. Joe completed his under-

Surrounded by my siblings, c. 1929.
Standing: Joe, a high school student. *Seated, either side of me:*
Mary and Rob, grade school students. I am a college student.

graduate work at Johns Hopkins (where I had studied); he earned a Ph.D. in History from Brown University and, in the same year (1939), a Master of Library Science degree from Columbia University. At the time he was called up for Army service in 1943, he was working at the New York Public Library. When he was killed in 1944, he left behind a widow, Sara, a Columbia classmate from Nebraska whom he had married in 1940, and an eighteen-month-old daughter, Mary Jo. (Sara remarried and lives in Santa Barbara, California. With a little matchmaking help from me, Mary Jo met and married my former undergraduate student Jim Hartle, and now lives in Santa Barbara, too, where Jim is a professor of physics.)

Rob, the most flamboyant among us, followed Joe to Johns Hopkins, and

My brother Joe with his bride-to-be,
Sara Hutchings, c. 1940.

went on to earn a Ph.D. in geology from Harvard. He spent most of his career in Texas, first in the petroleum industry, then teaching in several colleges. He married a fellow geologist, Marjorie Woodberry, and had two daughters, Robin and Bethel. Rob was a heavy smoker and, for some part of his life, was too fond of alcohol. Although we became fellow Texans when I went to Austin in 1976, our two-hundred-mile separation—Rob then taught at Lamar University in Beaumont—still kept us from seeing much of each other. In 1977, at sixty, Rob was diagnosed with a carotid artery obstruction. Janette and I drove to Beaumont and had a good visit with him three days before he went into the hospital for an operation, which he did not survive.

Mary completed her undergraduate work at Goucher College and afterward studied library science at the University of Denver in Colorado. Her first serious romance, with a Benson, Vermont, boy, was squelched by our par-

My brother Rob, c. 1950.

ents. The man she married, Alfred Beavin, was, like Mary, from Maryland, but oddly enough, they met in Midland, Texas, where Mary had gone to visit Rob and Marjorie and where her husband-to-be was teaching bomb-site maintenance at an Army Air Corps base. After their marriage in Baltimore, she was able to secure a library job in Midland, where they stayed just three years. Mary's affection for Vermont won over her husband. In 1946, they moved to Benson, where she bore and raised four children at The Green Door, a combination shop, store, and home where they sold small things that Al made on the premises. Mary lives now in Benson, in the same house to which my parents retired. Al was tall, slender, and good-looking, pleasant socially but a tyrant at home. Mary left him after twenty years of marriage and they were eventually divorced. During her years as what we now call a "single mom," Mary worked as a librarian in Wisconsin and then in Rutland, Vermont. Of her four children—Daniel, Joseph, Barbara, and Lee— three are still in Vermont, along with most of her grandchildren.

My sister Mary, 1969.

The second long excursion from Youngstown was to our farm near Benson. We had made brief visits to the farm before, but in the spring of 1921 my father decided that he needed a sabbatical to build up his health and chose the Benson farm as the place to do it. He and my mother and the four children—aged two to nine—piled into the family Model T and headed for Vermont. We stayed there from April 1921 to October 1922, and there I had my tenth and eleventh birthdays.

The farm had belonged to my great-uncle Oscar Hilon Bump, who was trained as a lawyer and had held a judgeship in Rutland, Vermont, but also worked as a farmer, a woodsmith, and an insurance salesman. Uncle Oscar had married my grandfather Wheeler's sister Emma. As Emma lay dying in 1919, her unmarried sister Hannah (whom we knew always as Annie) came to care for her. Two years later, she and Oscar were married, so he was my great-uncle twice over. Oscar was a tall, slender man with a large mustache, a bald head, and a forthright gaze. Annie, who weighed well under 100 pounds,

Uncle Oscar and Aunt Annie Bump in Benson, Vermont.

moved fast. In the days before skim milk could be purchased in the grocery store, she carefully skimmed her own milk, which she poured over her shredded wheat with a raw egg on top. I was close to them in my boyhood. My father had spent his first year after college living with Oscar and Emma to console them after their only child, Frank, died of pneumonia at thirty-seven.

Soon after we moved to Youngstown, my father took it into his head that he wanted to buy Uncle Oscar's farm, even though he could scarcely afford it. My practical mother was dubious, but she relented. Uncle Oscar set a fair price and allowed my father to pay it off a little at a time over many years, so my father got his prize. Much later, knowing how much both my parents loved Benson, Oscar bequeathed his house in town to them. That is where they lived in retirement and where my sister Mary still lives.

Some of my most pleasant childhood memories center on the farm and Benson. During our long sojourn there in 1921–1922, my father ran the farm himself, with the help of a hired hand and occasionally, in busy periods, two or three other hired helpers. Before and after that time, he rented it out to local people, who took over everything—fields, barn, equipment, livestock, and house. So, on most of our visits, we lived in a tent erected on one corner of the property, cooking over wood fires or cans of Sterno and getting our water from a spring and the nearby brook. The farm was a substantial operation: about 600 acres with a herd of some twenty-five to thirty dairy cows and about twenty pigs. The principal crop was hay for the cows. Milk from the farm made its way to New York City. (We were west of the "New England Divide" that separates milk flowing to Boston and milk flowing to New York.)

Eventually, my parents built a shack to replace the tent, and the shack, in turn, gave way to a larger cabin. In 1937, the farm was sold to Edward Wiskowski, who had been renting it. He couldn't make a go of it and defaulted on payments to my father. The farm ended up in the hands of a wonderful Vermont neighbor, Almon Charlton, whose speech I could listen to all day.

I was a fourth-grader in that spring of 1921, and attended the one-room Bump School, whose name honored my Uncle Oscar because he had provided the land on which it was built. My teacher, Mary Donovan, walked every day from the nearby farm where she lived, and taught about thirty-five children in eight grades. I don't remember being considered especially precocious, and I don't remember getting any special attention from Mary Donovan, but somehow, after a little more than one school year in Vermont, I moved into the eighth grade back in Youngstown, four grades beyond the one I had left. Part of the reason, I think, is that I could listen in on the teacher's instruction of the older children and quietly work along with them. Also, I had time during the day to move at my own pace through the available books, and I did as much mathematics as I could. Since the first grade, when my grandfather Archibald had introduced me to mathematics, I had loved it and found that it came naturally to me. Later, in the tenth or eleventh grade, I taught myself calculus.

Even though I knew I learned more quickly than most of my peers, I do not remember thinking of myself as special, either in Benson or in Youngstown. It was an era when children's character and intellect were supposed to be developed through discipline and hard work, not through rewards and flattery. Yet my parents could hardly conceal their conviction that I had unusual promise intellectually. I remember overhearing them one night after I had gone to bed as they talked about the problem my ability to soak up learning was presenting to them. They were wondering what more they could do to support my education.

Fortunately, several of my teachers at Youngstown's Rayen High School also went out of their way to encourage my progress. They were not the common variety of teacher who treats a fast learner as someone who can safely be ignored or even as someone who is a nuisance. The inspiration provided by Lida F. Baldwin, my mathematics teacher, is easy to remember but hard to define. She gave me extra work, extra reading, extra encouragement. Above all, she was *interested*. One afternoon she took herself down to the library to call on my father—to make sure, I suspect, that my parent's commitment to my education matched her own.

Lida Baldwin was slender, with sharp features, already in her middle years when I was her student. She was, in the terminology of that period, an old maid, dedicated to her students and her garden. She taught both English and mathematics at Rayen (I studied only mathematics with her), and, as I was to learn later, she wrote like an angel. I treasure a slim volume of her essays, published in 1941 after her death. Her loving essay on Abraham Lincoln's writing reveals much about her own greatness as a teacher, as she ties together the character of words and the character of the person uttering those words. Her essay on gardening (in which she applauds the men in the story who, limited to three books on a desert island, chose the Bible, Shakespeare, and a seed catalogue) is a masterpiece of humor and passion. In this essay she admits that although she easily obeys the Biblical Commandments against coveting her neighbor's ox, ass, or manservant, she is powerless not to covet his sandy loam, barnyard manure, and hillside leaf mold.

I did well in most subjects in high school, and won an essay prize sponsored by a local newspaper. But I always had trouble with music. I don't know if anyone is really completely tone deaf, but I think I come close. Perhaps this affliction would be less pronounced if I had not missed out on music in the fifth, sixth, and seventh grades.

During our extended stay in Vermont, my inquisitiveness almost cost me a finger. Back in Youngstown, I had learned how to make gunpowder with my chemistry set and also how to blow caps off bottles with acetylene and water. In Vermont, I was reading everything I could find on explosives. At the time, my father and his neighbors were using dynamite to blast holes for poles that would bring electrical lines, for the first time, to our farm and neighboring farms. I didn't want to set off a whole stick of dynamite, but I knew where some dynamite caps were stored in the pig barn. These little copper cartridges, containing mercury fulminate, were slid over the fuse, crimped with pliers, and then slid into a hole in the dynamite. When the fuse was lit at its far end, the flame ran down the fuse and ignited the dynamite cap, which in turn set off the dynamite. This meant that a flame, such as from a match or candle, could directly ignite a cap.

I couldn't resist the temptation. I took two or three dynamite caps from the pig barn and went across the road to a secluded spot in the vegetable garden. I stuck a match in the ground, lit it, and then dropped caps onto it. I kept missing, so I got lower and lower before I released the caps, in hopes of scoring a bull's-eye. Finally, my point of release was only an inch or two above the match flame. With a mighty bang, the cap exploded before I had even let go of it. For weeks afterwards, I was digging little pieces of copper out of my chest and arms and legs. By great good fortune, none of them landed in my eyes. My parents hurried me into our car, my mother climbed in beside me, and my father drove as fast as he could to a doctor in the village of Fair Haven, 10 miles away. The doctor inspected my mangled forefinger and thumb and decided that I was in no mortal danger. He sat down and ate his supper, then proceeded to snip off bits of excess flesh, soak the hand in a bowl of hot antiseptic solution, and dress the wound.

This experience should have cured me of an interest in explosions, but it didn't. There is still nothing I like better than a noisy Fourth of July. At our summer home on High Island, South Bristol, Maine, we now have a four-foot cannon from the Spanish-American War, with which we can mark important events with a blast that echoes across the water.

With librarians for parents, it is not surprising that I grew up among books. In Youngstown, my father used to bring home books so that the family could help evaluate them for possible purchase by the library. He and my mother would sit around after supper talking over prospective purchases. We children were allowed to participate too, as soon as we were old enough. I can't remember when I learned to read, probably around age five or six—not unusually early. One of my favorite books as a child was Howard Pyle's *The Wonder Clock.* I also remember *The Swiss Family Robinson* and *Robinson Crusoe.* I had an endless appetite for technical books. One of my favorites was Franklin Jones's *Mechanisms and Mechanical Movements*, which my father brought home from the library. It was for adult readers. With its guidance, I built a combination lock, a repeating pistol, and an adding machine—all from wood. The book that introduced me to science is one I still remember fondly: *Introduction to Science*, by the British biologist and writer J. (later Sir John) Arthur Thomson.

I was experimenting even before I started to go to school. At the age of three or four, I learned that if I put a marble inside an empty light socket, it shot out with a pop when I turned the switch on. Somewhat less hazardously, I loved to use the metal strips, girders, brackets, wheels, axles, gears, and pulleys of my Meccano set to build devices. I am told that once, when I was walking with my mother in Youngstown, we passed a site where some workmen were connecting pipes in an excavation. I looked down and said, "They are connecting it wrong. It won't work that way." Sure enough, they recognized their mistake and connected it correctly. (Things like this make one a hero in one's own family.)

America's first commercial radio station, KDKA in Pittsburgh, went on the air in 1920, the year I turned nine. My friend Rushworth Steckel and I made little crystal radios, as so many people were doing then, and we were able to pick up KDKA—Youngstown is less than 60 miles from Pittsburgh. Each week a special section of the local newspaper showed circuit diagrams and gave instructions on how to build radios. I remember the puzzlement on the face of the clerk at the hardware store when I went to buy a rheostat and asked for a "rohesat." Rushworth and I also built a telegraph set and strung wires between our houses, about half a block apart.

In high school, another friend, Burdette Moke, joined me in making things out of wood. Since our products included combination locks and wooden guns that shot wooden bullets, we formed the Wheeler-Moke Safe & Gun Company. (I can't help wondering why I turned into a theoretical physicist instead of an experimenter.) My friendship with Burdette was encouraged by his parents, who thought associating with me would enlarge his vocabulary. He later became a geologist and a college teacher.

I knew by then that I wanted to go to college and study a technical subject. I also knew that my parents had limited means. I confess that I sometimes got out of chores, at home or at the Library on a Saturday, by saying, "Well, I could do that, but it might not be the best way for me to spend my time if I am going to earn a scholarship for college."

I studied French in high school in Youngstown, and took it up again at Johns Hopkins, where I dove into a third-year course in French history taught in French. I was afraid I was going to sink instead of swim, but somehow I survived it. Much later, when I was with my family in France, and Janette and I were conversing in French with local people, our children would stand well apart from us so as not to be associated with our accents. I took up German as a high school senior in Baltimore—where the family had just moved—learning from a wonderful teacher, Konrad Uhlig. I pursued it further in college, for a physicist then had to be able to read German, the language in which most of the best work was being published. German also served me well in my postdoctoral year in Copenhagen. Inspired by Uhlig's good teaching, I won a prize for German in high school.

Perhaps because of the books he had written, my father became known outside of Youngstown, and known as a doer. In 1926, the city of Baltimore induced him to become director of its Enoch Pratt Free Library. His work at the Philadelphia World's Fair for the American Library Association came between the Youngstown and Baltimore jobs. He was horrified by the meager support that the Pratt Library received, but "problems exist to be overcome." He remained in Baltimore until his retirement in 1945, becoming an accomplished lobbyist and gaining a national reputation as both the public's use of the library and the size of its collections grew. In 1944, H. L. Mencken wrote approvingly of him as a "competent and diligent man."

This move meant that I would spend my last high school year in Baltimore.

It also meant that Johns Hopkins became the logical university to attend. I would be able to live at home and minimize expenses.

At fifteen, I enrolled as a senior in a public high school with the imposing name Baltimore City College. That's where I had the good course in German. It's also where I was introduced to debating. There is no better training for public speaking—and for thinking on your feet! At a typical debate, the student would climb up to the stage and be handed a slip of paper. On the paper was written the topic and which side to take, pro or con. I found that if I started by saying, "I shall discuss the reasons to restrict immigration [or whatever the topic] under the social, economic, and political categories," it gave me a moment to think. I was a member of the Bancroft Literary Society, one of two debating societies at Baltimore City College. The other was the Carrollton-Wight Literary Society. Our arch rival students were from the Baltimore Polytechnic Institute, another public high school with an imposing name. The climax of the year was a public debate between these schools.

Anyone who expects to create, be it as scientist or artist, scholar or writer, needs self-confidence, even bravado. How else can one dare to imagine understanding what no one else has understood, discovering what no one else has discovered? Where does this confidence come from? Fortunately, every young person is blessed with some if it. It is part of human character. What of the girl or boy who reads about Newton and Maxwell and Bohr and Einstein and says, "I want to build on what they have built; I want to add to the sum of knowledge about the most basic laws of nature"? That child has gained somewhere, somehow, even more than average confidence in his or her own capabilities. As I look at my own childhood and wonder what made me think I could grapple with nature's greatest mysteries, I have to give credit to a few teachers who saw some potential in me, and most of all to my father, for whom no mountain was insurmountable. He was no scholar, but he knew how to make his visions come true. Perhaps because I grew up in an environment where problem solving and achievement (as well as service) were the respected virtues, where the mind was supposed to *do* something, not just *know* something, I did not distinguish early between science and technology, between mathematics and devices. Although I gravitated in college to science and mathematics, I never lost my fascination with technology and devices.

I like to say, when asked why I pursue science, that it is to satisfy my curiosity, that I am by nature a searcher, trying to understand. Now, in my eighties, I am still searching. Yet I know that the pursuit of science is more than the pursuit of understanding. It is driven by the creative urge, the urge to construct a vision, a map, a picture of the world that gives the world a little more beauty and coherence than it had before. Somewhere in the child that urge is born.

4

I BECOME A PHYSICIST

MY PARENTS believed in owning, not renting. They bought a four-bedroom house at 5726 Uffington Road in the hilly, leafy Mt. Washington section of northern Baltimore, and stayed in it for nineteen years. Mary had her own room, Joe and Rob shared a room, and I had a small room of my own in which I built a desk and bookcase. We children stayed there until we were children no longer.

In both Youngstown and Baltimore, I worked as a paper delivery boy, and held other part-time jobs. I had saved a little money. Even with that small reserve and even with Johns Hopkins located in my new hometown, I needed a scholarship to go to college. My family had four children to support, and my father's salary was not princely. (Later, I learned that he chose to make his take-home pay even less princely. Out of his own pocket, he augmented the salary of one of the specialists he hired at the library, whom he otherwise could not have induced to come. His father before him had dug into his own pocket to support needy parishioners, over the objections of my grandmother. Now my dear wife has to cope with this genetic predisposition by restraining my impulses to treat money too casually.)

The scholarship that I secured was from the state of Maryland. It was controlled by a local political leader, who lived in our part of Baltimore. My father took me to see him. I think this rough-hewn politician sized up my father as different from the people on whom he depended for political support. Nevertheless, he gave his OK, so the next step was an interview for

admission to Johns Hopkins. The interviewer told me that I was not as careful about my appearance as I ought to be. "You don't even have your necktie
tied properly," he said. Well, he was right, and I didn't make that mistake
again. He also asked me what field I was interested in. I said engineering. At
the time, I was as intrigued by mechanical devices as by atomic theory. I had
devoured books about both. Engineering seemed like the only way to translate
my general interest in science and technology into a living. As a fifteen-year-
old bent on making my own way in the world, saying "physics" would have
been like saying "pottery making."

In the fall of 1927, now sixteen, I enrolled as a freshman in engineering. I
enjoyed the courses on strength of materials, electrical circuits, and alternat-
ing- and direct-current machinery. But the engineering library was my undo-
ing as an engineering major, for it was shared with physics. When I went there
to browse through engineering books, I found copies of *Zeitschrift für Physik*
(a leading physics journal of the time), with the latest reports applying the
newly discovered quantum mechanics to the behavior of electrons within
atoms. Moreover, my chemistry teacher, H. M. Smallwood, was gripped by the
excitement of the newest in atomic physics, the Schrödinger equation, which
assigned wave properties to electrons. It was easy to absorb his excitement.

My physics teacher, John C. Hubbard, was conscientious and likable. He
stuck to tried-and-true physics, not sharing with us any insights about the revolution then occurring in physics. But even the classical physics that he
taught was more inspiring to me than what I was getting in my engineering
courses. I was enchanted by the way in which physics united mathematics and
natural phenomena.

One day, in the waiting room of an optometrist's office, I encountered Hubbard's assistant R. Bowling Brown, who led the discussion sections that supplemented the main lectures in the physics course. He told me more about
the new ways in which quantum mechanics was illuminating atomic structure. I think it was in that quarter of an hour that the practicalities of making
a living suddenly seemed secondary. I wanted to be a physicist. With John
Hubbard's encouragement, I changed direction. My parents raised no objection. (I suppose they thought that their son would succeed, no matter what
he did.)

When I entered Johns Hopkins, it still bore the strong imprint of its first
president, Daniel Coit Gilman, who had helped organize it as America's first
research university in 1876. In Gilman's vision, research and graduate education were central to the university. (Contrary to some accounts, he also
made sure there was a strong undergraduate program.) Part of the Gilman
legacy was the option for nonstop flight from freshman to Ph.D. Choosing this
track, I was able to complete my combined undergraduate and doctoral work

in six years. I was sixteen when I started and not quite twenty-two when I finished. So I have no bachelor's or master's degree to hang on the wall.

The transition from undergraduate to graduate work was gradual, not marked by any special point. Almost all of my courses after the second year were in physics and mathematics. An exception was Medieval and Renaissance French History, a course I took in my fifth year. History had gripped my imagination since childhood and became a lifelong passion. Even as I try to grapple with questions at the frontier of science, I am always conscious of how my work is imbedded in history, how what has gone before shapes my outlook and my style. And several times in my life, I have looked to history for guidance as I wrestled with the contrary tugs of public service and the pursuit of science.

A first-year English course taught by Wardlaw Miles, an injured veteran of World War I, made Shakespeare come alive for me. Once, in his excitement over a passage from Shakespeare, he tumbled from his crutches and we students had to rush forward to pick him up off the floor. I continued in German, and, as mentioned before, I dived into a sink-or-swim French Literature course taught in French. I even found time for a little social life. But more of that later.

Scholarships provided support in every year but one while I was at Johns Hopkins. Starting in my third year, I served as an assistant, helping to guide beginning students in their laboratory work. In addition, I did some tutoring on the side, arranged privately, and worked many weekends at the Enoch Pratt Free Library. Thanks to my gifted, inspiring mathematics teacher, Irish-born Francis Murnaghan—who liked to tell us in class while teaching us new mathematical tricks that an Irishman gets over an obstacle by going around it—I also found part-time work for a nascent brewery. A friend of Murnaghan's, anticipating the repeal of Prohibition, planned to set up a brewery to make beer in Baltimore. He hired me to carry out calculations to determine the number and size of pipes required for the heat exchangers. I dug into some technical books that I found in the library and got the job done. Curiously, the experience came in handy more than a decade later as I worked with Du Pont engineers on the design of the water-cooling system for the Hanford reactors.

Earlier, in the summer of 1928, following my freshman year, my uncle John Archibald (one of my mother's two mining engineering brothers) offered me a job at his silver mine in Zacatecas, Mexico. This sounded like an exciting way to earn a little money, and I was ready for a break from studies. I got it into my head that the best way to get there would be by motorcycle, but my parents would have none of that. I went by train.

Everything about that summer still stands out clearly in my mind. I learned about mining, about electrical machinery, about management and employer-employee relations, about corrupt officials, and about being different—I was an ethnic minority in that setting, as well as a part of the power structure. It was a lot for a seventeen-year-old to absorb. As it turned out, the train fare and miscellaneous expenses just about used up all that I earned, but I never regretted the experience.

On the way to Zacatecas, I got off the train at El Paso, Texas, and carried my bags across the border to board the Mexican train in Juarez. The Mexican customs official asked me if I had anything to declare. I didn't know what he meant. So he opened my suitcase, and I blushed mightily when he found in it half a dozen pairs of ladies' stockings that my mother was sending to my aunt. "Are those for you?" he asked with a roar of laughter. Anyway, he didn't extract any import duty from me. He probably thought I was too young to understand *mordida,* the small bribe expected by officials who serve you.

Meals for the rest of the journey were served not on the train, but at stations where we stopped. At these stops, soldiers got off the train and paraded up and down with their guns. Despite such precautions, I learned later, the locomotive of a train on the same route the previous day had been dynamited off the track at Aguascalientes. I arrived in Zacatecas around midnight and was more than ready for the bed offered in Uncle John's house. There I lived for the summer as part of the family, working around the mine and playing with my uncle's children.

My uncle had witnessed some of the brutality and killing that was then common in Mexico as various factions struggled for control, but he had fortunately escaped it himself. He had learned how to function as a *patron,* making the right political contacts, paying the right bribes, and being benevolent to his workers. I had been there only a few weeks when he took me aside to explain the importance of courtesy and graciousness in dealing with his Mexican assistant Federico. "You don't just say 'good morning,' " he explained. "You chat a little more." As to the workers down in the mine, I didn't know enough Spanish to converse much. They were not disagreeable, but sometimes looked at me in a way that made me uneasy.

"What if I am tormented by a worker?" I asked my uncle.

"Either do nothing," he answered, "or knock him down. Don't do anything in between." My apprehension may have had no real basis. I never had to confront an unfriendly worker.

My job at the mine was rewinding wire on electric motors that drove pumps to bring excess water out of the deep mine. In the grit and moisture where these motors operated, they failed often, and needed rewinding. It was honest labor, not unlike hoeing a corn field or feeding the pigs at the family farm in Vermont. It was work that had to be done to make a larger enterprise succeed. Nonscientists probably think that scientists in their research are always using their higher mental faculties, always dealing with deep questions, always confronting the excitement of new knowledge. If only it were so! A lot of what the researcher does is not unlike winding motors, or hoeing corn, or feeding pigs. Laboratory apparatus has to be built and checked and often rebuilt; computer programs have to be debugged; a theoretical idea has to be cranked through tedious mathematics. There is some drudgery in every line of work, and I am glad that I wound motors and hoed corn and fed pigs as a boy. In some way, these activities helped prepare me for research.

By the time I was seventeen, I should have outgrown the foolishness that nearly cost me a finger when I was ten. But something, even at this older age, drove me to test my limits. Once, leading my younger cousins in follow-the-leader, I wanted to see what it would be like to touch an electrical line. The one I chose to touch (strung too low to meet any current safety standards!) was an 11,000-volt line. Fortunately, the instant muscle contraction threw me away from the line, not onto it. I survived, shaken. My cousins chose not to follow my lead. Another time, they did follow, and a guardian angel saved us all. We walked across the length of a pipe over a vat of cyanide solution that was used to dissolve out the gold and silver from crushed ore. Later that summer, I was told that when that tank had once been drained, buttons from the clothing of someone unknown had turned up. We had nothing to say about our exploit.

Having survived the summer, I came home via Mexico City, which my train ticket permitted without extra cost. A friend of my uncle showed me the sights. It was a glorious experience. I have delighted in returning there several times since.

I had the good fortune to be a student during a time of startling progress in physics. Some historians of science see the late 1920s and early 1930s as its golden age. In those few years, physicists discovered quantum mechanics; proposed the uncertainty principle; predicted and observed antimatter (in the form of the positron); discovered the neutron (thus clarifying nuclear structure); united quantum theory, relativity, and electromagnetism in electrodynamics; invented the cyclotron; and found the universe to be expanding. Despite all that—and even though it was more than enough to inspire a beginning physicist—I see an even broader panorama of progress in physics over this century. It was not only in my student days that momentous discoveries were being made.

What we call modern physics began in the last decade of the nineteenth century. In 1895, Wilhelm Roentgen, in Germany, discovered X rays (and immediately recognized their potential for medical diagnostics). In 1896, Antoine Henri Becquerel, in France, discovered radioactivity in a sample of uranium. In 1897, Joseph J. Thomson, in England, discovered the electron, a charged particle far lighter and smaller than a single atom. In 1898, Marie and Pierre Curie discovered the new elements polonium and radium. As the new century began, in 1900, Max Planck, in Germany, discovered the quantum principle of granularity in nature, and identified a constant—we now call it Planck's constant—that determines the scale of this granularity in the world of the very small. Eventually Planck's work explained not only why matter, divided finely enough, consists of grains called atoms, but also why light, examined closely enough, consists of chunks called photons.

All this happened at a time when many physicists were feeling glum about their field. Back in the 1870s, the distinguished British scientist James Clerk Maxwell had noted the limited vision of those of his colleagues who believed "that, in a few years, . . . the only occupation which will be left to men of science will be to carry [their] measurements to another place of decimals."[1] Such pessimism about the field of physics, persisting until the turn of the twentieth century, turned out to be far off the mark, as Maxwell foresaw. The discoveries of Roentgen, Becquerel, Thomson, the Curies, Planck—and others soon to follow—gave physics a new vitality. In my lifetime, the flood of discovery, invention, and insight has not ceased.

In 1994, Steven Weinberg, a leader in particle physics and cosmology—and, for a time, my colleague at the University of Texas—published a book called *Dreams of a Final Theory*. He is among a group of notable physicists at the end of this century who believe that we are within shouting distance of a "theory of everything," on which will be based all the laws of the physical world. I, too, dream of an all-encompassing theory, but my dream has quite a different shape than the dream of these particle physicists. In my dream, overarching principles of the universe at large impose order "from above" on the world we experience. In Weinberg's dream, laws governing the tiniest particles act "from below" to give structure to the world. But who can predict? Most likely, discoveries of the twenty-first century will astound us, or our heirs, with their novelty. Twentieth-century physics is unlikely to be the last word. There are too many loose ends, too many questions still needing answers, too many reasons to go on searching.

The startling march of physics in the first decade of this century made it seem that the discoveries of the preceding decade had only taken the cap off the bottle. Albert Einstein, in Switzerland, created the special theory of rela-

---

[1] From Alan L. Mackay, Ed., *Dictionary of Scientific Quotations* (Bristol, England: Institute of Physics Publishing, 1991), p. 167.

tivity, postulated the existence of a particle of light that we now call the photon, and enunciated what has become the world's most famous formula, $E = mc^2$, expressing the equivalence of mass and energy. The New Zealander Ernest Rutherford and his British colleague Frederick Soddy, working in Canada, discovered that radioactivity is accompanied by transmutation, the change of one element into another (the dream of the alchemists, we are reminded); and that the energy emitted per atom in radioactive change is vastly greater than in chemical change. The Curies, in France, along with Rutherford and others, unraveled the complex chains of transmutation in the natural radioactive decays of the heavy elements.

In 1911, the year I was born, Rutherford and his colleagues in Manchester, England, discovered that most of the mass of an atom resides in a tiny, positively charged core at the atom's center—the atomic nucleus. In the same year, Heike Kamerlingh Onnes, in Leiden, Netherlands, discovered the strange phenomenon of superconductivity—the ability of certain materials at low temperature to carry electric current without any resistance. In 1913, Niels Bohr, later to be my mentor, united Planck's quantum idea with Rutherford's nuclear atom in a successful but unsettling theory of atomic structure. His revolutionary ideas included the idea of a quantum jump (now part of everyone's vocabulary) and the idea that the frequency of light emitted by an atom need not match any mechanical frequency within the atom.

In 1915, Einstein, then in Germany, published his theory of general relativity, according to which spacetime is "warped" (deformed, or curved) by massive objects, and this warped spacetime steers the motion of particles and even of light. (Einstein's fuller, more pedagogical paper, came the next year, in 1916.) In his hands, gravity ceased to exist as a separate force, becoming just another property of spacetime. As the decade of the teens drew to a close in 1919, Einstein was propelled to international stardom when astronomers, taking advantage of a solar eclipse for their observations, found that starlight was deflected by the Sun, as Einstein's theory predicted.

In 1924, Louis de Broglie, in his dissertation research at the Sorbonne in Paris, postulated that material particles have wave properties, complementary to the particle properties exhibited by light waves. In 1925, which was to be a banner year for the physics of the very small, George Thomson in England as well as Clinton Davisson and Lester Germer in the United States verified that matter in the form of electrons indeed has wave properties. The Dutch physicists Samuel Goudsmit and George Uhlenbeck discovered that the electron spins and that the magnitude of its spin is only half as great as what had until then been assumed to be the smallest possible quantum unit of spin. (Later, I was to cross swords, in friendly fashion, with Sam Goudsmit over the publication of my papers when he was editor of the leading American physics journal, *Physical Review*.) In the same year, 1925, the twenty-four-year-old Werner Heisenberg, in Germany, pulled together the various threads of quantum theory that had been tantalizing physicists since the turn of the century. From them he constructed a beautiful mathematical theory that he called "matrix mechanics" (mathematically equivalent to what we now call quantum mechanics), the first theory capable of making truly accurate predictions of atomic phenomena. In Switzerland, Wolfgang Pauli advanced the exclusion principle, stating that no two electrons can overlap perfectly in the same state of motion. If one is moving in a certain way, all others are excluded from moving in exactly the same way. (I mentioned in Chapter 1 that this same principle applies to the neutrons and protons within nuclei, and helps to dictate how nuclei are built and how they behave.)

In the next year, 1926, the German physicist Erwin Schrödinger (later to emigrate to Ireland) developed what we now call the Schrödinger equation, whose predictions proved to be the same as those of Heisenberg's matrix mechanics. Shortly, their countryman Max Born (later to emigrate to England), gave to quantum mechanics the probability interpretation that has been its central feature ever since. In 1927, the year in which I turned sixteen and entered Johns Hopkins, Born and Pascual Jordan in Germany proved the equivalence of Heisenberg's and Schrödinger's theories; Heisenberg advanced the uncertainty principle, still today an idea that powerfully grips the mind; and Bohr generalized the uncertainty principle into a broader principle of complementarity. In 1928, Paul Dirac, in England, merged relativity and quantum mechanics into what we now call the Dirac equation, a beautiful synthesis that accounted for the electron's spin and led to the prediction of the electron's antiparticle, the positron.

It was not only in the world of the small that startling discoveries were being made. The American astronomer Edwin Hubble showed in 1924 that some of what is seen in the heavens is not within our own galaxy but lies far outside it. In 1929, by correlating the speed at which remote galaxies are receding from us with their distance from us, he revealed that we live in an expanding universe. Einstein's 1915 equations of general relativity *implied* that our universe is dynamic, not static, but neither Einstein nor anyone else who studied the theory was willing to accept that revolutionary concept until the evidence forced it.

Even if it is stretching a point to call the late 1920s the Golden Age of Physics, it is no exaggeration to call that period a watershed. Classical ideas about solidity, certainty, stability, and permanence were being abandoned. They were being replaced by *quantum* ideas of uncertainty and granularity and the duality of waves and particles; by *relativistic* ideas of spacetime as cosmic actor, not merely cosmic stage; and by *astronomical* ideas (backed up by relativity) of a universe that is expanding, not static, and of finite age, not eternal. I could not have picked a more exciting time in which to become a physicist.

To what extent did physics as it was then developing get into the curriculum? I mentioned earlier the excitement that my teacher of beginning chemistry, H. M. Smallwood, engendered over the Schrödinger wave equation and its application to atomic structure. Up-to-the-minute physics may not have greatly penetrated the standard physics courses, but it didn't matter, because so much of my learning was through apprenticeship. From about the third year, I felt very much a part of the department's research enterprise. I sat in on advanced seminars and attended colloquia at which visitors often spoke about their research. And, now and then, we students were asked to study and report on some recently published paper.

It fell to my lot in 1930 to report on a paper by the German physicists Walther Bothe[2] and H. Becker. They discovered that beryllium, when bombarded by alpha particles, produced a highly penetrating radiation, which they interpreted to be gamma rays (photons), although the apparent energy of the gamma rays was greater than one would expect from the nuclear process they were observing. I duly reported on their work, but could offer no more insight on the puzzling feature of the radiation than they did. The "radiation," it turned out, was a brand new particle, the neutron. They missed the chance to identify a new particle, and so did I. James Chadwick in England gets the credit for discovering the neutron. In 1932, Irène and Frédéric Joliot-Curie in France got close to the answer when they showed that this new radiation could kick protons out of hydrogen in a way that gamma rays would not be expected to do. Chadwick, after conducting similar experiments, boldly suggested that it took a new neutral particle to explain the results.

The neutron's discovery had several profound effects. It offered immediate clarification of nuclear structure. Suddenly every nucleus could be understood as a collection of protons and neutrons. The neutron also offered a powerful new tool for nuclear exploration. Being uncharged, and therefore immune to the positive charge carried by every nucleus, it easily penetrates nuclei. This property paved the way to nuclear fission. And the neutron put an end to the cozy world of protons and electrons as nature's only building blocks. It gave birth to the subject of elementary-particle physics.

---

[2] It was Bothe who later measured incorrectly the cross section for absorption of neutrons by carbon nuclei. As a consequence of his experimental error, German nuclear scientists concluded that graphite would not be a suitable moderator for a reactor.

An educational innovation in the Johns Hopkins Physics Department was to have each advanced physics student work one month with one professor, one month with another, and so on. In this way, I learned physics firsthand from masters, and learned laboratory techniques as well as theoretical techniques. From Augustus Pfund I learned physical optics; from Gerhard Dieke, the application of the new quantum mechanics to the characteristic light emitted by atoms and molecules (their spectra); and from Joyce Bearden, the study of X-ray spectra. All of these men were internationally recognized researchers. One of my first papers was prepared jointly with Bearden for presentation at a Washington meeting of the American Physical Society in April 1933.

My introduction to nuclear physics came from Norman Feather, who moved from Ernest Rutherford's Cavendish Laboratory in Cambridge, England, to Johns Hopkins while I was a student. Robert Wood, the great teacher of optics on the Hopkins faculty, being a member of England's Royal Society, went periodically to England. On one of these trips, he was charged by the department to find a new staff member in nuclear physics, clearly an important field for the future. Feather, the person recommended by Rutherford, accepted the invitation to join the Hopkins department. During my stint with Feather, I learned the tedious, exacting work of counting the individual dim flashes of light made by alpha particles hitting a zinc sulphide screen—a task that could be undertaken only after sitting half an hour in the dark. This was the same detection method used in Rutherford's lab when he discovered the atomic nucleus in 1911. Much of the advance in nuclear physics and particle physics since then has been made possible by ever better detectors. The human eye looking at a screen in a darkened room has long since been superseded—first by cloud chambers, then by photographic emulsions, then by bubble chambers and spark chambers and photomultipliers, culminating (so far) in complex arrays of plates and chambers and electronics weighing thousands of tons and costing millions of dollars.

Feather introduced us to the newly published book *Radiation from Radioactive Substances*, by Rutherford, Chadwick, and Ellis. This wonderful book helped to convince me that nuclear physics was a frontier field worth exploring. Even though nuclear physics was advancing at breakneck speed, this book remained a useful reference for years.

One of my fellow students was John Mauchly, later to become famous as a computer pioneer. Our personalities didn't mesh, for he was a pessimist and I an optimist, but I recognized his talent. Graduating in the depths of the Depression, he could get a job only at a small college, Ursinus, near Philadelphia. From there he went to the University of Pennsylvania, where, during World War II, he was a principal designer of the first true computer, the

Karl Ferdinand Herzfeld.
(Courtesy of AIP Emilio Segrè Visual Archives,
Physics Today *Collection.)*

ENIAC (electronic numerical integrator and computer). After the war, he joined with J. Presber Eckert, Jr., to develop the Univac computer and founded a company to manufacture it. For a period of time, the Univac was the best computer in the world.

The leading professor of theoretical physics at Hopkins was Karl Herzfeld, an Austrian who had emigrated to America in 1926 after studying in Vienna, Zurich (Switzerland), and Göttingen (Germany). He was noted as the author of an enormous article on the spectra and structure of solid matter that appeared in the definitive, multivolume German reference work, *Handbook of Physics*. Although not a prolific publisher of original research, he kept thoroughly up to date on physics and brought to the classroom not just knowledge but passion. I sat in on his graduate seminars beginning in my sophomore year and came away excited. Herzfeld began every course, whatever the topic, with a broad overview of physics and a discussion of the work of leading physicists. Then he showed how the particular subject matter of the course fit into

Maria and Joe Mayer, 1930s.
*(Photograph by Francis Simon, courtesy of AIP Emilio
Segrè Visual Archives.)*

the bigger picture. His lectures were not prepolished, gift-wrapped offerings; they were works in progress. How wonderful to see him think on his feet, take issue with himself, backtrack, and go forward again.

Since I found myself gravitating more and more toward theory, I rather naturally assumed that I would do my dissertation work under Herzfeld's direction. At one point, I considered working with Aurel Wintner in the Mathematics Department (since he had invited me to do so), but I decided that Wintner's work in mathematical physics was a bit too abstract and too removed from contact with experiment for my taste. Hubbard put in a good word for me, and Herzfeld accepted me as a student. I could not have wished for a better guide.

Another theorist at Hopkins at that time was Maria Goeppert Mayer, later to win a Nobel Prize for her work on the structure of nuclei. She and Herzfeld ran a seminar course in which a few of us sat around a table going chapter by chapter through the new German-language book of Max Born and Pascual Jordan on quantum theory. It was an exciting way to learn the subject. Maria's husband, Joe Mayer, was a professor of chemistry at Hopkins. She came to Hopkins as a newlywed with a fresh Ph.D. from Göttingen in 1930, when she was just twenty-four. Suffering the fate of many women academics, she had no proper position, being treated as a guest, with little if any salary. Herzfeld had the good sense to recognize her ability and brought her into the workings of the department.

Maria Mayer's situation got worse after the retirement of Joseph Ames as president of Hopkins in 1935. In the eyes of the new president, Isaiah Bowman, she suffered by being not only female but foreign. This made it easy for her husband, Joe, to accept a position at Columbia University in 1939. Even there, she was not appointed to a regular faculty position. She taught at Sarah Lawrence College while conducting her research at Columbia. During World War II, I had the chance to admire her abilities again when we worked together at the Met Lab in Chicago. After the war, while her husband, Joe, held a professorship in chemistry at the University of Chicago, she held half-time, nontenured research positions at the university and at Argonne National Laboratory. Not until 1960 did she get a professorship herself, when she and her husband moved to the University of California, San Diego. Her Nobel Prize was awarded in 1963. Despite the injustices she suffered for most of her professional life, she remained always cheerful and always vigorously active in theoretical physics. To her colleagues she was a valued full partner, whatever status she might be assigned by local administrations. A heavy smoker, Maria Mayer died in 1972 at the age of sixty-five.

I was nineteen when the first paper appeared with my name on it as an author, and I was justly proud. In the summer of 1930, after my third year at Johns Hopkins, I got a summer job at the National Bureau of Standards in Washington, working with the noted spectroscopist William F. Meggers. I still remember the stipend—thirty dollars per month. The heat and humidity were, I suppose, oppressive, but I have forgotten that. Following that summer's work, we wrote a paper together with the title "The Band Spectra of Scandium-, Yttrium-, and Lanthanum Monoxides." Since it was my first paper, let me explain it.

A "band spectrum" is observed in the light emitted by a molecule, to be distinguished from the "line spectrum" of light emitted by an atom. The atom's line spectrum is the simpler of the two types. In an atom, electrons can jump from one quantum state to another, emitting photons of light in the process. The frequencies (or wavelengths) of the emitted light are sufficiently well separated that they register as lines in a spectroscope. The motions within a molecule that give rise to radiation are more complex. The electrons in molecules occupy energy states, just as in atoms. But in addition, the atoms that make up the molecule can vibrate relative to one another, or the molecule as a whole can rotate. These vibrational and rotational motions give rise to much more closely spaced energy states. The frequencies (or wavelengths) of the light emitted by molecules are so fine-grained that they register as bands in a spectroscope (even though, on close enough inspection, the bands are made up of lines).

Among molecular spectra, the simplest—and therefore the best candidates to understand in detail—are those of diatomic molecules. A diatomic molecule is one containing only two atoms, such as carbon monoxide (CO) or hydrogen fluoride (HF). What I helped Meggers do was measure and analyze the spectra of light emitted by diatomic molecules consisting of one rare-earth atom (scandium, yttrium, or lanthanum) and one oxygen atom. This was the era of spectroscopy in which the accumulation of accurate data on spectral wavelengths was helping to identify the quantum energy states that exist within atoms and molecules. Our work that summer was devoted mostly to gathering extensive data, but it gave me an opportunity to study and learn about the theory of rotational and vibrational energy states according to the new quantum mechanics. An organized summary of that theory had been published only the preceding year.

When I told my father proudly about my work with Meggers, he said, "Yes, you measure the wavelengths, but what about the relative brightness of different parts of the spectrum? What do you do about that?" It was an astute question, and typical of my father's effort to steer me toward pioneering. The relative brightness is indeed important, but measurements of it came later.

Meggers was kind enough to invite me back for two more summers. The work provided a good exposure to research in which experiment and theory have to go hand in hand. It also provided a few helpful dollars. Still, it couldn't match Zacatecas, Mexico, for opening new vistas.

My more nearly independent research started in 1931 under Herzfeld's direction. He fit exactly the image of a scientist as portrayed in modern movies: a bit unkempt, of above-average height, slightly stooped, with an abstracted air, a cluttered desk, and a strong German accent. He had seen action on three fronts with the Austro-Hungarian army in World War I, including a stand against the last charge of the famous Cossack cavalry with its lances and flying banners. Herzfeld had two religions, Catholicism and physics. He told me once that he attended Mass every day because he had promised to do so in a prayer that he offered up when his unit was surrounded and in danger of annihilation in the Alps north of Venice.

During these student years, I had a chance to hear some of the greats of physics arguing with each other (in friendly fashion). I remember a meeting in Atlantic City where Nobelists Arthur Compton and Robert Millikan debated the nature of the primary cosmic rays. ("Primary" refers to the radiation arriving from outer space, before it creates new particles and stimulates other secondary effects within the atmosphere.) Compton argued for particles of matter such as electrons or protons, Millikan for particles of light (photons). Not long after, Compton won the argument—not by persuasive force on the lecture platform, but by measurements that he and others made of cosmic-ray intensities at various places on Earth. These measurements confirmed the so-called latitude effect. Because the incoming particles are positively charged (being mostly protons), they are deflected by Earth's magnetic field and, as a result, strike the earth more intensely near the Poles than near the Equator. Gamma-ray photons, being uncharged, do not "feel" Earth's magnetic field, and so exhibit no latitude effect.

It is interesting that Millikan's argument in the Atlantic City debate built on earlier work by Compton. Until the early twenties, the very existence of the photon was in some doubt, and even those who believed in it considered a "corpuscle" of light to be different in character from a "real" particle such as the electron. It was an experiment by Compton in 1923 that gave the first strong evidence that the photon deserves to be called a particle. He found that when X rays are deflected by atoms, the angles of deflection and changes of wavelength can be precisely and simply explained in terms of "billiard-ball" collisions. One set of billiard balls (the X-ray photons) rebound from another set of billiard balls (electrons in the atoms).

Nowadays, we tell our students that the Compton experiment "proved" that the photon is a particle. The acceptance of a new idea, especially a radical idea, is not always so swift. Planck, in 1900, had shown that when a solid emits or absorbs electromagnetic radiation, energy is transferred between the solid and the radiation in quantized bundles. Einstein, in 1905, had argued that the radiation itself behaves as if made up of a swarm of particles. Compton's 1923 experiment had seemingly revealed particle-like collisions of single photons. Yet general acceptance of the photon as a particle had to await the development of the quantum theory of radiation in the early 1930s. That theory offered the beautiful simplicity of treating electrons and photons as particles on an equal footing. Either kind could be created and destroyed (emitted and absorbed). One happened to be a charged particle with mass, the other happened to be an uncharged particle without mass. (The theory developed in my student days did not instantly clarify all aspects of light. Not until the late 1940s was the theory that we now call quantum electrodynamics fully developed.)

Another friendly argument took place between my teacher Karl Herzfeld and the eminent theorist Paul Ehrenfest, visiting from the Netherlands. After Herzfeld explained some mathematical aspect of Schrödinger's wave mechanics, Ehrenfest rose to say, "My dear Herzfeld, you are completely crazy." Such open, friendly argumentation is good for science, and it is good for students.

What better way to learn how physics is done than to see masters of the subject jousting with each other?

Herzfeld suggested for my doctoral research that I study the absorption and scattering of light by the helium atom, using quantum-mechanical methods. The helium atom, with two electrons, is much more complicated than the hydrogen atom, with its one electron. On the other hand, it is the simplest of all atoms heavier than hydrogen, and so lends itself to reasonably accurate theoretical calculations. (Helium, with its one nucleus and two electrons, constitutes what is called a "three-body problem." In both quantum and classical mechanics, the three-body problem is far more complex than the two-body problem. Earth, Moon, and the *Apollo* spacecraft, to give an example, constitute a three-body problem. Without the Moon, the spacecraft would follow (almost!) a simple elliptical orbit whose properties could be worked out on a piece of paper by hand. With the Moon, high-speed computers are required to calculate the trajectory of the spacecraft. A similar leap in complexity occurs from the hydrogen atom to the helium atom.)

This work led to my first solo-authored paper, "Theory of the Dispersion and Absorption of Helium," submitted to *Physical Review* in January 1933. In that paper, I was able to calculate the refractive index of helium to within 3 percent of its measured value and show how the "virtual transitions" to very high-lying energy states in helium contribute to its ability to influence light. (Refractive index is a combined measure of the absorption and slowing down of light passing through matter. Lenses and prisms offer everyday applications of refraction.)

As I look back now at that paper written when I was a twenty-one-year-old student, I am startled to find in it approaches to physics that have appeared again and again in my work throughout the rest of my career. First is my way of tackling problems (the practical doer in me). Second is my way of thinking about nature (the dreamer and searcher in me). I fearlessly jumped into mathematical analysis—and surely must have had to learn much of the needed mathematics as I went along. Equally fearlessly, I jumped into numerical calculation. There was, of course, no such thing as a computer at that time, nor even an electrically driven calculator. I used a hand-cranked mechanical calculator.

The problem suggested by Herzfeld had a special charm. It brought out the beautiful connection that exists in physics between absorption and scattering. If a photon—or any other particle—hits a target, it may be absorbed, or it may be scattered. (Scattering is defined as change of direction and/or change of energy. A bullet ricocheting from a metal fence is "scattered" because it changes both its direction and its energy. We also say it is scattered if it passes through a thin sheet of wood, losing some energy but not changing direction.) The relationship between these two processes, absorption and scattering, exists not only for light hitting helium atoms, but for *any* projectile particles hitting *any* target. If you know the chance that the projectile particle is absorbed at all energies, then you can calculate how it will be scattered at any energy. If you know how it is scattered at all energies, you can calculate its absorption at any energy. An equation that ties scattering and absorption together is called a "dispersion relation." In the few years following my work at Johns Hopkins, I applied dispersion relations to several problems in atomic physics. Later, my student John Toll—destined to serve as the president of three institutions of higher learning—extended dispersion relations to the relativistic domain and showed what general power they have as a tool for understanding electrodynamics.

Dispersion relations are among my long list of topics for unwritten books! Nuclear physics is another. In 1961, my colleague F. Bary Malik, from Dacca, Bangladesh, and I sketched out the contents of a book on nuclear theory, and he drafted a great deal of material for it. Then circumstances intervened. He moved to Yale University, and my attention was diverted as I got deeper and deeper into the physics of gravitation and spacetime. Our monograph on nuclear physics was placed on my virtual shelf of virtual books, right next to *Modern Mechanics, A New Look at Quantum Theory, Applications of Dispersion Relations,* and *The Acapulco Effect and Other Interesting Physical Effects.* (The "Acapulco effect" concerns great sloshing waves in a harbor like the waves in a saucepan that is jiggled from side to side.) Had I but world enough and time.

Although I was enthralled with physics at Johns Hopkins, I found time for some social life. I tried not to let living at home get in the way of my involvement in campus activities. After a few years, my fellow students elected me to manage the graduate student dances (their choice being based on my organizational skill, not my dancing ability). One of the young ladies I took dancing was from France. Another was from Denmark. These girlfriends from foreign countries seemed more exotic and charmed me more than most of the American girls I knew.

My best friend at Johns Hopkins was Bob Murray—Robert Taylor Keys Murray—who was also studying physics. While the rest of us kept our slide rules handy to make calculations, he refused to use one. The slide rule's limited accuracy wasn't good enough for Bob. He preferred to do the calculations by hand, and was a whiz. After getting a Ph.D. in physics, he taught at Brooklyn Polytechnic Institute, and later turned into a recluse, living alone on a boat based on Long Island. As far as I know, he never managed to fill his lifelong ambition to sail around the world.

On my first "date" with Janette Hegner (now my wife of more than sixty years), Bob Murray was along. He knew her better than I did, and got us together when he saw how I was inclined.

I knew Janette's younger sister, Isabel, from the young people's group at the Unitarian Church in Baltimore. My family had embraced Unitarianism in Youngstown when I made that church my choice for Sunday School. In Baltimore we attended regularly, usually sitting near the Hegner family, who were also Unitarian regulars. Isabel and I went often to meetings of the young people's group, but Janette did not. She had been away from home for most of the previous six years—as a student at Radcliffe College for four years, holding a job in Washington, D.C., for a year, and studying history on a fellowship in Italy for a year. At this time, she was back in Baltimore, living at home again while taking graduate courses in history at Johns Hopkins—and, as I learned later, keeping busy with her own social life and her own boyfriends.

In the spring of 1933, Isabel invited me to a dance at her house. There I became intrigued with her sister Janette. She looked me straight in the eye. No fluttering eyelashes for Janette. Quickly discovering that I wasn't much of a dancer, she suggested that we sit and talk. Janette says now that she doesn't especially remember this encounter at her house, but it stands out clearly in my mind. I was attracted to her quick wit, her obvious intelligence, and her commonsense approach to matters we talked about.

I told Bob Murray afterward about being attracted to this young woman. "Well," said Bob on the day I passed my final oral exam for the Ph.D. later that spring, "Why don't we invite Janette Hegner for a walk?" So I called her up, and the three of us went walking in Druid Hill Park. At the time, I must have been more intrigued than passionate, since I didn't pursue Janette right away. Then, a few months later, when I was working at New York University, I discovered that she was teaching at the Rye Country Day School near New York. Three dates later, we were engaged. More of this in the next chapter.

When Janette is asked now when she first met me, her answer is, "I didn't *meet* Johnny. I *knew* Johnny." Despite the fact that we often sat near each other in church on Sunday mornings, I had somehow not taken careful notice of Janette until that dance at her house and the walk in Druid Hill Park. How fortunate that I finally did take notice.

5

·

I TRY MY WINGS

WHAT DOES a young researcher need at the beginning of a career? Perhaps, most of all, a good mentor. (Einstein was an exception to the rule. He did brilliant work in isolation.) And freedom—freedom to experiment with ideas, freedom to try new directions, freedom to make mistakes, freedom to think without distraction.

In two postdoctoral years, I was blessed with two wonderfully strong mentors, Gregory Breit and Niels Bohr. In this chapter and the next, I will speak about these two remarkable scientists, so different from each other in so many ways, yet both brilliant, both possessed of a burning commitment to understanding the basic laws of nature. Each of them provided for me an almost ideal work environment for my early development. They provided inspiration and leadership without the hierarchical differentiation of master and student. With each of them, I felt like a colleague from the first. And they provided freedom without neglect. They were accessible, always ready to talk, yet did not try to direct details of my work. I don't think I could have built a better base for a career in theoretical physics than I did at New York University with Breit and at the University Institute for Theoretical Physics in Copenhagen with Bohr.

Money was not easy to come by in 1933 and 1934. Probably neither Breit nor Bohr would have been able to pay me a salary if I had come without support of my own. What made it possible for me to work with them was a generous and farsighted fellowship program of the National Research Council

(NRC), funded by the Rockefeller Foundation. This program had been start-ed in 1919 and continued for thirty years, making it possible for many a young scientist to get a career launched or redirected. Robert Oppenheimer, for example, used NRC Fellowships to pursue his postdoctoral studies at Har-vard and the California Institute of Technology (Caltech) in 1927–1928. Arthur Compton, going to Cambridge, England, in 1919–1920, and Grego-ry Breit, going to Leiden, Netherlands, in 1921–1922 and to Harvard in 1922–1923, were among other NRC Fellows helped toward notable later careers.

As I was nearing completion of my doctoral research on the helium atom in the spring of 1933, I was perfectly clear in my mind about the path I want-ed to follow. I wanted to do theoretical physics research, and I wanted—but not immediately—an academic career, which seemed to me to provide the best environment for lifelong learning and fundamental research. In the immediate future, for at least one year, I wanted a chance to learn more by doing more research, unfettered by teaching or other obligations, as part of a group centered on a leading theoretical physicist. I was advised by Herzfeld to apply for an NRC Fellowship, which I did. But before filling out the applica-tion, I had two decisions to make. What kind of physics did I want to do? With whom did I want to work?

The first question required very little thought. My path in physics came nat-urally. From my earliest student days, I was most intrigued by questions about fundamentals. What are the basic laws that govern the physical world? How is the world, at the deepest level, put together? What are the constituents? How do they interact? What are the unifying themes? In short, what makes this world we live in tick? Quantum mechanics was the new theory of the very small. I had applied it, in my doctoral work, to the atom. I wanted to go deep-er, to atomic nuclei, to the individual particles within atoms and nuclei, and to their interactions with electromagnetic radiation. At the time, that seemed to define the frontier for me, the places where new insights were most likely to arise. (General relativity and gravitation, the physics of spacetime itself, had not yet fired my imagination. That came quite a bit later.)

Then the question With whom? That was harder. It was natural to think of picking someone in Europe. There were leading centers of physics in Copenhagen, Göttingen, Leipzig, Zurich, Cambridge. Many a young Amer-ican physicist went to Europe for advanced training, bringing back to Ameri-ca the latest from one of those centers. Werner Heisenberg's recent pathbreaking work on the atomic nucleus so inspired me that I formed a plan to join him in Leipzig a year later. But I decided to try my wings first in the United States. There were various good possibilities: Robert Oppen-heimer at Berkeley, John ("Van") Van Vleck at the University of Minnesota,

Gregory Breit at New York University, Eugene Wigner at Princeton.

Van Vleck, thirty-four, was a leader in the new field of solid-state physics. He applied quantum mechanics to the electrical, magnetic, and other properties of bulk matter. I recognized the challenge of that work and its importance, but it did not intrigue me in the way that nuclear physics and electrodynamics did. I did not see answers to the deepest questions about nature coming from the study of solids. (In a roundabout way, solid-state physics has proved critically important to particle physics and cosmology. Without the deep understanding of solids provided by Van Vleck and his followers, we would not have the sophisticated electronics that makes possible the extraordinary measurements that have been essential to progress in these fields.) The year after I considered Van Vleck for postdoctoral work, he moved from Minnesota to Harvard. He spent the rest of a very productive career at Harvard, and received the Nobel Prize for his work on solids in 1977.

Wigner had come to Princeton in 1930, along with his boyhood friend from Budapest, John von Neumann. Both had studied and worked in Germany, and both had made early reputations—Wigner in physics, von Neumann in mathematics. (As it turned out, Wigner made some significant contributions to mathematics, and von Neumann, to physics.) Despite his recognized excellence and his mastery of the new quantum mechanics, Wigner was not well known to Herzfeld and others at Johns Hopkins, and I had not yet met him. He worked mostly alone and had not built a group at Princeton. For me, looking about the physics firmament, he was not as visible a star as Oppenheimer, Van Vleck, or Breit, and I did not give serious consideration to a postdoctoral year with him. In later years, I have had no more valued colleague and friend than Eugene Wigner. My admiration for his talent and good sense is boundless. We also turned out to be soul mates in our thinking about defense issues and public service. His death in 1995 was a great personal loss for me. Nevertheless, as I look back, I see that he would not, in fact, have been the best choice as an early mentor. Wigner was more a worker—a producer—than a teacher. His counsel meant more to me later than it might have in a first postdoctoral year.

My serious choice came down to Oppenheimer or Breit. Oppenheimer, then thirty, was already well known as the brilliant, charismatic leader of groups in both Berkeley and Pasadena, California. He was "Mr. Theoretical Physics" on the West Coast, traveling back and forth between the University of California campus in the north and the Caltech campus in the south. Oppenheimer had worked in pair theory (the electrodynamics of electrons, positrons, and photons—"pair" refers to the electron and positron, a particle-antiparticle pair). He was interested in cosmic rays. And, through the Oppen-

Robert Oppenheimer and Gregory Breit, 1946.
(Courtesy of AIP Emilio Segrè Visual Archives,
Physics Today *Collection.)*

heimer-Phillips effect, his name was associated with nuclear physics. He and
Melba Phillips had analyzed so-called (d,p) reactions in which a deuteron (d)
strikes a nucleus, a neutron is stripped from the deuteron, and the leftover proton (p) emerges, leaving an excited nucleus behind. There was no doubt
about his stature in physics or about his abilities as a teacher. Yet there was
something about Oppenheimer's personality that did not appeal to me. He
seemed to enjoy putting his own brilliance on display—showing off, to put it
bluntly. He did not convey humility or a sense of wonder or of puzzlement. As
I look back now on my many later interactions with Oppenheimer—in the
Manhattan Project, in the thermonuclear weapons project, and doing physics
in Princeton (he at the Institute for Advanced Study, I at the university)—my

feelings toward him remain as ambivalent as they were more than sixty years ago. Oppenheimer was a complex human being. I never felt really close to him. I never felt that I really understood him. I always felt that I had to have my guard up.

Gregory Breit, thirty-three years old at the time, was also a broad-based theoretical physicist, known for his work in both pair theory and nuclear physics (and known, too, for his prickly personality and flashes of temper). Although I scarcely knew Breit at the time, I had formed a good opinion of him from hearing him speak at Physical Society meetings. I resonated with his style. Like me, he seemed to be always puzzling and was not afraid to let his puzzlement show. Breit had done his doctoral work at Johns Hopkins and was well known to Herzfeld and to others there. Herzfeld thought that Breit would be right for me.

So I contacted Breit and got his approval to come the next year to New York University (NYU) to join his group if I could find support elsewhere. I sent in my application for an NRC Fellowship, explaining that I wanted to work with Breit on problems in nuclear physics and electrodynamics. I was successful. Fourteen NRC Fellowships were awarded in physics that year, and I had the good fortune to get one of them, a $1,600 stipend. I am sure that Herzfeld's letter of support was critical.

The story of Breit's emigration to America from Russia I learned not directly from him, but from his sister, Lubov, years later. Lubov was probably a year or two younger than Gregory, and they had an older brother, Leo. Their mother, Alexandra, died prematurely in 1911 when Gregory was twelve. The next year, their father, Alfred, set off from Odessa to establish a new home in America, leaving his three children in the care of a German nanny. He picked Baltimore. Not until 1915 were the children able to follow. Gregory, Lubov, and Frau Schneider traveled by train to Archangel and by ship to New York, where Alfred met them. Leo made his way separately, via Turkey.

Gregory, who turned sixteen that summer, had only one good outfit, a sailor suit with short pants acquired for the trip. He must have demonstrated his intellectual promise at once, for he was admitted to Johns Hopkins, where he completed his undergraduate work in three years and his doctoral work in three more, finishing while he was still twenty-one. To his mortification, he had only the sailor suit to wear to classes at first, and his fellow students may have teased him cruelly, as young people can sometimes do—especially when the victim not only dresses differently but is younger, of small stature, and speaks differently. Whether this contributed to Breit's later sensitivity to criticism and his belligerent defense of his own ideas is hard to say.

Showing great promise in physics, Breit moved swiftly toward his Ph.D.

While a student, he worked at the National Bureau of Standards. There and at Hopkins he concentrated on the properties of radio-frequency circuits and radio propagation in the atmosphere. His postdoctoral year with Paul Ehrenfest in Leiden ignited his interest in atoms and quantum phenomena. In a second postdoctoral year as an NRC Fellow in 1922–1923, he went to Harvard, where he became acquainted with another young postdoc, John Van Vleck. The two of them moved together to faculty positions at the University of Minnesota. As Breit later put it, "Van enjoyed teaching (and I not especially)."[1] After a year in Minneapolis, Breit accepted a job offer from the Department of Terrestrial Magnetism of the Carnegie Institution in Washington, D.C., and went back to that familiar territory, where he could pursue his dual interests in radio propagation and the acceleration of particles for nuclear research. Deciding, perhaps, that formal teaching responsibilities were not so bad after all, Breit moved to the Bronx campus of New York University in 1929 and spent the rest of his career in universities.

Breit was short, intense, sometimes pugnacious. He had a high forehead and wore small circular eyeglasses. Although he was stubborn and difficult with some of his colleagues, that was a side of him that his research students did not see. The brood that he presided over like a mother hen consisted of several graduate students—I remember Hugh Wolfe—and several postdoctoral fellows. One of the postdocs was Jenny Rosenthal, scarcely older than I, who was one of the first women to earn a Ph.D. in physics in the United States. Another was Montgomery Johnson, whom I was to encounter years later in the aerospace industry in southern California. I struck up a friendship with Norman Heydenburg, like me an NRC Fellow, and, to stretch our limited budgets, we shared a room in the Bronx. We ate many of our meals together in local restaurants and talked physics as we ate. (At one restaurant where we ate often, I regularly ordered Apple Brown Betty for dessert. Finally, one evening, the waitress said, "Betty? Always Betty. Isn't it time you took someone else out?")

Breit went at his recreation with the same compulsive determination he brought to his work. On many Saturday afternoons, he and his research entourage would board a train at Grand Central Station, ride to a New York suburb, walk for miles in the woods, and take a train back. I don't think we felt we had any choice in this matter, but we would certainly have had no inclination to excuse ourselves from the outings. We saw them as a privilege, not a duty. They provided a wonderful opportunity to get to know Breit as a per-

---

[1] The remark is from a letter that Breit wrote to the Berkeley physicist Edwin McMillan on June 5, 1977, printed in Roger H. Stuewer, Ed., *Nuclear Physics in Retrospect: Symposium on the 1930s* (Minneapolis: University of Minnesota Press, 1979), p. 138.

son, and they knitted us together as a group. Needless to say, physics was not entirely forgotten as we marched through the woods.

Breit had wide-ranging interests in physics, including both theory and experiment. The students and postdocs who worked with him were doing theoretical work, but he himself was also supervising some experimental work in nuclear physics at NYU. In addition, he continued his collaboration with researchers at the Carnegie Institution in Washington, helping them to design and build better accelerators to fire energetic protons at nuclei and stimulate nuclear reactions. (Accelerators were still a novelty. In the late 1920s and early 1930s, Breit and Merle Tuve in Washington, Robert J. Van de Graaff in Princeton, and John Cockcroft and Ernest T. S. Walton in England had constructed devices that used high voltage to accelerate charged particles. In 1932, Ernest Lawrence in Berkeley, California, had invented the cyclotron, which held the prospect of accelerating particles to energies much greater than could be achieved with the high-voltage machines.)

Physicists like to seclude themselves in their laboratories and offices to work. But they also like to talk. Without the give and take of face-to-face communication, progress would be much slower, and too many researchers would struggle up blind alleys. To tell the truth, most physicists are gossips. They like not only to share ideas with one another, but also to talk about what other physicists in other places are up to.

Breit was easy to talk to. And there was no problem with accessibility: we shared an office. He had allocated so much space on the first floor of the small building to experimental work on accelerators that office space on the second floor was quite cramped. Regrettably, he was a heavy smoker, but in those days one put up with it. He also had a habit of snorting every once in a while like a bull about to charge. Yet we didn't grate on one another as office mates sometimes do. We worked away on opposite sides of the room as if in private, and talked when we needed to.

Through visiting speakers and regular seminars, we kept up to date on work going on elsewhere. Among the exciting news that came to us that year was word on the work of Fermi in Rome on neutron-induced nuclear reactions. In theoretical physics, "barrier penetration" was a hot topic. This is a quantum phenomenon in which a particle can sneak through an energy barrier that, according to classical physics, is impenetrable. This mechanism was advanced by George Gamow and, independently, by Edward Condon and Ronald Gurney in 1929 to explain alpha-particle radioactivity. We heard about the suggestion by Robert d'Escourt Atkinson and Fritz Houtermans that the sun's energy is generated by nuclear-fusion reactions, made possible by barrier penetration. Not everyone accepted this idea right away. Arthur Eddington, the noted British astrophysicist, did accept it, and reportedly told

I. I. Rabi and Ed Condon, 1930s.
(*Courtesy of AIP Emilio Segrè Visual Archives.*)

one doubter, "If you don't think the center of the sun is hot enough, go find a hotter place."[2] (Five years later, Hans Bethe refined this idea in work that was eventually recognized with a Nobel Prize.)

In one memorable talk during that year, Oppenheimer made a prediction that turned out, fortunately, *not* to be true. Just after Christmas 1933, Breit's student Hugh Wolfe (later to become director of publishing at the American Institute of Physics) drove me to a meeting of the American Physical Society in Boston. I now remember the drive better than I remember most of the meeting. A mixture of snow and sleet was falling, and we had to stop frequently to clear the windshield. A trip now completed easily in under five hours took us nearly twice that time. Oppenheimer's evening lecture, delivered to a large and enthralled audience, dealt with quantum electrodynamics, the theory of electrons, positrons, and photons. In that lecture and in a subsequent paper, he suggested that the then current theory of electrodynamics had a limited range of validity, and failed above a certain energy corresponding to about 70 million volts. If that were true, what happens in the cosmic radiation would be beyond understanding until some new theory was developed. The next year, while I was in Copenhagen, this matter sorted itself out, to everyone's relief, through work of Niels Bohr and E. J. Williams. Quantum electrodynamics seems to be valid at all energies without limit.

It was a special pleasure, during my year with Breit, to get to know I. I. Rabi,

---

[2] Another famous Eddington quote: When someone told him that only three people in the world understood Einstein's theory of general relativity, he is said to have asked, "Who is the third person?"

an experimental physicist at Columbia who had recently (in 1931) launched a series of measurements with atomic beams destined to illuminate subtle features of atoms, notably properties of their nuclei. Breit, often instigator of, and almost always allied with, great physics, made essential contributions to the theory of Rabi's measurements, which would bring Rabi the Nobel Prize in 1944. Rabi was one of the leaders of a New York seminar managed jointly by the uptown and downtown campuses of NYU and Columbia University. The uptown campus, where I worked with Breit, was in the Bronx. The downtown campus was in Greenwich Village. Columbia was between the two. For convenience, the evening seminars were often held at Columbia. Frequently, after a meeting, the participants repaired to Rabi's apartment for refreshments and more talk. Rabi's attractive wife, Helen, always made us feel welcome.

Throughout his long life, Rabi was beloved by nearly everyone who came in contact with him. He was a born-and-bred New Yorker with the accent to prove it. He had rock-solid self-confidence without a trace of arrogance. He established easy rapport with everyone he dealt with, be it a laboratory technician, beginning graduate student, or distinguished colleague. His down-to-earth candor showed itself years later, in 1954, when he testified on behalf of Oppenheimer before the special board convened to consider whether Oppenheimer's security clearance should be revoked. After ticking off Oppenheimer's many contributions to the nation, he rhetorically asked the board, "What more do you want, mermaids?" Rabi built Columbia's Physics Department into one of the finest in the world. Thanks to his leadership and skill in identifying talent, his department became something of a Nobel Prize mill. Developments that were germinated at Columbia—and based, to a large extent, on Rabi's own early work—included the maser, the laser, the atomic clock, and magnetic resonance imaging (MRI).[3]

In October 1933, only a month or so after I started to work with Breit, Albert Einstein came to America. Then fifty-four, Einstein was so famous that his first scheduled talk in Princeton that fall was not publicly announced, for fear of an overflow crowd. Eugene Wigner called his friend Breit to invite him to attend. Breit accepted and took me with him. That is how I first met Einstein. It was my impression, based on listening to him that day, that he was

---

[3] The maser (for microwave amplification by stimulated emission of radiation) uses a process first predicted by Einstein in which one photon stimulates the emission of other identical photons to achieve intense concentrations of microwave energy. The now-familiar laser applies the same principle to light, producing narrow, intense beams. Atomic clocks couple radio waves or microwaves to the reliably constant vibrations of atoms to keep precise time. For an engaging account of Rabi's life, see John S. Rigden, *Rabi: Scientist and Citizen* (New York: Basic Books, 1987).

no longer at the frontier. He seemed to be pursuing his own idiosyncratic track in a direction veering away from the main stream of physics. How strange that I, a youngster, should so cavalierly dismiss the greatest physicist of this century. Later, when I, too, lived in Princeton, and even before general relativity became my passion, I became friends with Einstein and had some wonderful conversations with him in his house on Mercer Street. He was kind enough also to welcome my students to his home. With his marvelous insight, he could invariably illuminate corners of the subject of gravitation where we were having trouble seeing clearly.

The image of Einstein with his slippers and baggy pants and sweatshirt, his long hair flying every which way, is familiar to millions. He stands in the public mind for a perfect blend of disheveled innocence and towering intellect. Even those of us who knew him well sometimes had trouble separating Einstein the icon from Einstein the man. He was actually a very down-to-earth person, warm and charming, with a puckish sense of humor.

That first visit to Princeton was memorable not just for meeting Einstein. I also met some members of the university faculty, including Ed Condon, who was then a thirty-one-year-old associate professor. We talked about an idea I had come up with the previous year, at Johns Hopkins, to apply methods of molecular physics to nuclear structure. My thought was that the nucleons (protons and neutrons) in the nucleus could group and regroup themselves into substructures, constantly changing from one grouping to another. It was as if, at a party, all the tall people clustered together at one moment, with all the short people in another cluster; then at the next moment, all the women clustered together and all the men; then at the next moment, four groups formed, consisting of guests from the north, east, west, and south parts of the city; and so on, an endless flux of groupings, each dependent on some attraction bringing sets of people together. In the nucleus, a proton might join one moment with another proton and two neutrons to form a temporary alpha particle; at another moment, it might join a neutron to form a temporary deuteron; at another moment it might be solitary. "That's the idea," I said to Condon, "but I need a name for it."

"How about 'resonating group structure,' " he suggested.

So that was it. A little cumbersome, but I couldn't think of a better name. I didn't actually use the term in print until 1937, when I tried to pull together this line of thinking about nuclei in a pair of papers published in *Physical Review*.

Ed Condon, a talented and imaginative physicist (and a very likable one), did not remain at Princeton. After completing a masterful book with George Shortley on methods of treating atomic spectra—a bible on the subject from the moment of its 1935 publication—he moved to Westinghouse Electric

Company in Pittsburgh as director of research. After World War II, he was
appointed director of the National Bureau of Standards in Washington, D.C.,
and revitalized the Bureau's research efforts. Born in Alamogordo, New Mexico, Condon was a Quaker with liberal leanings, and was forced from government service in the crossfire of the McCarthy period. (I'm told that he
once said privately, "I join every organization that seems to have noble goals.
I don't ask whether it contains Communists.") The Corning Glass Works,
not intimidated, invited Condon to head its research laboratory. He pitched
into the physics of glass with gusto and enjoyed success there until he
returned to academia in 1956.

I described Breit earlier as a compulsive worker. Like many others in love
with their subject, he worked most of the time. He was neat, highly organized,
and scrupulous in his attention to small details. I recall watching him mark up
a manuscript being prepared for submission to *Physical Review*. In the margins, he carefully redrew every Greek character and wrote its name with a col-
ored pencil. This scrupulousness extended to all parts of his research. He
wanted everything right, checked and double-checked. It was good training.

With me and other members of his group, Breit was generous with his time
and warmly supportive of what we were doing. He could be critical when we
deserved it, but never with an unfriendly edge, never in a way to deflate one's
enthusiasm. When I heard about run-ins he had with others, I was puzzled,
because I did not see that side of him. Once, when I was at NYU, according to
a reliable report he actually came to blows with a faculty colleague over some
disagreement. In the heat of anger, they ended up wrestling on the laboratory floor.

Only years later, in 1940, did I encounter some of Breit's stubborn inflexi-
bility. He asked me (and others) to stop publishing anything about uranium,
to avoid the possibility that our results might help an enemy. I was open-mind-
ed on the question, and wanted to discuss it with him. To whom would I be
making the pledge? And under what future circumstances could I be released
from the pledge? Breit was not in a mood to discuss such questions, and
showed his anger, since he felt so strongly about it. As it turned out, American physicists—I among them—did agree to a self-imposed moratorium on
publishing anything that might be related to nuclear weapons, although we
made no formal pledges. Breit, Wigner, and Szilard were principal organizers
and proponents of the moratorium. We learned later from the Soviet physicist
Georgii Flerov that the sudden cessation of uranium physics in U.S. journals
was itself a piece of significant information for scientists in his country. Flerov,
with his colleague Konstantin Petrzhak, had discovered spontaneous fission in
uranium and published the result in the American journal *Physical Review*
in July 1940. He followed the literature, and when suddenly the flow of relat-

ed papers ceased, he drew it to his colleagues' attention and pushed for acceleration of uranium research in the USSR. (One of the last published fission papers by American authors appeared in *Physical Review* a month before the Flerov-Petrzhak paper. In it, Edwin McMillan and Philip Abelson announced the discovery of neptunium, element number 93 and a stepping stone to plutonium.)

I had hardly showed myself at Breit's door before I was drawn into research with him. Our first joint paper was a short one completed in about two months. It was submitted as a Letter to *Physical Review* on November 15 and published on December 1 (a speed hard to match today). I brought to the Letter a knowledge of two-electron systems such as in the helium atom. Breit was interested in understanding better the quantum wave functions of the innermost two electrons close to the nucleus for heavier atoms, where there are tiny effects of nuclear magnetism on the electron energies. We took a first step in that direction by calculating the overlap of electrons with the nucleus in an ionized lithium atom. (Lithium is the third element in the periodic table. Its atom normally contains three electrons, but only two remain when it is ionized.) This was an atomic-physics calculation with a nuclear-physics goal: to learn more about nuclear magnetism through its tiny effect on the energies of atomic electrons.

I had arrived at NYU carrying some unfinished work in my bag. During the previous year, while completing my Ph.D. work on the helium atom, I had also been working with Joyce Bearden on the propagation of X rays through matter. In that problem, as in the helium problem, the mathematical link between scattering and absorption was important. Bearden and I had submitted one abstract and presented a paper at a meeting while I was still at Johns Hopkins. During the year at NYU, I continued to work on the subject. Bearden and I completed a more extensive paper in the fall of 1934. We improved on previous calculations of X-ray absorption by taking account of the important "exclusion principle" that had been discovered nearly ten years earlier by Wolfgang Pauli in Switzerland. As mentioned in Chapter 4, Pauli deduced that no two electrons could be in the same state of motion at the same time. His discovery is a pillar of quantum theory for particles such as electrons that have one-half unit of spin. (The photon, a different sort of particle, has one unit of spin and does not obey the Pauli exclusion principle. The unit in which spin is measured is a fundamental quantum constant written $\hbar$ and pronounced "h-bar.")

Imagine a heavy atom containing lots of electrons—every one, according to the Pauli exclusion principle, occupying a different state of motion. An X-ray photon arriving on the scene cannot be absorbed by causing an electron in,

Fellow physicists recognize the meaning of this physicist's license plate.
Others have wondered if the driver is a lawyer, a rancher, or the proprietor
of a drinking establishment.
(Photograph by Robert Matthews, courtesy of Princeton University.)

say, the first state, to jump up into, say, the tenth state if that tenth state is
already occupied by another electron. The exclusion principle *inhibits* the
absorption of this photon. A photon may, however, be absorbed by causing the
electron in the tenth state to jump clear out of the atom. A not-so-obvious consequence of the exclusion principle, as Breit and I applied it, is that this jump
to freedom *enhances* the absorption of the photon. Since the electron in the
tenth state can't fall down into the already occupied first state, it actually
jumps with greater zest (greater probability) into a free state. Our calculations
provided numerical estimates of these inhibiting and enhancing effects of
the Pauli exclusion principle on X-ray absorption.

The important new physics that Breit taught me was pair theory—the theory of interacting electrons, positrons, and photons. In a beautiful piece of
work in 1928 (mentioned in the previous chapter), Paul Dirac in England
brought together Einstein's special theory of relativity and the new quantum
mechanics of Heisenberg and Schrödinger—that is, he created a relativistic
quantum mechanics—and out of it fell, like magic, three conclusions about
electrons: that they have one-half quantum unit of spin, that they obey the
Pauli exclusion principle, and that they are accompanied by a positively
charged sister particle, now called the positron.

The first two of these conclusions were already known to be true; Dirac's
theory provided a satisfying theoretical explanation for them. The third was

revolutionary, and physicists at first didn't know how to interpret it. Dirac's theory predicted that electrons could exist with either positive or negative energy. Since the electron has mass, it surely has positive energy. What could negative energy mean? The only interpretation that seemed to make sense was that these predicted negative-energy states are completely filled except for the occasional absence of a filled state, which "looks like" a positively charged electron, or positron. In this "hole theory," the sea of negative-energy electrons is unobservable except when one electron is missing. The negative-energy sea can be compared to the ocean, the positive-energy region to the air above the ocean. A positron is analogous to an air bubble in the ocean (the absence of water), while an electron is analogous to a water droplet in the air above.

If a drop of water fills a bubble, the drop disappears, the bubble disappears, and only the smooth ocean remains. In a similar way, in Dirac's theory, an electron can "fall into" a hole, causing the electron and the hole (the positron) both to disappear. Likewise, a negative-energy electron can be elevated to positive energy, leaving a hole behind. This constitutes the effective creation of two particles. To say that the hole theory with its infinity of negative-energy electrons filling all of space made physicists uneasy is an understatement. The unease was alleviated, at least partially, when physicists in Europe showed that the same theory can be alternatively formulated with the positron as a real, positively charged particle, not a hole. In this reformulation, electron-positron pairs can be created if energy is added in the right way, and electrons and positrons can be annihilated with the release of energy when they meet. The positron is called the antiparticle of the electron (and the electron the antiparticle of the positron). We now know that for every particle in nature there is an antiparticle.[4]

The antiparticle theory, with its prediction of annihilation and creation of pairs, was easier to digest than the hole theory—even though the theories differ only in their imagery, not their mathematics. But where was the positron? It had not shown itself in any experiment. The only fundamental particle with positive charge known at that time was the proton. To be sure, the proton has one-half unit of spin, but it is nearly 2,000 times heavier than the electron, whereas Dirac's theory predicted equal mass for the positron and electron. Theorists are ingenious creatures. Some of them, including Dirac himself, set about trying to rationalize the proton as the sister particle to the electron. Fortunately, an experimental physicist, Carl Anderson at Caltech, stepped into the picture. In 1932, he found in his cloud chamber, in the debris from cosmic radiation, a positive particle with a mass apparently the same as—or at least close to—the mass of the electron. Here was Dirac's positron. Soon

[4] The photon is special. It is its own antiparticle.

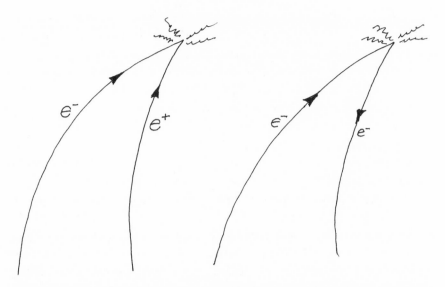

Two ways to look at positron-electron annihilation: as two particles encountering
and annihilating each other, with release of electromagnetic radiation, and as a
single particle moving forward, then backward, in time.
The latter viewpoint is incorporated in Feynman diagrams.
(Drawing by John Wheeler.)

other examples were found, in which high-energy photons created positron-
electron pairs. (When Dirac was asked years later why he had not followed the
dictates of his own theory and predicted a positive particle having the same
mass as the electron, he reportedly answered, "Pure cowardice!")

Some years later, in 1940 or 1941, I came up with yet another way to look
at the positron (also without changing the mathematics of Dirac's theory). Sitting at home in Princeton one evening, it occurred to me that a positron
could be interpreted as an electron moving backward in time. I was excited
enough about the idea to phone my graduate student Richard Feynman at
once at the Graduate College, the on-campus residence where he lived.
"Dick," I said, "I know why all electrons and all positrons have the same mass
and the same charge. They are the same particle!" What had sprung into my
head was a vision of a single electron, tracing its world line — its path in spacetime — first forward in time, then reversing to go backward in time, then
reversing again to go forward in time, and so on. At any given moment of
time — that is to say, on a single slice of spacetime — one sees numerous electrons and numerous positrons, not knowing that at various future times and
past times, their world lines are joined into a single tangled thread. I knew,
of course, that, at least in our corner of the universe, there are lots more electrons than positrons, but I still found it an exciting idea to think of trajecto-

ries in spacetime that could go unrestricted in any direction—forward in time, backward in time, up, down, left, or right. Feynman later incorporated this idea into the diagrams that now bear his name and that have become powerful tools of calculation in quantum electrodynamics and elementary-particle physics. The lines in a Feynman diagram are pieces of world lines that one can take hold of and twist into any direction in spacetime.

Even before Carl Anderson's discovery of the positron, some theorists were making calculations that assumed its existence. What Breit and I did, the year after the positron was discovered, was to calculate the probability that two photons, in colliding, would create an electron-positron pair. We gave thought, even then, to a process in which a high-energy photon passing through interstellar space would interact with one of the lower-energy photons in the hypothetical radiation filling empty space. Based on the assumption that this "interstellar temperature radiation," as we called it, was made of starlight crisscrossing the galaxy, we calculated that the photon-photon collisions would be too infrequent to observe. Had we been astute enough to think of the cold background radiation now known to fill all of space as a relic of the Big Bang, we would still have concluded that the process we were studying would not show itself in space. (Gamow first predicted such radiation in 1948. When Arno Penzias and Robert Wilson at AT&T Bell Laboratories in New Jersey picked up puzzling "static" from space in 1964, my Princeton colleague Bob Dicke could at once offer an explanation, for he had just set two of his students to the task of finding and measuring this radiation.)

Breit and I not only abandoned hope of seeing the effects of photons hitting photons in outer space; we also concluded that no source of radiation would be strong enough to make pair creation by photons visible in the laboratory either. We weren't quite right. Sixty-three years later, in 1997, scientists at Stanford Linear Accelerator Center announced the observation of electron-positron pairs created by photon-photon collisions, conforming to the prediction that Breit and I had made. This achievement, a real tour de force, was made possible by stunning developments in both accelerator and laser technology, far beyond anything we could have imagined in 1934.

However, immediately following our announced result, E. J. Williams, in England, realized that one did not have to wait for super-intense light beams. The idea could be applied to a photon interacting with the strong electric field near an atomic nucleus. To the incoming photon, the electric field "looks like" a cloud of photons. So our photon-photon collision could occur after all, deep within an atom. This is, in fact, a common process in cosmic radiation. It leads to a "shower" in which a photon creates a pair of particles, one or both of these particles create more photons, which in turn create more particles, and so on— a proliferation that ends only when the energy that started the process is spent.

Some of the work I was doing with Breit involved extensive numerical calculation, at that time a laborious activity involving many hours at mechanical calculators. Breit hired some students for this work, and also had some help from a man paid by the Works Progress Administration (WPA). I remember this older gentleman, who worked with such care and concentration, because it was one of my assignments to supervise him. The WPA is often associated with road building and other construction projects during the Depression, but, in seeking to put people back to work, it also supported the arts and sciences.

Out of this productive nine months at NYU came some ideas that haunted me for many years. I was so enchanted with the electron, with its beautiful, exact Dirac theory and its ultimate simplicity, that I couldn't help wondering: Is everything made out of electrons? Isn't there some way to tie together the electron and its antiparticle, the positron—perhaps with the help of the photon—to build all other particles? This kind of thinking, more dreaming than based on solid ground, was not congenial to Breit. He was a powerful theorist with broad interests, but he liked to keep close to what was being measured, or could be measured, in the laboratory. On my own, I kept trying this and that as I worked away at my desk, but I never came up with a plausible way to build the world from positive and negative electrons. Now we know that there are indeed simpler units of which protons and neutrons and many other particles are built, but they are quarks, not electrons.

The paper on polyelectrons that I submitted to the New York Academy of Sciences around the end of World War II was much more modest in its scope than some of my earlier dreaming about a world made of electrons. I just explored some of the atoms and molecules that could be constructed from electrons and positrons alone, and calculated their properties. The simplest such atom, made of one electron and one positron, I called a bi-electron. I learned as my paper was going to press that such an entity had been conceived also by Arthur Ruark in the United States and by Lev Landau in the Soviet Union. It is now called positronium, and its properties have since been extensively studied.[5] It has the "purity" of my early dreaming. Unadulterated by quarks or anything else, its properties can be wholly understood in terms of the electron, the positron, and the photon.

Later, I went further, calculating how a large collection of positronium atoms might behave. Liquid positronium should be superconducting. But will

---

[5] Positronium was first observed in the laboratory by Martin Deutsch at MIT in 1951. Later, in 1981, the three-particle system, composed of two electrons and one positron, was created and detected by Allen Mills at Bell Labs in New Jersey.

we ever see a drop of it? It would be worth striving to demonstrate its existence, for it would be a new extreme form of matter. And it would make nitroglycerin seem as tame as tap water. Positronium matter is, in effect, pure energy. Whereas nitroglycerin, when it explodes, converts only about one-billionth of its mass to energy, positronium matter would disintegrate with the conversion of 100 percent of its mass to energy.

The musing in my first postdoctoral year about a world made of electrons set me to thinking, equally radically, of a world *without* photons—a world of particles without fields. So was launched what I call the Everything Is Particles period of my life. One of the problems with a composite particle made, say, of an electron and a positron bound together in a very small volume (smaller than the volume of a proton) and having a large mass, is that this entity would radiate away its extra mass energy extremely quickly. It would be highly unstable. So, I asked myself, grasping for a way to stabilize it, is there any way to sidestep the inevitability of this radiation? Can radiation be eliminated from electrodynamics and replaced by action at a distance between particles, without losing all that we know to be successful in the predictions of electrodynamics?

Not many years later, when I started working with Dick Feynman at Princeton, we found the answer to this question to be yes. We found a self-consistent formulation in which forces (actions) propagate through space without any need for electric or magnetic fields, either static or in the form of radiation. This was a satisfying result, given that it came at a time when the thinking of all physicists was embedded in fields. However, it did nothing to resolve my old puzzle about neutrons or other particles made of electrons. Nor, as it turned out, did it contribute insights that greatly advanced particle physics or electrodynamics. It did, however, make a most remarkable prediction about a hypothetical world containing only few particles rather than the nearly limitless number of particles in our world. In such a simpler world, the future would affect the past! We concluded that the one-way flow of time that we observe in our world is attributable, in part, to the existence of all the other matter in the universe. So our excursion into physics without fields provided an unexpected bonus. It taught us more about the mathematical structure of electrodynamics and about hypothetical other worlds, but—at least so far—has not advanced further our understanding of the way the world we live in is put together.

Back in Baltimore in the spring of 1933, after Janette and I had gone walking in Druid Hill Park and she learned that I was going to be in New York that fall, she said, "You must look up my friend Elsie Field Doob." Janette did not yet know that she herself would be teaching in the New York area that fall or that much of anything would come of our friendship. Elsie had been her good friend at Radcliffe College. Elsie had been a regular at Boston Symphony concerts. For three years she sat next to a young man without speaking. In the fourth year, they spoke, and before long they were married. The young man was Joseph Doob, a mathematician, later to be a professor at the University of Illinois and a trustee of the Institute for Advanced Study at Princeton. Elsie, like Janette, was interested in history and culture. She became a physician and was influential in the organization Planned Parenthood.

Finally, after a few months in New York, I followed up on Janette's suggestion and contacted the Doobs. They invited me over for a relaxed evening. I enjoyed their company and returned several times. They were a cheerful, carefree young couple, fun to be with. The Doobs introduced me to alcohol and got me back in touch with Janette Hegner (no connection). Until I had an after-dinner drink pressed upon me by the Doobs, I had never tried anything alcoholic. I decided that it was not likely to be fatal, and thereafter indulged from time to time, always in moderation. I have heard that some authors find that strong drink stimulates their creative faculties. I fear it does not work that way for me. I have always partaken only when my brain is not going to be called on for work soon thereafter.

The Doobs did not actually play matchmaker and bring Janette and me back together. They only let me know that she was in the neighborhood. That was enough. She was teaching at the Rye Country Day School in Westchester County and, after a while, I got in touch with her. On our first date, we went to see a play—I think it was *Farewell to Arms*. On our third date, we became engaged. That was in the spring of 1934. In May I sent flowers to Janette and they were delivered, by chance, on Mother's Day. Her fellow teachers wondered if that wasn't a bit premature.

By then my plan to spend the next year in Europe—although in Copenhagen rather than Leipzig—had crystallized. (The reality had to await renewal of my NRC Fellowship.) Many times since, I have thought that I should have tried, one way or another, to marry that summer and take Janette with me. It would have been a much better experience for both of us than spending a year apart. And, as I realized once I was actually in Copenhagen, Janette and I together would have been able to forge closer ties with the Bohr family than I was able to do alone. Bohr's remarkable wife, Margrethe, took an interest in all that he did and was an important advisor to him. (Back in 1911–1912, they, too, postponed marriage for a year while Bohr studied in England.)

But with no money and no certainty of any support in the fall, I could see no way to get married at once. Both Janette and I assumed as a matter of course that our wedding would wait until the summer of 1935. (Being wiser than I, Janette may have felt that a year for reflection was not a bad idea.) She did urge me, if not to study in Italy, at least to visit it while I was in Europe, because she had so much loved her year there working on the history of the Italian unification movement (the *resorgimento*).

In the hope that I would be in Copenhagen that fall, I found a young Dane who worked at the Danish consulate in New York and arranged for him to tutor me in Danish. I must have talked to him about Janette. He said to me once, "You ought to marry that girl and take her to Denmark."

"Where would I get the money?" I asked.

"Well, borrow it from her father" was his answer. A bold suggestion, but not one compatible with my Yankee pride.

That summer (1934), Janette went off with her father on an automobile trip across the country and I went to my family's farm in Vermont. By careful prearrangement, I talked to her once by phone. She was in La Jolla, California. I phoned at the designated time from a shop in the nearby village of Fair Haven, Vermont. That was the last time we spoke for a year. But the letters went back and forth during that year, at least one a week each way. Since it took, typically, ten to twenty days for a letter to cross the Atlantic, there were always two or three en route each way. (Years later, deciding that my letters to her from Copenhagen were just too sentimental and—to use her word— jejune, Janette destroyed them. Somehow I eventually lost her letters to me. So the written record of that transatlantic romance is gone from my archives.) I arrived back in the United States on June 5, 1935, and we were married in Baltimore five days later. More about the wedding and our early married life is to come in Chapter 7.

6
·

AN INTERNATIONAL FAMILY

ENCOURAGED BY Breit, I shifted my sights from Leipzig to Copenhagen for a second postdoctoral year. Breit knew both Heisenberg and Bohr, and thought that Bohr was right for me. Moreover, as he pointed out, Copenhagen was a crossroads where I would meet all the leading physicists, including Heisenberg, when they came to visit. When I applied to the National Research Council in the spring of 1934 for a fellowship renewal, I said that Bohr "is the best man under whom to investigate the nucleus. He is the man with the great mind and imagination who stimulates and foresees all the others." At age forty-eight, Bohr was already a "senior citizen" of physics, a sort of father figure. His most famous work, applying quantum theory to the structure of the hydrogen atom, had been carried out in 1912–1913, when he was twenty-seven. Since then, he had contributed to the further development of quantum theory and the understanding of atomic structure. His University Institute for Theoretical Physics in Copenhagen attracted physicists from all parts of the world for stays short and long.

Under Bohr's leadership, the "Copenhagen interpretation" of the new quantum mechanics was hammered out in the late 1920s. It was that interpretation that Bohr defended in his famous on-again, off-again debates with Einstein. It is that interpretation that has withstood every onslaught since, but which is still debated—and is still a source of uneasiness for many physicists, myself included. Some are uneasy because the Copenhagen interpretation sweeps away certainty at the core of physics, replacing it with uncertainty. That's what Einstein couldn't accept. I am uneasy for a different reason. I see no bedrock of logic on which quantum mechanics is founded. What is the underlying *reason* for quantum mechanics? I keep asking myself. It has to flow from something else, and that something else remains to be found.

Bohr was noted at the time for his idea of complementarity, enunciated in 1927. It is the idea that there are mutually exclusive ways of describing an event or a process, such that a description based on one kind of measurement may make it impossible to offer a description based on a different kind of measurement. Different ways of examining the system under observation are "complementary." Think of the blind men who reached different conclusions about the nature of an elephant depending on what part of the beast they touched. Quantum physicists examining a system not only reach different conclusions depending on how they "touch" the system; they find that the very act of making one measurement rules out making a different kind of measurement at the same time. It is as if when one blind man touches the elephant's trunk, the elephant's leg disappears into a kind of fog, escaping the touch of another blind man. Yet if the blind men return to the "quantum elephant" later, they can touch the leg—provided they don't try to grasp the trunk at the same time.

Bohr introduced complementarity into the framework of quantum physics as a generalization of Heisenberg's uncertainty principle, which puts limits on the precision with which certain specific variables, such as energy and time, can be measured simultaneously. Later he came to see it as an idea of broader scope, with application outside of physics—in biology and psychology, for instance. I even heard him refer once, perhaps tongue in cheek, to the complementarity of love and justice in dealing with children. Nowadays, one can find references to a "principle of complementarity" in the Baha'i view of marriage, in a judicial opinion on the separate roles of the U.N. Security Council and the courts, and in a recommended balanced diet for vegetarians (to name a few).

Most of the friends I talked with about my wish to work with Bohr encouraged me enthusiastically. At the same time, they were a little awestruck. It was as if a little-known, fledgling sculptor in 1523 had stated his intention to go and work with Michelangelo. Once, when I was talking with Harvey Hall, an older postdoc in Breit's group at NYU, about a subtlety in the theory of absorption of radiation by atoms, I said, "Wouldn't it be great to be able to talk over this point with Niels Bohr!"

"You have to be a mature physicist before you can get much out of being with Bohr," Harvey answered.

I didn't let Harvey Hall's little put-down deflect me. Copenhagen seemed

Niels Bohr's University Institute for Theoretical Physics, early 1940s. Renamed the Niels Bohr Institute in 1965, it remains one of the world's important physics centers. *(Courtesy of Niels Bohr Archive, Copenhagen.)*

the right place, and Bohr the right person, if I was to get deeply into the mysteries of the quantum. That was what excited me. It appeared to be the path to the fundamental laws of nature. Breit encouraged me and put in a good word with Bohr, who accepted me. Most important, the NRC renewed my fellowship for a second year.[1] My study of the Danish language was not to be in vain. (In Copenhagen, it was needed only *outside* the institute. Within the institute, German and English were the principal languages.)

Bohr lost his oldest son, Christian, seventeen, in a boating accident shortly before I arrived in Copenhagen. It was a blow from which he had great difficulty recovering. He kept working, and kept everyone at the institute on their toes, but his manner was abstracted and subdued for some months. Flashes of humor and exuberance were rare. Bohr was a thinker—a meditator, really—but he had a strong physical presence, the legacy, perhaps, of his early athletic prowess. To many Danes, Bohr was as well known for his exploits on the soccer field (or football field, as it is called there) as for his achievements in physics. Not much taller than I, he was still muscular although no longer slender. He often rode his bicycle to work—not so unusual, really, in

[1] Breit, in supporting my renewal application, wrote favorably of my abilities in physics and mathematics, and added that I didn't show the "nervous irritability frequently met with in men of high theoretical ability."

a city filled with bicycles. One day, soon after I arrived in Copenhagen, as I was cycling toward the institute I saw a worker pulling down vines from the front of the building. To my astonishment, as I got closer, I saw that the worker was Bohr. This may have been part of his personal therapy after his son's death.

Bohr was famous in laboratories and institutes all over the world for his special mannerisms. There was the invariable pipe, being incessantly lit but never staying lit. There was the mumbling speech that everyone strained to hear. Bohr gave always the appearance of a man thinking deeply, very deeply, with his deep thoughts struggling to find expression. There was the slow pacing and turning at the front of the room. (It was said of him when he visited Los Alamos later that his talk ended when the microphone cord was wound around him as many times as it would go so that he could rotate no longer.)

When another person was speaking at a seminar, Bohr would sit quietly for perhaps fifteen minutes. Then, if the subject engaged his attention, he would shift gradually from passive listener to active participant. First a question. Then rising to make a longer point. The end of the hour might find Bohr at the board with the original speaker listening and trying to get in a few words. In private discussions with me, he was somewhat the same way. I would start talking about what I was working on, and Bohr would say, as if his mind were elsewhere, "That's beautiful" or "very interesting." (It was always necessary to "renormalize" Bohr's comments. "Beautiful" meant "probably correct, even if not significant." "Interesting" meant "not quite entirely trivial.") Then, at some point, Bohr might catch, in what I was saying, one thing that intrigued him. On that point we could have an animated discussion for a while, but it would die down if I did not happen to say something else to spark his interest.

Dialogue was Bohr's mode of making progress in physics, be it with a lone postdoc in his office or with a large group in a seminar room. He always liked to have at least one other person present, even if he were lost in his own thoughts. When the moment came that he wanted to pull forth an idea and examine it, he needed a foil, someone with whom he could toss the idea back and forth. Léon Rosenfeld filled this role for some years. So did Bohr's son Aage. (These "foils" were first-rate physicists themselves. Rosenfeld made notable contributions in theoretical physics, and Aage Bohr won a Nobel Prize for his work on nuclear structure.) One morning, on my way to work, I saw Niels and Aage in a car, also headed for the institute. The older Bohr, at the wheel, was in animated conversation with his son, complete with gestures. Knowing Bohr, I felt that he had forgotten all about the fact that he was driving a car. I prayed that he would reach the institute in one piece. (He did.)

Although Bohr never had a hard word for any other person and could

Niels Bohr, 1934.
*(Photograph by John Wheeler, courtesy of Niels Bohr Archive,
Copenhagen.)*

always find something good to say about anybody's work, he did expect visitors
to Copenhagen to adhere to the institute's culture of questioning, searching
dialogue—of full respect for the other person's point of view. During the year
I was there, I saw two instances in which a visitor was too self-assured, too pontifical for Bohr's taste. In each case, Bohr quietly arranged for the person to
take a position at another institution. I never knew for sure, but always suspected that Bohr talked over these troubling cases with Margrethe, and relied
on her advice on how to deal with them. She was a very sharp observer of the
people at the institute and their interactions with one another. Margrethe
stood about my height with the bearing of a princess. She was slender, with a
lovely face, her brown hair parted in the middle. She was born a Nørlund; her
close relatives included explorers of Greenland, the Arctic, and the Antarc-

Margrethe Bohr, 1934.
*(Courtesy of Niels Bohr Archive,
Copenhagen.)*

tic, educators, and government ministers. As matron of the House of Honor, the mansion provided to Bohr by the Royal Danish Academy's Carlsberg Foundation, she enriched the life of the institute, and she was kind enough to show a personal interest in me and my engagement. Later, Janette and I got to know her much better, and gained even greater respect for her judgment and thoughtfulness.

It would be hard to name any physicist of note in the 1930s who did *not* spend some time at Bohr's institute. Paul Dirac from Cambridge was among those who visited. He came on the recommendation of Bohr's old mentor, Ernest Rutherford. Dirac was as famous for his economy of speech as for his brilliant physics. After Dirac had been in Copenhagen for a while, Bohr, as he told me later, ran into Rutherford at a meeting, and said, "This Dirac, he seems to know a lot of physics, but he never says anything."

"Let me tell you a story about a parrot," replied Rutherford. "Once a man bought a parrot from a pet store and tried to teach it to talk. The parrot

Paul Dirac with Werner Heisenberg, Brussels,
1933.
*(Courtesy of Niels Bohr Archive,
Copenhagen.)*

wouldn't talk. So the man took the parrot back to the store and asked for
another, explaining to the store manager that he wanted a parrot that talked.
He took the new parrot home and had no better luck. It wouldn't talk either.
So he went back angrily to the store manager and said, 'You promised me a
parrot that talks, but this one doesn't say anything.' The store manager paused
a moment, then struck his head with his hand, and said, 'Oh, that's right!
You wanted a parrot that talks. Please forgive me. I gave you the parrot that
thinks.' "

Because of Bohr's style and his special role as stimulator and guide in
Copenhagen, it is difficult to know how many significant achievements by
others were built on his original ideas. Many, I suspect. Neither then nor at
any time later did I ever hear Bohr breathe a word of regret that someone
else got credit for an idea of his. In his mature years, at least, he seemed gen-
uinely indifferent to personal credit. Margrethe was more sensitive to the
issue, and sometimes hinted that others had borrowed her husband's ideas
without giving credit. There was, for me, a somewhat regrettable side to Bohr's
unconcern for priority. It meant he wanted to ruminate on a topic at length,
patiently polishing its details, when the typical physicist would have said, "Let's
publish what we've got, and deal with further details in a later publication."

Once, during that year in Copenhagen, Bohr suggested that Milton Plesset and I not publish work we had done on the interaction of gamma rays (high-energy photons) with atomic nuclei. We had shown, using what is called the "causality principle," that from a knowledge of the *absorption* of photons in a heavy nucleus, one could deduce a limit—that is, a maximum magnitude—for the *scattering* of photons by that nucleus. Although we were happy about our work, we respected Bohr's judgment that it was incomplete, and we set about refining it. As it turned out, we never published this piece of research. The year ended. Plesset and I underwent our own "scattering" (to other places) and our own "absorption" (in other projects). The reader might wonder whether we felt even a trace of bitterness toward Bohr over this incident. The answer is not just no; it is more than no. It didn't even cross our minds that there might be anything to be bitter about.

I have long since lost track of how many times I have traveled to Europe. But the first impressions from the first trip remain clear and special. For $55 in that September of 1934, a Diamond Line freighter carried me from New York to the Belgian port of Antwerp. Among my four cabin mates was a young German who was part of a small reverse migration of that time—he was abandoning his farm job in America to "join Hitler." When the freighter landed in Antwerp, I had only half a day to explore a city whose art, culture, and stock market existed before the first European had set foot in America. There I discovered the remarkable Plantin-Moretus Museum. In a building that matched my image of what a college in Oxford or Cambridge must look like, Christophe Plantin had created one of the great printing and publishing houses of the sixteenth century. When the city of Antwerp acquired the property and turned it into a museum, it had been in his family for over three hundred years.

The next morning, after an overnight train ride to Cologne, Germany, I encountered another young German, this one in swastika-decorated black shirt. Apart from a waiter, he and I were the only people in the cavernous railroad station restaurant. He marched ostentatiously and noisily back and forth at his end of the room, giving me occasional hostile glances. I was happy to board my train and be on my way. The impression of a depressed economy conveyed by the deserted restaurant was reinforced that day as the train wound through the Ruhr Valley, once (and again later) the industrial heart of Germany. Through the train window I could see factories that were shut down and people ambling in the streets, seemingly without purpose.

In Warnemünde, Germany, the train and its passengers were loaded onto a ferry for the hour-and-a-half trip to Gedser, Denmark. As we rolled on through Denmark's lush farming country, a Dane sitting next to me asked why

I was going to Copenhagen. "To work with Niels Bohr," I answered. "Aha," he responded, "the great football player." My first impression of Copenhagen: throngs of people, about as many on bicycles as on foot. After a night in a hotel, I set about looking for a room. All I could find in the historic downtown district was a room with only a skylight, no window, and with a price beyond my budget. So I turned for help to Frøken [Miss] Betty Schultz, helpful friend to every institute visitor. Years later she became *Fru* [Mrs.] Schultz, apparently as a result of the passage of time, not because of marriage. With her help, I landed in a pleasant boardinghouse at Strandvaenget 8, half a block from the main coastal artery, Strandvej, and a block from the Øresund, the sound separating Denmark and Sweden. The boardinghouse provided breakfasts and dinners, and in due course I adjusted to buttermilk soup. Lunches I took at the institute. All that remained was to buy a used bicycle, which I did at once. By my second day in Copenhagen I had met Bohr and was doing physics.

But my head emerged from physics now and then. As always in my life, I took time to read, and sometimes I went to the theater and the opera, with the Plessets or other institute friends. Milton Plesset, from Oppenheimer's group in California, was also an NRC Fellow. His wife was the daughter of a California specialist in mental illness, and once I went along as she and Milton tracked down the identical twin of a mentally ill patient of her father. Astonishingly, we found that the twin in Denmark had contracted mental illness of the same kind and at about the same time as the twin in California. A Baltimore Unitarian friend, Fred Bang, brought me into contact with his relatives in Copenhagen, who were kind enough to invite me several times for dinners and excursions. Through them I got to see the historical treasures of Copenhagen's Rosenberg Castle as well as castles elsewhere in the country. And then there was my correspondence with Janette, which kept us close during the year. I must have written some fifty letters to her.

I had been in Copenhagen no more than a couple of weeks when, with Bohr's encouragement, I boarded a train for England to attend the London International Conference on Physics. As was typical for those times, the conference covered all of physics. Conferences of that generality are rare today. Yet there was a principal focus of excitement—cosmic rays.

One cannot pinpoint a single date when cosmic rays—energetic radiation from outer space—were discovered. Bits and pieces of evidence accumulated over several decades. At the beginning of the century, scientists noticed that gases in containers were being ionized apparently by some agent entering the containers from outside. In 1911–1912, the Austrian Victor Hess pioneered balloon flights to study this ionization, work that eventually—in

1936—earned him the Nobel Prize. He and others after him found that what-ever was causing the ionization became more intense at higher altitude. Then came studies in deep lakes, and observations of individual tracks in cloud chambers. But only after the discovery of the latitude effect (see page 99) in 1927 and its interpretation in 1929 as arising from the deflection of charged particles by Earth's magnetic field did the modern picture of the primary cosmic radiation emerge. While I was a student, evidence accumulated that particles hurtling toward Earth from outer space are positively charged and extremely energetic. By the time of the London conference, cosmic rays had moved, so to speak, front and center in physics.

Cosmic radiation had yielded up one new particle, the positron, in 1932, and it was natural to wonder whether there might be others. Experimenters, such as Bruno Rossi and his group in Italy, were turning up new information about the altitude effect, the latitude effect, and the fantastic penetrating power of cosmic rays. Theorists were hard at work trying to understand the complex progress of cosmic rays passing through the atmosphere—complex because air is not transparent to cosmic rays, as it is to sunlight. The particles from space strike atoms and nuclei in the atmosphere, and, as a result of the collisions, are slowed and deflected and may create new particles. What is observed by experimenters at the surface of the Earth—or even aloft in bal-loons—is quite different from what irradiates the top of the atmosphere.

In 1934 there were still more puzzles than answers surrounding the cos-mic rays. Just what charged particles were coursing through the vacuum of space beyond the atmosphere was not known (we now know it is mostly pro-tons, with some nuclei of heavier atoms). Just what new particles were being created in the maelstrom of collisions in the atmosphere was not known. (Electrons, positrons, and photons were recognized; we now know that many other particles are also created.) To me, cosmic radiation was a romantic sub-ject. Part of the romance was in the experiments, conducted often on icy mountaintops or in instrument packages carried aloft by balloons and borne by winds to unknown destinations. For me as a theorist, there was romance, too, in the contemplation of particle collisions at energies far higher than any accelerator could reach, energies sufficient to test the limits of electrody-namic theory, and sufficient, through conversion of energy to mass, to make new particles never seen before.

In December 1933, on my trip with Hugh Wolfe to Boston, I had heard Robert Oppenheimer speculate that some of the perplexing results of cos-mic-ray experiments might mean that electrons, positrons, and photons at very high energy do not follow the same laws that we had come to trust at low ener-gy. The energy he cited where trouble might begin was tied to a famous numerical constant in physics called the "fine-structure constant." This con-

Hans Bethe, Ann Arbor, Michigan, 1935.
(Courtesy of AIP Emilio Segrè Visual Archives,
Goudsmit Collection.)

stant, which measures the strength of interaction between electrons and photons, is small. Its magnitude is $\frac{1}{137}$. Oppenheimer suggested that the inverse of this constant—that is, 137—multiplied by the rest mass of the electron might set the energy at which the familiar laws of pair theory break down. (I won't try here to explain his reasoning.) This energy, some 70 MeV (million electron volts), although larger than any energy then available in accelerators, was much smaller than typical cosmic-ray energies. I didn't care for Oppenheimer's idea. I had no more valid reasoning for opposing it than he had for proposing it. Both of us were working more from intuition than from logic or mathematics. To me, Dirac's relativistic quantum theory (pair theory) was just too beautiful to be wrong.

At the London conference, Hans Bethe presented his impressive calculation of what came to be called an electron shower: A high-energy photon is created somewhere, perhaps by the deflection high in the atmosphere of a heavy charged particle such as a proton. That photon, passing close to an atomic nucleus, can create an electron-positron pair—an example of matter being made from energy. The electron and positron fly forward until one or both are deflected in an encounter with an atomic nucleus. In such a deflection, more photons are emitted. Some of these photons may have enough energy to make more electron-positron pairs. Each of these, in turn, can make new photons. And so the process continues until all the original energy is dissipated in the masses of electrons and positrons, the energies of the residual photons, and the recoil energy of nuclei. Showers in which a single photon gives rise to a cascade of hundreds or even thousands of particles are not uncommon.

A shower is "soft." When it hits solid matter, it is like spray from a garden hose hitting a piece of stretched wet cloth: some of it gets absorbed; what gets through is a dribble made up of numerous but less energetic droplets. But observations showed that the cosmic radiation has a "hard" component that passes through a plate of solid matter as an arrow would pass through the stretched cloth. Either there is more in the cosmic radiation than electrons, positrons, and photons, or the theory worked out by Bethe is wrong. The total energy in a large shower is much greater than the 70 MeV ($137mc^2$) that Oppenheimer suggested might set a limit on the validity of pair theory.

Long before the London conference, Bohr had developed a method of treating the motion of very energetic charged particles passing through matter, his "equivalent field" method. By applying this method to the analysis of the penetrating cosmic rays, E. J. Williams of Manchester University provided what might be called a climate of opinion—an intellectual background—for the discovery of a new particle. It took theory to clarify what the observations were revealing.

Williams avoided any direct appeal to the theory of high-energy electron processes, which might conceivably hold surprises (as Oppenheimer had suggested). Instead he looked at the interaction of a fast electron with a block of lead not in the "laboratory frame of reference" (electrons flying toward a motionless lead block) but in the electron's frame of reference (a swarm of lead nuclei hurtling past a motionless electron). From this perspective, the electric field surrounding a moving nucleus is equivalent in its effect to a beam of radiation consisting of photons of relatively low energy. This trick of looking at the process from a new vantage point made it suddenly clear that no new physics needed to be invoked. The effects of these equivalent low-energy photons were well established. The inescapable conclusion was that it is totally impossible for a fast electron to penetrate inches of lead. Whatever was getting through the lead blocks as the "hard" component of the cosmic rays could not have been electrons. It must have been something else.

Over the next couple of years, evidence gradually accumulated for a new particle intermediate in mass between the electron and the proton. By 1936, especially through the measurements of Carl Anderson and Seth Neddermeyer at Caltech, it was clear that there existed in the cosmic radiation near the Earth particles some two hundred times more massive than the electron (about one-ninth the mass of a proton). These particles were dubbed mesons, after a Greek word meaning "intermediate." According to the Bohr-Williams theory (and Bethe's shower theory), these particles would not be stopped in lead the way electrons are. They are the hard component of the cosmic radiation. (What was not yet known at that time is that there are in fact several different kinds of mesons. The picture that we now have, briefly stated, is this: Protons from outer space strike atoms and nuclei high in the atmosphere. These interactions produce various kinds of particles, but mainly two kinds: photons—which go on to make showers—and a kind of meson called the pion. The pions have lifetimes so short that most of them do not reach the surface of the earth. They decay into another kind of meson called a muon. Many of the muons do reach the surface of the earth, thanks to their "long" lifetime of two-millionths of a second, and thanks also to relativity, which tells us that for particles in high-speed motion, time is "dilated," making the particles live longer, according to Earth-based clocks. These muons are the principal part of the hard cosmic rays. Hundreds of muons strike each of us every second.)

In Copenhagen, I made my first acquaintance with Hungarian scientists, who have been showing up in my life ever since. Edward Teller came along on the boat train to London with his bride of a few weeks, Mici. Edward was black-haired and bushy-browed, with intense dark eyes. Mici was slender, lively, and likable. I struck up a friendship with the Tellers, which has endured, through thick and thin, to the present day. (My first scientific collaboration with Edward came in 1938, when we wrote a paper on rotational energy states of nuclei.)

Also on the boat train was George Hevesy, an affable aristocrat from Budapest and a member of Bohr's institute. He told me of his interest in heavy water, a substance that had been identified only two years earlier by Harold Urey at Columbia University. With a flair for the dramatic, Hevesy had drunk some heavy water, without ill effect, a stunt that got good press coverage in Denmark. At forty-nine, Hevesy was older than most of the rest of us. He had served with the Austro-Hungarian Army in World War I and, like many other scientists of that period, had moved around a good deal among major centers of science in Europe. Even today, physicists often feel an allegiance to their colleagues throughout the world and to their discipline that is stronger than their allegiance to any one institution. As for me, institutions—North Carolina, Princeton, and Texas—have provided what I need most, students and stimulating colleagues. But their walls have not been confining. Colleagueship extended around the globe has been critical to my professional life.

George Hevesy, 1944.
(Courtesy of Niels Bohr Archive, Copenhagen.)

One of the reports that caught my attention at the London conference dealt with the scattering of gamma rays by solid lead, by the experimenters Louis Gray and Gerald Tarrant from Cambridge, England. The process they studied seemed to me to involve fundamental physics, and it gave a puzzling result, a strong scattering backward. The gamma ray (a high-energy photon) interacts with the strong electric field near the lead nucleus. Among the possible outcomes is the creation of electron-positron pairs. The gamma ray can also create a so-called virtual pair, an electron and positron that exist only momentarily and annihilate to create another gamma-ray photon. The subject intrigued me also because of my interest in the relation between scattering and absorption. Milton Plesset and I decided to work together to see if we could generate any new insights into the processes at work when a gamma ray flies past a lead nucleus. (This was not the work mentioned earlier in this

chapter that Bohr had trouble accepting; that work concerned a photon interacting within the nucleus.) We completed one paper the following spring on
the process in which the incoming high-energy photon creates three particles:
an electron, a positron, and a lower-energy photon. This was of some interest
in analyzing cosmic-ray events. We undertook another, more ambitious paper
in which we envisioned a mini-shower, a whole cascade of events, all within
one atom. In the shower in air, as studied by Bethe, no more than one event
occurs near any given nitrogen or oxygen nucleus. But in the much stronger
electric field near a highly charged lead nucleus, several events might occur
in sequence. This paper, like the one that Bohr found only "interesting," got
shelved and was never published. The amount of calculating that would
have been necessary to push the idea to meaningful conclusions was beyond
our power then.

It was during my year in Copenhagen that Bohr developed the "compound-nucleus model," and I participated in many discussions about it. Bohr
realized that what a nucleus did after it got excited (had energy added to it)
was largely disconnected from how it got excited. A nucleus might, for example, absorb a proton and thereby gain 8 or 10 or 12 MeV of energy (some of
it from the energy of the bombarding proton, some contributed by the attractive force holding the proton within the nucleus). This energy is distributed
quickly among all the particles in the nucleus. Sitting there at the center of
the atom is then an excited nucleus that has "forgotten" how it got excited. It
proceeds to dissipate its excess energy in accordance with its excited state,
unrelated to how that state got formed. It is rather like a patron at a bar who,
after being needled by the person on the next barstool, flies into an agitated
state of high energy, and then, forgetting what caused his excitement, unleash-
es his energy on others around him.

As an embodiment of the compound-nucleus model, Bohr soon adopted
the "liquid-droplet model." Like many other ideas, it was "in the air" in
Copenhagen that year. I recall discussions of the liquid-droplet model in the
spring of 1935. According to Rudolph Peierls, who headed the British dele-
gation to Los Alamos and who later edited the volume of Bohr's collected
works devoted to nuclear physics, this model took shape gradually in Bohr's
mind from the fall of 1934 to the spring of 1936. I extended and applied the
model in the years that followed, and it proved crucial to the work that Bohr
and I did on fission in 1939. George Gamow's earlier suggestion that a nucleus behaves like a liquid droplet came out of his effort to understand nuclear
masses rather than nuclear reactions.

Suppose that the nucleus behaves like a solid. Then a particle fired at it
might crack it or might bounce off of it, and, if it bounced off, might leave
some energy behind, recoiling with less than its initial energy, like a "dead"

Lev Landau, George Gamow, and Edward Teller *(adults, left to right)*,
with Niels Bohr's sons Aage and Ernest *(children, left to right)*,
in the yard of Bohr's institute, 1931.
(Photograph probably by H.B.G. Casimir, courtesy of Niels Bohr Archive, Copenhagen.)

tennis ball bouncing weakly from the floor. But the incoming particle would
not easily lose its identity by amalgamating with the nucleus it hit. Now suppose that the nucleus behaves like a gas. Then a particle fired at it might fly
right through it, or might bounce off one of the nucleons within the nucleus. But again, the incoming particle would not easily be slowed by multiple
collisions to become itself a part of the gas, because there is too much empty
space between the nucleons and not a very large number of them. (Even in a
uranium nucleus containing 238 nucleons, only about 10 nucleons stretch
across a diameter from one side of the nucleus to the other.)

Finally, suppose that the nucleus behaves like a liquid. Then it is easy to
imagine that a particle fired at it splashes into the nucleus, sharing its energy
with all the particles, and itself becoming part of the nucleus. The liquid-
droplet model turns out to work well in certain applications, although the

gas model also describes features of the nucleus. What I later called the "collective model," very similar to what Aage Bohr and Ben Mottelson have called the "unified model," incorporates both liquid-like and gas-like features of the nucleus. To oversimplify a bit, nucleons well inside the nucleus move about freely, like gas molecules, but are constrained by a surface that behaves much like the surface of a liquid drop. It has surface tension and it can deform and rotate.

Working with Bohr and working with Breit were complementary experiences. Breit and I dived into problems, did calculations, and wrote papers. We didn't spend much time pondering the deepest questions of physics. With Bohr, I spent more time thinking, less time calculating and writing. My work in Copenhagen generated only one paper, and that was with Plesset, not Bohr. (As I mentioned, two other papers with Plesset never made it to published versions. I also used some time in Copenhagen to complete two papers with Breit, work left over from the previous year at NYU.)

Breit taught me new mathematical and calculational techniques. Bohr taught me a new way of looking at the world, a new way of raising questions. Undoubtedly, my style of doing physics owes something to both Breit and Bohr. It might be better to say *styles*. I have always loved pushing the mathematics beyond formalism, to get numerical results that can be turned into pictures and compared with experiments. At the same time, I have had a lifelong fascination with the meaning of the quantum and the urge always to think about what physics might be like twenty years hence, not just the day after tomorrow.

Another benefit of being in Copenhagen was the chance to get acquainted with many of the outstanding physicists of the time. One after another, they passed through (except for those from the Soviet Union, who were rare visitors). Physics, not politics, dominated our conversations, but we could hardly ignore the political currents of the time. Hitler was newly in power in Germany. I took special pains to talk with the Germans who visited the institute.

Heisenberg, boyish at thirty-three with short-cut, blond hair, was the "good German." In our talks, he tried to steer clear of politics. Always circumspect, he neither praised nor condemned Hitler. He was a patriot. My impression, both then and later, was that he wanted to keep his country and its people separated in his mind from the leaders and their policies. Even though Heisenberg and I shared no intimacies, I felt sympathy for him. I could understand his commitment to his country. In science, Germany led the world. In culture and the arts, it had a centuries-long record of achievement. I was inclined to believe, as he no doubt did, that an immoral dictatorship was a transitory evil, something a great nation could endure without

Werner Heisenberg and Niels Bohr, Copenhagen, 1934.
(Photograph by Paul Ehrenfest, Jr., courtesy of AIP Emilio Segrè Visual Archives.)

lasting harm. Of course, I was wrong. So was Heisenberg, who never openly opposed the Hitler regime.

Max Delbrück, an assistant to Lise Meitner in Berlin, visited Copenhagen to work both on quantum theory and on his emerging interest in biological physics. He was carefree and witty, and we never talked politics. A few years later, in 1937, he left Berlin for what was to be a temporary stay at Caltech as a research fellow. When war came, he accepted a position at Vanderbilt University in Tennessee and quickly gained a reputation for brilliance in his new field of biological physics. He brought to biology just what biology needed at that time, a deep knowledge of the principles of physics. Bohr, sensing the role that fundamental physics could play in molecular biology and genetics, had encouraged this new direction for Delbrück, even though it was far removed from the rest of the work at his Copenhagen institute. Lesser people than Bohr might have regarded biological physics as a backwater. After World War II, Delbrück returned to Caltech and stayed there until his retirement in 1977, being honored with the Nobel Prize in 1969.

The noted atomic physicist James Franck, a Jew, showed up in Copenhagen while I was there, remaining a year until he emigrated to America and a professorship at Johns Hopkins. After Hitler came to power, he felt vulnera-

James Franck, Copenhagen, 1934.
(Courtesy of Niels Bohr Archive, Copenhagen.)

ble despite his service in the German Army in World War I and despite the Nobel Prize he won in 1925. He resigned his position in Göttingen and, before coming to Copenhagen, managed to find positions outside Germany for all the staff members in his laboratory who lost their jobs with him. "It will take my country fifty years to recover from this," he told me. I never forgot those words, yet nearly ten of those fifty years passed before I fully embraced his fears. Franck was later my valued colleague at the Met Lab in Chicago. Following the defeat of Germany in May 1945 but before the end of

Wolfgang Pauli and Niels Bohr, with a "tippy top" in Lund, Sweden, 1954.
(Courtesy of AIP Emilio Segrè Visual Archives, Margrethe Bohr Collection.)

the war with Japan, he headed a committee whose Franck Report[2] advocated that the atomic bomb be demonstrated in an uninhabited area to representatives of all nations before it was used in war. I myself doubted the wisdom of this recommendation, but, in any event, it was rejected by America's military and political leaders. Just as Franck had seen very early the threat posed by Hitler, he saw sooner and more clearly than most the likelihood of a massive nuclear arms race after World War II, and warned of it in the Franck Report.

During that memorable year, I also came to know Otto Frisch, Lise Meitner's nephew, another German Jew who was in Copenhagen because he was leaving Germany for good. In Chapter 1, I told the story of how he and Meitner came to understand nuclear fission.

Bohr's institute has been called the Vatican of physics. To be sure, it drew pilgrims from every land. And, to be sure, Bohr presided as a benevolent, fatherly figure. (In a 1932 skit based on Goethe's *Faust*, written by Delbrück

---

[2] The Franck Report, dealing with social and political implications of the atomic bomb, was the sixth and last of a set of panel reports requested by the Secretary of War. Officially entitled "A Report to the Secretary of War—June 1945," it appears as an appendix in Alice Kimball Smith, *A Peril and a Hope* (Chicago: University of Chicago Press, 1965).

Drawings by George Gamow of Niels Bohr, Albert Einstein, and
Wolfgang Pauli, from the 1932 *Faust* parody, which appears as an
appendix in Gamow's *Thirty Years That Shook Physics.*
(Courtesy of Dover Publications.)

and illustrated by Gamow, Bohr appeared not as the Pope but as God himself,
constantly bedeviled by the fallen angel Wolfgang Pauli.)[3] Yet, despite the
touch of formality that did enter into my relationship with Bohr, I never cared
much for the Vatican metaphor. Here was an institute free of hierarchy, free
of pomp, nearly free of ceremony. Bohr presided with a very loose rein. People came to Copenhagen not just to interact with Bohr, but to form and
renew friendships and professional bonds with fellow physicists from everywhere. To me, the best metaphor for Copenhagen is family. That is where I
joined the international family of physics.

[3] This hilarious parody was performed by physicists to wrap up the spring 1932 conference in Copenhagen. Among its characters are the British astronomer Arthur Eddington
as an archangel, the Dutch theorist Paul Ehrenfest as Faust, and the newly postulated neutrino as Gretchen.

7
·
SETTLING DOWN

WHILE IN London for the International Conference on Physics in October 1934, I bought two green Chinese vases at Liberty of London for Janette and had them shipped to her, without bothering to check on what duty might be imposed on them. It turned out to be hefty, but Janette was kind enough not to mention it at the time. It was the first, but not the last, time she would conclude that I did not have a good head for practical detail.

In those days, engaged young ladies didn't pursue professional careers. But Janette was not one to sit around for a year planning a wedding and dreaming of married life. She moved back in with her family in Baltimore and found a job in the Pediatrics Department of Johns Hopkins Hospital. When spring rolled around, she and her mother set about planning a wedding. They did such a good job that all I had to do was buy some suitable clothing, enjoy the prewedding reception at Joe and Maria Mayer's, and get myself to the Hegners' house on time. I have made a few mistakes in my life, but marrying Janette wasn't one of them. Maybe it was blind luck. But it's a marriage that has worked for more than sixty years.

And in those days, sensible young men didn't get married without firm employment prospects. In 1935, the depth of the Depression, the job market in physics, as in so many other fields, was dismal. But I was not one to worry. I had been lucky so far in my young life. Everything had clicked. I was a natural Pollyanna, assuming that I would find a good academic job at the end of my fellowship year in Copenhagen. That spring, I pulled out my battered portable typewriter and wrote letters to Harvard and Chicago, as I remember, and perhaps one or two other places. No job offers came back. Then, out of the blue, I heard from Arthur Ruark at the University of North Carolina in Chapel Hill. He offered me an assistant professorship there. He had heard me give a talk at a meeting of the American Physical Society a year earlier (before I went to Bohr's institute), on the subject of scattering and laws of force between particles. He must have liked what he heard.

UNC was a good university with a strong physics program—and its offer was the only one I had. I accepted. My first-year salary was to be $2,300 (a little less than what Janette was offered to go back to Rye Country Day School).

Our wedding day was Monday, June 10, 1935. We were married at about 10:00 A.M. so that my sister, Mary, could take an exam at Goucher College later that morning. Only she and Joe and Rob, the Hegners, and a few friends were present. My parents were in England. My father had been invited to a library meeting in Spain, all expenses paid. It was too good an opportunity to pass up. Neither he nor my mother had ever been to Europe, so they decided to make a summer of it—although, unfortunately, she missed the chance to practice her Spanish. He went ahead to the meeting in Spain, then met my mother in England, where they toured all the places they had read about in the English literature they both loved.

Between Spain and England, my father paid me a surprise visit in Le Havre, France. The *Europa*, which I had boarded in a north German port, docked for a few hours in Le Havre to pick up additional passengers. My father, after sitting up all night on a train from the south of France, came to the dock and spotted me standing at the rail. With his limited French but considerable sales ability, he talked his way onto the ship and appeared, to my surprise and delight, at my side for a short visit. Then, on the Atlantic crossing, I was able to exchange radiotelegrams with my mother, whose ship was steaming toward England at the same time.

After the ceremony, Janette and I climbed into my parents' old Dodge sedan and drove to Charlottesville, Virginia, for our wedding night. The next day, after visiting the University of Virginia campus so beautifully laid out by Thomas Jefferson, we drove on to Chapel Hill, spending several nights at the Carolina Inn while house hunting and getting acquainted with my future colleagues. We found a pleasant house to rent at 416 Pittsboro Street. It served us well for the time we spent in Chapel Hill, as our family grew from two to four.

Most of that first summer of our married life we spent in Benson, Vermont. Thanks to my parents' extended stay in England, we had the use of their Benson house (the one that had belonged to my great-uncle Oscar

Janette, early 1936.
*(Courtesy of AIP Emilio Segrè Visual Archives,
Wheeler Collection.)*

Bump) as well as their car. My three unmarried siblings were in the neighborhood at the same time, staying at the family's hillside camp on our farm. To Janette's delight, the house was full of books. It was here that we started our lifelong practice of reading aloud to each other. (On our first night in the Benson house, we left the light on late and the town's telephone operator called to ask, "Is everything all right?") I spent a lot of my time doing physics — more time, no doubt, than Janette had foreseen. I also had the task of building a fire in the woodstove in the morning and keeping it going all day.

As the end of the summer neared, we returned the Dodge to my parents in Baltimore and bought a used black Ford roadster, a two-seater with a rumble seat, for our trip south. Everything about the Chapel Hill community and the university appealed to us. I had a stimulating environment in which to teach and do research. We made good friends and enjoyed the town and the surrounding countryside. Our first two children were born in Duke University's hospital in neighboring Durham. When, after three years, I accepted an offer from Princeton University and we prepared to leave, Janette wept. I could understand.

Arthur Ruark, who had recruited me to come to UNC, was the best known of my fellow physics colleagues. He and Harold Urey had written the definitive book on modern physics of the time, *Atoms, Molecules, and Quanta*, published in 1930. (Not long after its publication, I had won a debating prize at Johns Hopkins and was offered a choice of books. That's the one I chose, and I proceeded to devour it.) Ruark, besides being a master of the new physics, was a forceful department chair whose aggressive tactics disconcerted a few people at the university as he worked at building a first-class department.

While in the Pittsboro Street house, I lovingly constructed a mini-greenhouse from zinc, wood, and glass, and mounted it in a window. I was so proud of it that when I visited my old Chapel Hill Physics Department in 1993, I asked my friend Jimmy York to drive me by the house to see if my handiwork had survived the fifty-five years since we moved out. Well, it wasn't there. I had survived longer than the window-box greenhouse.

When I returned from Copenhagen and settled in at North Carolina, I was wonderfully, gloriously free. As an academic scientist, I was privileged, and I knew it. I had teaching duties, to be sure, but I was free to pursue research of my own choosing, without specific obligation to the university, or to any funding agency, or to particular colleagues. (I had just finished my joint paper with Milton Plesset on creation of positron-electron pairs by gamma rays; a few loose ends of work with Gregory Breit on quantum wave functions would soon be wrapped up.) As a new junior professor, I also had that special freedom that goes with being recently hired: few committees, little outside correspondence, no calls for service to professional societies and government agencies, no requests to address general audiences or review books, no letters of recommendation for former students. With time, such tasks accumulate. Although they are gladly undertaken, they add baggage to the senior professor as barnacles add baggage to an old ship.

Working with Herzfeld and Breit and Bohr—and some of the young scientists gathered around them—I had learned a lot of physics and armed myself with techniques of theoretical research. I had also developed a certain "taste" in physics—the acquired layer of intuition (or prejudice!) that helps the practicing scientist pick out of all the unanswered questions the ones that seem most exciting, most worthy of attention at the moment. As the Danish writer Piet Hein put it in one of his pithy "grooks,"

> Problems worthy
> of attack
> prove their worth
> by hitting back.

Piet Hein[1]

I had tried my wings in independent research and found that I could fly. I was twenty-four years old, ready to chart my own course.

Although this was a critical moment in my career, I did not give labored thought to my options. I had too much momentum and was too buoyantly optimistic—and confident—to ponder for long the choice of a research path. I had priorities, and I held opinions about productive and nonproductive directions, but I acquired these almost unconsciously, not as the result of careful, ordered thought. I was driven, above all, by curiosity about the most fundamental laws, those governing the world of the very small, laws on which all else is built—"deep happy mysteries," I like to call them. Yes, there is happiness to be found in the mere contemplation of the deepest mysteries.

It is always difficult to analyze one's own motivations. Was I moved in part by a desire for recognition and applause by my peers and by a wider public? I wasn't aware of it, but perhaps there was a small driving force of this kind buried somewhere in my brain. In any case, gaining the approval of my university and my departmental colleagues was a practical necessity if I was to remain securely employed and was to feel that I was doing my part to repay an institution that granted me so much freedom.

I saw two directions of interest. One was electrodynamics—the study of electrons, positrons, and photons. (Sixty years later, these remain the simplest and the most fundamental particles of which we have any knowledge.) Dirac's quantum pair theory was a thing of mathematical beauty, yet it raised as many questions as it answered. It brought us the concept of virtual particles—particles that are created and annihilated in limitless number all the time and everywhere. It opened the door to a vibrant world of submicroscopic activity, in which even a perfect "vacuum" is filled with fluctuating quantum phenomena. As it turned out, I didn't actively pursue electrodynamics until a few years later, when I did some work on cosmic-ray showers with Willis Lamb and explored action at a distance with my brilliant student Richard Feynman. But I kept up to date on what others were doing, and I kept thinking about the question whether there is not some simpler approach that can skirt the complexities of the infinite jumble of virtual particles. Paradoxi-

---

[1] Piet Hein © Grooks. This grook and the one on page 357 are reprinted with the kind permission of Piet Hein a/s, DK-5500 Middelfart, Denmark.

cally, when I later became interested in gravitation and general relativity, I found myself forced to invent the idea of "quantum foam," made up not merely of particles popping into and out of existence without limit, but of space-time itself churned into a lather of distorted geometry.

The other interesting direction was nuclear forces. What held neutrons and protons together in the nucleus? Indeed, what were neutrons and protons? Were they elementary in the sense that electrons and photons are elementary, or were they composite, made of other particles and having some measurable physical extension? This direction intrigued me because it led "downward" (that is, toward smaller dimensions). Studying the nuclear force was peeling off one more layer of the onion, at the core of which might reside the most fundamental laws. This was the direction that absorbed most of my intellectual effort during my stay at UNC.

The nuclear problem had another appeal: it offered plenty of opportunity for confronting theory with experiment. In the 1930s, as accelerators were being constructed, and with neutrons as well as protons, deuterons, and alpha particles available as projectiles, a wealth of experimental data on nuclei was being accumulated. I chose to focus on mathematical techniques for describing nuclei even when the details of the force between the constituent nucleons were unknown. I looked at some very general possibilities for this force that depended on the relative motion, not just the relative position, of the interacting particles (unlike, say, the gravitational force in the solar system, which depends only on the relative positions of the sun and planets). I also developed the method of resonating group structure, which Ed Condon had helped me name. It took two years to pull these ideas together. In the summer of 1937, I submitted two related papers on nuclear structure and another paper on wave functions outside the nucleus to *Physical Review*. When they were published together later that year, they occupied forty-five densely packed pages of the journal. (Writing long papers became a lasting affliction, and later got me into hot water with some editors.)

In the second of these three papers, I introduced the scattering matrix, or S matrix, as it came to be called. Like most of what was in that paper, the S matrix was "methodological"—that is, it provided a formal way of describing events in the quantum world independent of particular details. It gave a before-and-after description of scattering and reaction events. Each of many initial states (such as a proton flying toward a nucleus) can give rise to various possible final states (for example, a neutron flying away from a nucleus). The S matrix is a package description of all of these possibilities. In the hands of Werner Heisenberg and others, it later proved a useful tool for analyzing elementary-particle events, not just nuclear events.

At UNC I taught courses at the advanced undergraduate and graduate levels, on topics such as classical mechanics, quantum mechanics, and nuclear physics. (Years later I discovered the pleasures and rewards of teaching freshmen.) Since I was younger than most of the graduate students, some of them started calling me Brother Wheeler.

Without delay, I got several advanced students going on research problems. Katharine (Kay) Way helped me in my work on nuclear structure and, for her Ph.D. dissertation, carried out her own independent research on the magnetic properties of rotating nuclei treated as liquid droplets. She was one of a tiny handful of women in physics at that time. (They are more numerous now, but still not nearly numerous enough.) She was tall and slender, blond with blue eyes, full of enthusiasm for physics, and not reluctant to express her definite opinions on people and events. She went on to a notable career in nuclear physics. During the war, she spent some time at the Met Lab in Chicago while I was there, and then moved to Los Alamos. She made the decision not to teach, electing instead to pursue research, first at Oak Ridge Laboratory in Tennessee, then at the National Bureau of Standards in Washington. She became known as the grand curator of nuclear data, and remained active as a journal editor until just before her eightieth birthday. Kay was eight years my senior. In 1993, when she was ninety, I visited her in her Chapel Hill retirement community, Carolina Meadows, where I found her as lively and opinionated as ever. She died two years later.

Hermon Parker, from Vass, North Carolina, worked with me on features of the model of resonating group structure. After his master's degree at North Carolina, he went on to Cornell, earning a Ph.D. with Hans Bethe. From these two students I discovered what I have applied ever since, that I learn best by teaching. In fact, as I got older, I concluded that I could learn *only* by teaching. Universities have students, I like to say, to teach the professors.

My introduction to Princeton came through a three-month visit to the Institute for Advanced Study, a young organization that had already attracted some leading intellects. For me, it was a mini-sabbatical, and it came early. I had been at UNC only a little over a year. I felt, already, the need for a little quiet time to work, away from classes, and I longed for the chance to be in daily contact with people like Eugene Wigner and the great mathematicians Hermann Weyl and John von Neumann. I also hoped to get better acquainted with Albert Einstein, but our interests were then so different that I didn't expect to learn very much from him. I applied for a visiting appointment at the Institute and was accepted. The University of North Carolina generously approved a three-month leave of absence for me, from December 1936 to March 1937. Janette and little Letitia lived with Janette's parents in Baltimore during this time, while I took a rented room in the home of a laborer on Patton Avenue in Princeton. That gave me some insight into a different kind of family. My landlord was so abusive to his wife that he felt called upon to explain himself to me. "That's the way my boss treats me," he said. "So I treat her the way my boss treats me. Otherwise I'd go crazy."

The driving force behind the Institute was its director, Abraham Flexner. Major gifts from the Bamberger and Fuld families, inspired by Flexner's writings, provided the financial stability that enabled him to set about creating a major research center unlike any other in the United States. He communicated his vision effectively, attracting leaders in mathematics and other fields. The fact that Einstein came to the new Institute in 1933 rankled Robert Millikan of Caltech, who was building a great physics center in the West and wanted Einstein to be part of it—indeed, *expected* Einstein to be part of it. During my short stay at the Institute, I met Flexner only in passing, but I got better acquainted with him later, when Janette and I bought a lot from the Institute and built a house on it. We couldn't help liking this small, bright-eyed man, always eager to make the younger members of the Institute family feel at home. At a formal dance once, he wore a business suit, knowing that the younger researchers would not have formal wear.

Although the Institute for Advanced Study had land, it did not yet have buildings. Its mathematicians and scientists were still in Fine Hall on the university campus. As a short-term visitor, I was assigned no office, only space and a file drawer in the Fine Hall Library. This gave me a chance to concentrate on my work when I wished to, but also to talk with the Fine Hall regulars. Afternoon tea was already a fixed tradition. I listened to a series of lectures by Weyl and got better acquainted with Wigner. Einstein I saw only from time to time.

I was asked, while there, to give some lectures on nuclear physics in the Physics Department, next door in Palmer Lab, which I was glad to do. Part of the reason for the invitation, I deduced later, was to size me up as a possible Princeton faculty member. Princeton wanted to strengthen its work in both theoretical and experimental nuclear physics. Wigner was probably my advocate. One day, he invited me and Harry Smyth, the department chair, to lunch at Lahière's, a good local restaurant (that is *still* a good local restaurant). Hints were dropped about my coming to Princeton, but nothing definite followed for some time. The person they had in mind for the experimental position was Milton White, a cyclotron expert from California. He accepted an offer to come and build a cyclotron at Princeton, and he headed a strong research group there for many years.

Soon after I got back to Chapel Hill in March 1937, I got the good news of a promotion to associate professor effective that fall; it was made clear that I could remain there with tenure. It was a good place, and Janette and I were

Some of the participants in the 1937 Theoretical Physics
Conference at George Washington University. *First row, left
to right:* Hans Bethe and Niels Bohr. *Second row:* I. I. Rabi
and George Gamow. *Third row:* Fritz Kalckar and I. *Fourth
row:* Edward Teller (partially hidden) and Gregory Breit.

happy, but I felt an urge to work, if possible, at an even stronger center of
research in nuclear and particle physics. (As I like to say, "Nobody can be anybody without somebodies around.") During the following year, I received a
tempting offer from my alma mater, Johns Hopkins. It was an associate professorship with tenure. Yet the possibility of Princeton appealed to me more.
I held off on responding to Johns Hopkins until Princeton could decide what
it wanted to do. In time, I did get a firm offer from Princeton for an assistant
professorship without tenure. Had I been security-minded, I would have gone
to Johns Hopkins or stayed at UNC. But rank and tenure status were not
important to me. I wanted to be in the most exciting environment for the kind
of research that interested me, and that was Princeton.

In 1937, the same year in which my long papers on nuclear structure and
scattering were completed and published, I got into some collaborative work
with Edward Teller—then at George Washington University in Washington,
D.C.—on nuclear rotation. It started as a conversation at some meeting during the year, and continued when he came later for a visit to nearby Duke
University in Durham.

The liquid-droplet model implied that nuclei can rotate, but the calculated quantum energies of rotation are smaller than any known excitation energies. Teller and I worked on the problem and developed a theory of nuclear rotational states that seemed to conform better to what was known experimentally. We showed how certain symmetry considerations forced the excitation energy to be higher than had been calculated before, and also showed how the rotational energy should depend on the mass of the nucleus. Our paper was published in the spring of 1938. Had we been just a little more clever and far-sighted, we might have hit on what later became the key to understanding nuclear rotation more fully—the fact that although the surface of the nucleus behaves like a liquid droplet, the nucleons in the nuclear interior behave more like the molecules of a gas. It was a decade later before insights connected with the independent-particle aspects of the nucleus came to the fore, and another few years after that before the independent-particle and liquid-droplet properties of the nucleus were successfully united in the unified, or collective, model. (Fortunately, a successful theory of fission did not require that unification. It could be based on the liquid-droplet properties alone.)

In the summer of 1938, Janette and I stuffed two-year-old Letitia and two-month-old Jamie, and such belongings as would fit, into our car and made the long drive from North Carolina to Salisbury Cove, Maine, to visit Janette's parents before settling in at Princeton. On the way through Princeton northbound, we stopped to look for a place to live. With the help of an agent, we found a house to rent and drove on, only to learn upon reaching Maine that it wasn't available after all. Another agent had rented it to someone else the same day we were there. We negotiated for another by mail, a small house on Murray Place, and it proved to be satisfactory.

Once back in Princeton that September, I settled happily into the department, and we felt quickly at home in the community. We made friends and we helped to found a Unitarian Church. Although I don't consider myself religious, churchgoing has been important to me during much of my life. There is something comforting about colleagueship based on shared values, and a good sermon stimulates my thinking—making me perhaps a better, more caring person. There are many modes of thinking about the world around us and our place in it. I like to consider all the angles from which we might gain perspective on our amazing universe and the nature of existence.

Everything in Princeton seemed so congenial that we decided to put down roots. (Some young assistant professors in today's more uncertain job market would scratch their heads about this.) The Institute for Advanced Study had some land available for sale to university faculty members as well as to members of the Institute. Abraham Flexner had in mind an intellectual ghetto that would cement good relations between the two institutions. With some critically important help from Janette's parents, we were able to buy a lot at 95 Battle Road, and we retained an architect and builder, Raymond Bowers, to help us plan and build a house. Flexner said to me, "Wise men buy houses; foolish men build them." But to paraphrase the poet A. E. Housman: I was seven-and-twenty; no use to talk to me. And how glad both Janette and I were over the years that we were not deflected from our purpose by Flexner's discouraging words. Our "foolishness" was wise.

Ray Bowers finished our Battle Road house on time, and we moved in on September 1, 1939. It proved to be an ideal home in a stimulating neighborhood. Across the street, Carl Ten Broeck, director of the Princeton branch of the Rockefeller Institute for Medical Research, and his wife, Janet, became good friends. Good neighbors around the corner were Vladimir and Katusha Zworykin. Russian-born Vladimir, a principal inventor of television technology, was a research director at nearby RCA Laboratories. Next door to the Zworykins, and separated from us only by a field in which the children played, lived the Institute's economics luminary, Winfield Riefler, and his wife, Dorothy. Riefler told us about his service as an advisor to the administration of President Roosevelt. It was a time, with America emerging from the Depression, that feelings about economics and about Roosevelt ran high. "Tax and tax; spend and spend," chanted Roosevelt's critics. (It's a durable idea. We hear still about "tax-and-spend liberals." Both then and now, I count myself in Riefler's camp, a liberal on economic issues.) On America's entry into World War II, Riefler went off to London to help advise on the allocation of scarce resources.

Just beyond the Rieflers lived the great mathematician Hermann Weyl. His extraordinary mind was matched by a nobility of character, and I treasured our friendship. As a nineteen-year-old sitting in a Vermont meadow working my way through his opus, *Theory of Groups and Quantum Mechanics*, in the original German, I could hardly have imagined that he would one day be my neighbor and friend. Like me, he learned by teaching. He told me once that he liked to teach the history of mathematics, because only by ranging over the whole subject and its historical development could he clearly see the gaps and the places where new work would pay off in new understanding.

Our immediate neighbor on Battle Road was the eminent art historian Erwin Panofsky, likable but prickly with the children. At a memorial service for him years later, I recall one speaker saying of him, "He hated children, grass, and birds, /And loved all dogs, a few friends, and words." His sons Wolfgang and Hans came to be known as the smart Panofsky and the dumb Panofsky because one of them finished first and the other second in the same Princeton class (1938). (Wolfgang, known as Pief to his friends in physics,

declines to identify which one of the two is the smart Panofsky but acknowl-
edges that he received one B in his academic career. Pief became the director of the Stanford Linear Accelerator Center, as well as a tireless and effective
worker for international arms control.)

We didn't suspect in those early Princeton years that we would have to
vacate our Battle Road home for more than three years for war work. Nevertheless, it is the home that our children most remember from their early years.
It was our family homestead.

Princeton's Palmer Physical Laboratory and the adjoining Fine Hall, together
forming a unit that housed the physics and mathematics departments, provided an astonishingly vibrant center of intellectual effort in 1938 when I
arrived. One could not walk to the tea room without bumping into outstanding people who were working at the frontiers of science. These encounters
included people dropping in from the Institute for Advanced Study, which
had only recently moved to its new quarters across town. It was a happy place
and an informal place. I could chat with Milton White about cyclotron work,
Eugene Wigner or Wolfgang Pauli about nuclear theory, Rudolf Ladenburg
about spectroscopy or Palmer Lab's new accelerator. With John von Neumann and Hermann Weyl I could explore subtle points in mathematics.

I was equally inspired by the undergraduate program. With a push from the
dean of the Graduate College, the mathematician Luther Eisenhart, Princeton had initiated a requirement that all seniors engage in research and prepare
a senior thesis. I have supervised many a senior thesis in my years at Princeton,
and some of them rate in quality and significance with Ph.D. dissertations. A
later requirement that juniors prepare a historical paper and a short independent paper added to the strength of the undergraduate program and helped
prepare students for more substantial senior projects. It added to the fun of
one of my first teaching assignments, an undergraduate course on mechanics,
that the students were casting about for research topics and I could help
them find subjects to investigate. One of my early students in this course was
Robert Dicke, later to become my colleague on the Princeton faculty. He
loved to tackle experimental problems that others thought too difficult, and
he was a master of theory as well.

Dick Feynman, who had earned his bachelor's degree at MIT, showed up
at my office door as a brash and appealing twenty-one-year-old in the fall of
1939 because, as a new student with a teaching assistantship, he had been
assigned to grade papers for me in my mechanics course. As we sat down to
talk about the course and his duties, I pulled out and placed on the table
between us a pocket watch. Inspired by my father's keenness for time-and-
motion studies, I was keeping track of how much time I spent on teaching and

Dick Feynman, Los Alamos Laboratory ID photo, 1940s.
*(Photograph from Los Alamos National Laboratory, courtesy of
AIP Emilio Segrè Visual Archives.)*

teaching-related activities, how much on research, and how much on departmental or university chores. This meeting was in the category of teaching-
related. Feynman may have been a little taken aback by the watch but he
was not one to be intimidated. He went out and bought a dollar watch (as I
learned later) so he would be ready for our next meeting. When we got together again, I pulled out my watch and put it on the table between us. Without
cracking a smile, Feynman pulled out his watch and put it on the table next
to mine. His theatrical sense was perfect. I broke down laughing, and soon
he was laughing as hard as I, until both of us had tears in our eyes. It took quite
a while for us to sober up and get on with our discussion. This set the tone
for a wonderful friendship that endured for the rest of his life.

My collaborative work with Willis Lamb started during my first semester
at Princeton. He came down often from Columbia. Willis was a tall, lanky
twenty-five-year-old, soft-spoken but confident. He could have played the
part of a cowboy in a Western. He had just moved from California to New
York. I. I. Rabi had recognized his talent in theoretical physics and invited
him to join the Columbia University department. As it turned out, Lamb's talent was not limited to theoretical physics. His 1955 Nobel Prize was awarded
for his experimental determination of a small correction to the Dirac theory
of the energy of the electron in a hydrogen atom—a correction attributable to

the swarm of virtual particles that make the vacuum such a lively place. Quantum mechanics just won't stand for quiescence. This "Lamb shift," as it is called, stands as a reminder of the complexity that rules in the world of the very small.

But my work with Lamb was pre–Lamb shift. We explored the passage of high-energy electrons through matter, and especially through air. We wanted to contribute something to the detailed understanding of showers initiated by the cosmic radiation, so we focused on "super-relativistic" electrons, those with energies of motion (kinetic energies) much greater than their rest energy (the energy equivalent of their mass). Such electrons, because of their great energy, can spawn swarms of new particles. This was a satisfying piece of work because it was one for which we could get reasonably accurate numerical answers (unlike some problems in nuclear structure), and one that related directly to experimental work then in progress.

Once Bohr arrived in January, I was caught up in fission studies, but Lamb and I had finished most of the fast-electron work by that time. I was able to draft the paper in odd hours sandwiched between those spent on fission, and we submitted it to *Physical Review* in early March 1939. At Lamb's generous insistence, my name appears first on the paper, a rarity for me on jointly authored papers, given where the letter W appears in the alphabet. (My later collaborator Wojciech Zurek finds himself even more often relegated to last place in a list of authors!)

Where was I headed when fission came along? I suspect that, without its sudden thrust into my life, I would have drifted away from nuclear physics sooner than I did. I entered physics in the early 1930s by applying the new quantum mechanics to atoms and radiation. Then, in the mid-1930s, I turned to nuclear physics because it seemed to hold the promise of revealing new and deeper laws of the submicroscopic world. After several years of intense work in the field, I saw that this promise was not being fulfilled. The nucleus is a system of great complexity that shields whatever simplicity might lie at deeper levels. The force that acts between neutrons and protons proved to be of far greater complexity—indeed obscurity—than the forces of electromagnetism that dominate atoms and radiation or the force of gravity that dominates the astronomical realm.

In the 1930s, Hideki Yukawa had advanced a theory of meson exchange as the basis of nuclear forces; Werner Heisenberg, Ettore Majorana, and others (including Eugene Wigner) had developed theories in which the particles interact by exchanging their position and their spin and possibly their charge, and I had explored forces that depend on momentum and angular momentum, not just relative position. The nuclear problem, as physicists like to say, is "messy." Working on it was fun and it was challenging, but I began to feel that it was not going to lead to more fundamental understanding of nature.

So my attention was turning back to electrons and photons and to the new particle, the meson, that had been found in the cosmic radiation. (This so-called meson turned out to be the muon, a fundamental particle that is cousin to the electron and is no longer called a meson. Other particles of intermediate mass were discovered, beginning soon after World War II. They *are* mesons and, like neutrons and protons, they exert and feel forces that are strong and complex.) With this new thinking, I was concluding that the route to more basic understanding was through the study of high-energy particles. This would lead me, immediately after the war, to campaign for, and to secure, a cosmic-ray laboratory at Princeton. The controlled energy provided by accelerators could be important, but I was in a hurry and I wanted to see particle research conducted on my doorstep. Accelerators were major installations and were clearly going to get larger and more expensive in the future. They took years to build and were not going to be suitable for a single campus. So cosmic rays caught my fancy.

An accelerator lab requires a large and expensive machine to push particles to high energy and various detectors to study them. A cosmic-ray lab requires only the detectors. The high-energy particles come free, streaming down from above. Cosmic rays don't arrive in intense controllable beams as particles from an accelerator do; they come helter-skelter in direction and energy. But the peak energies in the cosmic radiation are far greater than in any accelerator then envisioned (or ever constructed) and, as I reasoned in 1945, they were available *now*.

As I look back on my 1930s work, I see that I was perhaps too adept at mathematics and calculation. I absorbed complexities of quantum mechanics and group theory and matrix algebra like a sponge. I could roll up my sleeves, grind through complicated mathematical manipulations, and then, using only a slide rule or mechanical calculator, slog painstakingly through numerical calculations based on the mathematical results, ending finally with impressive tables and graphs. It gave me a sense of accomplishment, such as a mountain climber must feel, and it brought me to the attention of other physicists. But was it preventing me from spending more time in contemplative thinking about deeper problems?

It is paradoxical that, looking back now, I ask that question, because many of my colleagues, looking at my later career, have an opposite perspective about it. In the eyes of some of them, especially the younger ones, I have, in recent years, spent too much of my intellectual energy in "far-out" speculation and too little in the analysis of what one can get a firm grip on—that which the Germans call *handfest*. My kinder friends tell me that I have earned the right to do that. I am struggling now, as I have been throughout my career, to learn something new about what is most fundamental. I have not lost my faith in calculation nor my primary interest in being able to make theory and experiment confront one another, but I see more clearly now the need to look beyond the immediate, to follow a vision, to do some guessing in the hope that it will inspire me or others to construct a new theory that can indeed confront experiment.

My vision during the 1930s and 1940s was of a world made only of simple particles—perhaps only electrons and positrons—even though I saw no way to construct that world. A related vision was of a world in which particles could interact without the intermediary of fields, since all of the nasty mathematical difficulties of quantum theory seemed to be associated with fields. Dick Feynman and I pushed this latter vision to its logical limit, working in fits and starts as time permitted during World War II and completing the work after the war.

Fission, on the one hand, kept me in nuclear physics. On the other hand, in its wartime, applied manifestation, it became a job, not a calling. During the war, I found time to talk with colleagues such as Enrico Fermi about fundamental physics and I found time to think about my own directions in postwar physics. General relativity and gravitation, which became the love of the second half of my life, were still sitting hidden in the back of my mind. When they emerged, I finally had a calling.

8

·

PHYSICS AFTER FISSION

PURE CHANCE played a big role in my involvement in nuclear fission—chance that fission was discovered just as Niels Bohr was about to board a ship for New York, chance that he was headed for Princeton, chance that I was there and prepared to work with him on the theory of fission. Those circumstances led directly to my war work in Chicago, Wilmington (Delaware), and Richland (Washington). But whatever the circumstances, I would surely have joined the war effort in some capacity. There was little debate and little indecision among scientists in 1942. We signed on.

The three and a half years I spent working on reactor physics were not years lost to my career. I always kept my "Princeton physics" alive, and I had the frequent stimulus of conversations with people such as Fermi, Feynman, and Wigner. Moreover, engineering work can provide dividends to pure research in surprising ways. I have always found that my applied work and my basic research enrich each other. Some mathematical method applied to a practical problem or some way of looking at a machine or a factory can unexpectedly find application in the "purest" research. I feel like the artist who, after spending a summer doing farmwork, returns to his studio with a clearer vision and stronger technique.

But to members of Congress and the public, the question wasn't what war work had done for science; it was what science had done for the war. Everyone soon recognized that the Allied victory in 1945 was as much a victory of science as of military might. As a consequence, government dollars flowed into my field of physics, and indeed into all fields of science, at unprecedented levels for decades to come. Until my final retirement from the University of Texas at Austin, I never again lacked for research support (my needs as a theorist were modest).

The atomic bomb brought special fame to physicists—even though that achievement required the equally vital contributions of chemists, engineers, and industrial managers. Radar, another potent instrument of war, was likewise credited to physicists, who properly should share the credit with experts from other fields. These great developments—the bomb and radar—are what captured the public's imagination. But scientists made their mark in many another wartime development—including the proximity fuze, a blind landing system for aircraft, magnetic mines, and the operations research that helped protect convoys.

The rising star of science in the postwar firmament (perhaps it should be called the stardom of science) affected me personally, as it affected many of my friends. I received invitations to give talks, write articles for the general public, advise the government, and sit on boards. Although I accepted many of these invitations, my primary commitment to teaching and research never wavered. No other professional activities have ever been half so exciting—or half so satisfying.

As my duties in Richland and Hanford came to an end in the summer of 1945, Janette and I thought deeply about our future. We hardly thought of the wartime years as years of sacrifice, for they provided many friends and unforgettable memories. But the time had come to resume the life we had started before the war. We were eager to return to Princeton, rejoin the academic community there, settle into the house we had occupied so briefly after we built it, and raise our children in that house and that community. Our daughter Letitia turned nine that summer; our son, Jamie, was seven; Alison, our youngest, was three. We were lucky. Our vision of the life ahead largely came to pass, although we did not fully foresee the extent of my outside involvements, nor did we anticipate that five years later, my work on thermonuclear weapons would pull us away from Princeton again for a year.

Janette and the children left Richland a few weeks before I did. By July 1945, I had concluded that the war would end that summer or, at worst, in the fall, so we decided to get the children back to Princeton in time to enroll in school at the beginning of the fall term. With July temperatures in Richland running over 100 degrees day after day, Maine was looking more and more attractive, so we planned for Janette to take the two younger children for the last part of the summer to her family's summer home at Salisbury Cove on Mt. Desert Island. (Our oldest, Letitia, had gone earlier to stay with my parents in Baltimore.) I left work for a few days in order to accompany the family by train and boat via Seattle to Vancouver, British Columbia, where they could board a train for the trip to the East Coast. I saw them off in the early evening of August 6. Leaving the station, I encountered a newsboy calling "Extra! Extra!" There was the news of the bombing of Hiroshima.

Janette, relaxing in the train's observation car as it rolled eastward from Vancouver, picked up a paper and discovered the same news. She excitedly pointed it out to her fellow passengers. "Look!" she said, "This means the war will be over in a week or two!" To her surprise, the Canadians seemed unmoved. Perhaps they saw the atomic bombing as just one more piece of war news, of which they had had so much for so long. She didn't get more news until two days later when the train reached Winnipeg. A newspaper she bought there gave as much attention to women's clubs as to the atomic bomb.

I found a similar unconcern when I stopped on August 7 at the Canadian Astrophysics Observatory near Victoria on Vancouver Island. With the end of the war now clearly in sight and the exciting prospect of more time ahead to investigate fundamental questions, I wanted to take the opportunity to find out more about what was going on in astrophysical research. Soon I was into a discussion with a Canadian colleague on new observations of a supernova explosion. Not being able to contain my excitement over the previous day's headline, I said, "Yesterday's explosion in Japan is even more important for mankind." "Yes, that was interesting," he answered, and went back to telling me about the radiation from the supernova.

If there had been excitement or celebration at Hanford over the news of the bomb, I had missed it. Calm had returned by the time I got back from Canada. At Hanford, as at Los Alamos, scientists immediately began thinking of getting back to their prewar pursuits. I was eager to be back in Princeton in time for the start of classes. I didn't dare drive our ancient car across the country, so I sought a buyer for it. The Mormon worker who answered my ad told me how well he had done in real estate investment in Salt Lake City by following the advice of his church elders on when to buy, when to subdivide, when to sell. Since he tithed, his good fortune was also the church's good fortune. I was just as glad that he didn't consult the elders before buying my car. So I returned by train, while a moving van arranged by Du Pont took our slender belongings back across the country. I fetched the car we had left behind in Princeton and drove the more than one thousand miles to Maine and back to bring the family home.

Even after the war, fission occupied my attention from time to time, both in public service and in research. Fission is a good avenue to deeper insights about nuclear structure in general, and I exploited this later in work with my student David Hill and with Niels Bohr. But in the years immediately following the war, there were other interests I wanted to pursue. Among these was the challenge of understanding action at a distance.

To explain action at a distance, the idea that had intrigued me since the 1930s, I have to go back to Isaac Newton. In 1692, he wrote to his friend the Reverend Richard Bentley, Chaplain at Worcester, "that one body may act upon another at a distance through a vacuum, without the mediation of anything else, . . . is to me so great an absurdity, that I believe no man who has in philosophical matters a competent faculty for thinking, can ever fall into it." In other words, it makes no sense (said Newton) to imagine a truly empty vacuum with nothing in it. Something must be there, he thought, to transmit a force from one place to another, such as the force of the Sun on Earth. This something—which the Greeks, far earlier, had postulated—was called "the ether," and played a major role in thinking about the physical world up until the twentieth century.

It is to Aristotle, working in the fourth century B.C., that we owe the popular maxim that "nature abhors a vacuum." It is more accurate to say that *people* abhor a vacuum. Newton called it an absurdity. Scientists ever since have developed our picture of nature in terms of what I may call "local action," to distinguish it from "action at a distance." The idea of local action rests on the existence of "fields" that transmit actions from one place to another. The Sun, for instance, can be said to create a gravitational field, which spreads outward through space, its intensity diminishing as the inverse square of the distance from the Sun. Earth "feels" this gravitational field locally—right where Earth is—and reacts to it by accelerating toward the Sun. The Sun, according to this description, sends its attractive message to Earth via a field rather than reaching out to influence Earth at a distance through empty space. Earth doesn't have to "know" that there is a sun out there, 93 million miles distant. It only "knows" that there is a gravitational field at its own location. The field, although nearly as ethereal as the ether itself, can be said to have physical reality. It occupies space. It contains energy. Its presence eliminates a true vacuum. We must then be content to define the vacuum of everyday discourse as a region free of matter, but not free of field.

When the classical (prequantum) theory of electromagnetism was perfected in the latter half of the nineteenth century, the concept of fields came into its own. James Clerk Maxwell recognized that electric and magnetic fields are dynamic entities. They can oscillate and move through space, accounting for light and radio waves, among other forms of electromagnetic radiation. We now know that gravitational fields can propagate too—as gravitational waves—but at that time (around 1865), the Sun's gravitational field was regarded as static. It just sat there, to be experienced by the planets and comets and asteroids. In a similar way, the electric field of an atomic nucleus can be regarded as static. It occupies the space around the nucleus and influences the motion of electrons. But the electric field in a light wave is dynamic. Together with a companion magnetic field, it oscillates, and the two tango together as they rush through space. It is easy for us to recognize that the filament of a light bulb doesn't "reach out" to influence the retina of our eye through action at a distance. The actions are "local." A light wave is created locally at the filament, sent through space, and detected locally at the retina.

In the twentieth century, after physicists developed the quantum theory of electromagnetism and after they discovered numerous new particles and new interactions among them, physics was fully in the grip of the field point of view and the idea of local action. All interactions of all kinds could successfully be described as interactions happening at local points in space and time, with effects at other places and other times being the result of propagating fields. It was a powerful, unified vision of nature, with a web of interconnections that totally eliminated any need for action at a distance. Newton could smile in Heaven. No one in the twentieth century believed that one body could act upon another at a distance through a vacuum, without the mediation of anything else.

Well, almost nobody. I saw chinks in the armor of field theory, and thought that it was time for a new look at action at a distance. I wanted to explore the idea that one charged particle could act upon a distant charged particle without an intermediate field. In my vision of electromagnetism, an *effect* propagated from one point to another—for, according to well-established principles of relativity, no effect could act instantaneously elsewhere—but this transmitted effect lacked the physical reality of a field (in technical language, it had no "degrees of freedom" of its own). This concept of a transmitted effect was only a way of describing the literal action of one particle on another at a distance.

Was I just swimming upstream to be different? I didn't think so, because electromagnetic field theory does contain some notable internal difficulties. The worst of these difficulties is the prediction that the field in the immediate vicinity of a point particle is infinite in magnitude, resulting in "infinite self-energy" of the particle, which at first seems to imply infinite mass. Fortunately, this infinite self-energy can be swept under the rug—just ignored—without disturbing all of the successful predictions of the theory. I couldn't help thinking that some of the other infinities—all of them mathematical nuisances—that kept cropping up in quantum field theory in the 1930s and 1940s resulted from this self-energy problem left over from the classical theory.

I had another motivation as well for pursuing action at a distance, for I clung to my hope that all of the matter in the world could be reduced to electrons and positrons. Yet I knew that if an electron and a positron were to be crowded together in subnuclear dimensions, some way would have to be found to get around the prediction of conventional theory that they would quickly radiate away their energy in the form of electromagnetic fields. Per-

haps, I thought, an action-at-a-distance version of electromagnetic theory—
one without fields—might explain the suppression of such radiation and permit the particles to live happily in such a confined space.

I didn't invent the idea of action at a distance. It had been examined mathematically over the first several decades of this century by several theoretical
physicists, including Karl Schwarzschild (my friend Martin's father), H.
Tetrode, Adriaan Fokker, and Paul Dirac. What I found, however, when I
turned my attention to the subject in 1938, was that loose ends remained.
These earlier theorists had discovered the remarkable fact that to work without fields, it is necessary to consider "advanced" as well as "retarded" effects.
This means that some signals arrive elsewhere *before* the action that initiated
them—seemingly a nonsensical idea!—and some arrive after. But these workers weren't able to assign any physical meaning to the advanced effects. Moreover, using action at a distance, they found no way to account for "radiation
reaction." Let me try to explain what that is. If you push on someone, you
transfer energy to that person. You lose some energy and the person you push
gains energy, preserving the overall energy balance. In the process of gaining
the energy, the person being pushed pushes back: he or she "reacts." You can't
tranfer energy to something else unless that something else pushes back,
reacting to the transfer. It happens even with light. When the hot filament of
a lightbulb emits light, the light, as it leaves the filament, pushes back on the
filament. That is what is called radiation reaction. The problem in the early
theories of action at a distance was that there didn't seem to be any radiation
reaction, because there were no fields to do the reacting. So I saw that action-
at-a-distance theory was still incomplete and needed more work.

I want to tell the story of my work with Dick Feynman on action at a distance because it shows the remarkable—indeed almost miraculous—power of
mathematics in physics. The great unifying theories of physics—the mature
ones, at least—can be expressed very economically with just a few, decep-
tively simple equations. What flows from these equations can be quite startling, more than their discoverers imagined. One square inch of paper
comfortably holds the equations of general relativity, set down by Einstein in
1915. As I write, eighty years later, new insights are still flowing from these
equations. Einstein did not recognize when he first wrote them down that they
predicted an expanding universe, black holes, and gravitational radiation. Yet
he was quite convinced of the correctness of these equations, based as much
on their elegance as their application.

Other examples are not hard to find. When Dirac combined quantum
ideas and special relativity into a single, simple equation of the electron, it was
as if he had assembled two subcritical pieces of uranium to make a critical
mass. Bang! Out flew the spin of the electron and the prediction of antimatter. Back in the seventeenth century, Newton's few simple equations of

mechanics fueled two centuries of further developments as his successors found more and more treasure locked in these equations.

And then there is electromagnetism. Maxwell's synthesis of all the known laws of electricity and magnetism in 1865 can be condensed into either two or four equations, depending on whether one merges space and time into space-time or treats space and time separately. In either case, this grand unification has a very simple appearance. Again, the equations fit in a small corner of a piece of paper. Yet what power is locked within them! Maxwell himself derived the first great conclusion from these equations: that electromagnetic waves of all frequencies (or all wavelengths) propagate at the same fixed speed—186,000 miles per second. And he shed light on light, showing that what we see with our eyes is just a thin slice of a grand electromagnetic spectrum of radiation.

But the Maxwell equations held more. Forty years later, Einstein invented special relativity as a way—the only way—to assure the complete validity and self-consistency of Maxwell's equations. By taking them seriously and insisting on their universal validity, Einstein showed that space and time are relative concepts and that mass is energy! Hardly conclusions that a casual glance at Maxwell's equations would reveal. Later, scientists showed that the massless-ness of the photon follows from Maxwell's equations. And, as mentioned earlier, the early workers on action at a distance showed that among the solutions to these equations of electromagnetism are ones in which effects arrive at one place before they start from another.

I have guided the work of many graduate students. I try to adjust the suggested topic of their work to their interests and abilities. (Often the student pleases and excites me by going beyond my expectations.) Few come along with the astonishing skills of Richard Feynman. Not only was he bright; he also had a wide-ranging curiosity and took delight in looking at old problems in new ways. He had an infectious enthusiasm and sense of fun. I had no hesitation in suggesting to him that he work with me on this problem of action at a distance.

This idea of an effect that precedes its cause, fantastic though it seems, is an idea that Feynman and I found we had to accept as real—if we were to work *without* fields to duplicate all the successes of past workers *with* fields. Yet, in nature, we don't see any examples of effects that precede causes. The startling conclusion that Dick Feynman and I reached, which I still believe to be correct, is that if there were only a few chunks of matter in the universe—say only Earth and the Sun, or a limited number of other planets and stars—the future would, indeed, in reality, affect the past. What prevents this violation of common sense and experience is the presence in the universe of a nearly infinite number of other objects containing electric charge, all of which can participate in a grand symphony of absorption and reemission of signals going both forward and backward in time. The mathematics told the two of us that in the real world, the myriad of signals headed back in time, apparently to influence the past, miraculously cancel out, producing no net effect. Common sense and experience are saved.

Here, indeed, is the power of mathematics. Who, casually inspecting Maxwell's equations or using them in straightforward applications such as radio propagation, would have imagined that they contain this strange power to allow for signals that fly forward and backward in time but which, taken all together, keep the arrow of time running forward? I use the word "miraculously" above. Of course, no miracles are involved. It only seems that way. The conclusions are all there in the equations, waiting to be ferreted out.

In the end, our analysis was more than just an intriguing way to look at old phenomena in a new way. We concluded that the existence of vast realms of matter elsewhere in the universe crucially affects what happens here on Earth. The field description, by itself, does not lead to this conclusion. Nor does it reveal that in a hypothetical universe with only a limited amount of matter, the future would, in reality, affect the past. Nor that a charged particle alone in an otherwise empty universe could not radiate at all.

While working on our second action-at-a-distance paper (published in 1949), Feynman and I went to call on Einstein and see what he had to say. "Well," he said (I am paraphrasing), "I have always believed that electrodynamics is completely symmetric between events running forward and events running backward in time. There is nothing fundamental in the laws that makes things run in only one direction. The one-way flow of events that is observed is of statistical origin. It comes about because of the large number of particles in the universe that can interact with one another."[1] Once again, Einstein exhibited his astonishing intuition about the physical world. Without having done any of the calculations or analyses that Feynman and I had been through, he had reached the same general conclusion about the role of distant absorbers in influencing what happens here and now in small spaces and times.

This wasn't the first time that I had consulted Einstein about work that Feynman and I were doing. Back in 1940 or 1941, Feynman had come up with a new way to look at quantum phenomena that I called "sum over histories." The idea, in brief, is this: In quantum mechanics, if you want to find

---

[1] As early as 1909, Einstein had stated his belief that "irreversibility [i.e., the one-way flow of time] rests entirely on probability considerations." Einstein's prominent colleague Walter Ritz did not share this opinion. In an admirable show of colleagueship, they wrote a joint paper setting forth their respective views, in which this quotation appears ("On the Present Status of the Radiation Problem," *Physikalische Zeitschrift*, vol. 9, 1909, p. 323).

out how something at point A influences something at point B, you can get the answer by pretending that all of the ways that A *might* send a signal to B happen at once; the actual effect is then a particular sum of all the "virtual" effects from all the different paths. It is as if a baseball pitcher, instead of throwing a single ball toward the batter, could launch simultaneously a thousand balls that travel a thousand different paths through space and time on their way to the batter. Each of these thousand baseballs has a "history" as it flies from pitcher's mound to plate. What the batter sees and swings at is the result of all these histories combined. A mind-bending idea, to be sure, but it's just what happens in the quantum world.

I was excited by the idea. I went to Einstein at his home and sat down with him in his upstairs back room study, spending about twenty minutes telling him about Feynman's idea. "Professor Einstein," I concluded, "doesn't this new way of looking at quantum mechanics make you feel that it is complete-ly reasonable to accept the theory?"

He wasn't swallowing it. "I still can't believe that the good Lord plays dice," he answered. Then, after a pause, he added, "Maybe I have earned the right to make my mistakes."

Einstein, as he once said about himself, could be as stubborn as a mule. So, once again, there was no moving him from his conviction that quantum mechanics, at its core, is defective.

Soon after World War II, the American Philosophical Society invited me to contribute to a symposium on atomic energy and its implications. I chose to speak on "Problems and Prospects in Elementary Particle Research," deliv-ering my talk in Philadelphia on November 17, 1945. The point of my talk was to emphasize that the frontier of the very small had moved beyond atom-ic nuclei and atomic energy to the deeper world of particles. I wanted to speak about the relative roles that accelerators and cosmic radiation might play in future particle research, and I could not help speculating on directions that theoretical research might take in the quest to unify the descriptions of gravi-ty, electromagnetism, and electron-positron pairs. In short, I didn't really want to talk about atomic energy. I wanted to talk about what lay beyond it.

In the audience for my talk was Zay Jeffries, a vice president of the Gener-al Electric Company and one of the six trustees of the Battelle Memorial Insti-tute in Columbus, Ohio. My remarks struck so much resonance with him that years later, when a vacancy arose on the Battelle board of trustees, he was instrumental in getting me invited to serve. I accepted and was rewarded with thirty years of warm comradeship as a Battelle trustee, working on fasci-nating projects from 1960 to 1990.

Battelle, a contract research organization founded in 1929, takes on tasks for industry and the government, and has grown into nearly a billion-dollar enterprise, with operations in several countries. One of its major technology centers is at my old workplace, Hanford, Washington. When Du Pont left Hanford at the end of the war, GE took over its management. Eventually, GE wanted out too, and in 1965 Battelle was invited in. My own wartime experience and my nuclear background played a role in Battelle's decision to accept. Battelle Northwest, as it was then called, has become Pacific Northwest National Laboratory, now conducting research in many fields other than plutonium production and processing.

One famous story about Battelle concerns Xerox. The independent inventor Chester Carlson had an idea on how to make a copying machine using the process that we now call xerography. His first xerographic image is pinned to a specific date and place: October 22, 1938, Astoria, New York. In approaches to company after company, he struck out, until finally Battelle agreed to underwrite the development. Battelle found a small photographic-paper company in Rochester, New York, the Haloid Company, willing to undertake the commercialization of the process. The rest, as they say, is history. Haloid became the gigantic and successful Xerox Corporation, and Battelle became richer.

All of this happened before my time as a Battelle trustee, but when I came on the board, I found that Battelle had a disconcertingly large portion of its investment portfolio in Xerox stock. I remembered the hardship that Johns Hopkins University had suffered from having too big a portion of its assets invested in the Baltimore & Ohio Railroad. When that stock plummeted in value, Johns Hopkins lost money and lost academic programs. Not intimidated by the greater experience in money matters of my fellow trustees, I urged the board to sell off a major part of its Xerox stock and diversify. We did so. As it has turned out, Battelle would now be even better off if it had not followed my advice, but how were we to know?

In the 1930s and 1940s, cosmic rays, as I have emphasized already, stood at the frontier of the very small. Immediately after the war, I campaigned in my Princeton department for the creation of a cosmic-ray laboratory on the Princeton campus. Harry Smyth, the department chair, was sympathetic to the idea, and so were most of my other colleagues. With little opposition, the laboratory was approved. Of course, approval of an idea is the least of the necessary steps to get a new enterprise going. Space and money and a staff are required, too! Almost inevitably, the pusher of an idea gets tapped to implement the idea, so I had little choice but to accept the directorship of the cosmic-ray lab. Later, I was able to transfer this responsibility to George Reynolds, an experimental nuclear physicist in the department. He was an accomplished scientist and an affable administrator. Among the younger people at Princeton who were made available to the lab as their wartime duties ended were W. Y. Chang, from China, and Thorbjorn Sigurgeirsson, from Iceland. Both of them made important measurements that were related to my theoretical interests.

Chang's wife, also a physicist, worked at the University of Michigan in Ann Arbor. Like so many couples today, they had to cope with being separated most of the time. Later they returned to China, where she taught physics at Peking (now Beijing) University, and he worked at the Institute of Modern Physics in that city. Then he was assigned to the Joint Institute for Nuclear Research in Dubna, USSR, near Moscow, where he spent three years before returning to Beijing. That must have made Ann Arbor and Princeton seem almost next door. At Princeton, Chang discovered the pattern of gamma rays emitted when a muon (then called a mu meson) behaves like an electron in an atom, cascading through energy states around a nucleus (dropping from one "orbit" to another) on its way to the lowest-energy state, near the nucleus. It was an important discovery, and I referred to this radiation whenever I could as the "Chang radiation." Unfortunately, the name didn't catch on.

Sigurgeirsson also returned later to his native country to assist the development of physics there. John McPhee, in his book *The Control of Nature*, gives a lovely account of Sigurgeirsson's successful effort to save the only town on the island of Heimaey from lava flow by persuading fire fighters from all over Iceland to bring their equipment and spray water on the lava. Iceland's harsh environment played a direct role in Sigurgeirsson's gaining a permanent academic post. He was hiking one day on the slope of the volcanic Mt. Hekla with an older colleague when a minor eruption ejected a rock that landed on his colleague, killing him. This created a university vacancy.

As to space and money: Fortunately, an ancillary building at Princeton that Walker Bleakney had used for wartime shock-wave experiments was available. We established our cosmic-ray beachhead there. Most of the subsequent funding for the work of the laboratory came from the federal government. Some came also from the generous private contributions of many of my old Du Pont friends, including Crawford Greenewalt (who, as I noted before, became Du Pont's president), Dale Babcock, Lombard Squires, Charles Wende,[2] Hood Worthington, H. C. ("Ace") Vernon, and George Graves. They established a fund named the Friends of Elementary Particle Research, from which I was able to allocate expenditures, especially to support students. By drawing on it sparingly to meet special needs when other funds were not available, I made it last many years.

The cosmic rays that race toward the Earth from all directions in outer

---

[2] Wende had been the doggerel laureate of Hanford. In one of his verses, Lombard Squires was Lom (to rhyme with bomb), I was Johnny the Genie, and Crawford Greenewalt was Greenie (to rhyme with genie).

space, at speeds only infinitesimally less than the speed of light, are positively charged particles, some of enormous energy—energy that would be imparted by acceleration through a billion billion volts or more. As I mentioned in Chapter 6, most of these particles are protons (nuclei of hydrogen); some are nuclei of heavier elements—helium, lithium, and so on. When these submicroscopic missiles strike nuclei of nitrogen and oxygen high in the atmosphere, they not only disrupt the nuclei; they also create new particles of all kinds. The most common newly created particle, the pion, comes in positive, negative, and neutral varieties, and has a mass about 270 times the mass of the electron. The neutral pion, with a half-life less than a millionth of a billionth of a second, goes almost nowhere before decaying. The charged pions, which live on average more than a hundred-millionth of a second— and have this lifetime further extended by relativity's time dilation—penetrate farther into the atmosphere, although few of them reach the earth's surface. Some of the pions strike atomic nuclei, causing further disruption and creating other particles. Some undergo their natural radioactive decay. The pion's decay produces a muon and a neutrino.

The muon is lighter than the pion, as it must be since it is the product of the pion's decay. Its mass is some 207 times that of the electron. As elementary particles go, it has a long lifetime, 2 millionths of a second. This is long enough that many muons reach the surface of the earth. Some three-quarters of the cosmic-ray debris at the earth's surface consists of muons. Much of the rest consists of electrons, positrons, and photons.

Positive muons undergo natural radioactive decay. A positron and two neutrinos are the result. A negative muon can do the same thing (producing an electron instead of a positron), or it can be captured by an atom and cascade to lower energy states, giving rise to Chang radiation. Once in the lowest energy state in the atom, it can decay, just as it would have done if isolated, or it can be absorbed by the nucleus. In this absorption process, a proton is transformed to a neutron, and a neutrino is emitted. Apart from the facts that they are both radioactive and have masses that are not very dissimilar, the muon and the pion have nothing in common. The pion interacts strongly with nuclear matter.[3] The muon does not. The pion has zero spin. The muon, like the electron, has one-half quantum unit of spin. It is what is called a "weakly interacting particle." It belongs in the same family as electrons and neutrinos. It is more "elementary" than the pion.

It took some twenty years, roughly from 1930 to 1950, to reveal the facts related in the three paragraphs above (the quark composition of pions, men-

---

[3] We now know that the pion, like the neutron and proton, is composed of quarks. Quarks are the common constituents of all "strongly interacting particles."

tioned in a footnote, came later). It was an exciting quest, as, one by one, new facts came to light. In Chapter 4, I mentioned the latitude effect—the observation that cosmic rays are more intense at high latitudes (near the Poles) than at low latitudes (near the Equator). The latitude effect showed that the incoming particles from outer space must be charged, since they are deflected by the earth's magnetic field. An east-west asymmetry showed, further, that they are positively charged. The variation of intensity with altitude, with depth in lakes, and underground showed that the cosmic radiation is bombarding us from outer space, not created within the earth or the atmosphere. And, as mentioned earlier, we learned that the cosmic rays near the earth's surface contain a "hard" component, which can penetrate many inches of lead, and a "soft" component, easily stopped by a thin sheet of lead.

As late as the 1950s, physicists were still fussing about the name to be given to the intermediate-mass particles. During my term as a vice president of the International Union of Pure and Applied Physics (1951–1954), I found myself sitting on an international committee on symbols, units, and nomenclature (the SUN Committee). Robert Millikan of Caltech, although not a committee member, was promoting the term *mesotron*. Most of us didn't like it because we knew that the Greek root *-tron* meant "tube," not "particle." (*Electron* has a different root. It comes from the Greek word for amber.) We liked *meso-* because it implies "intermediate," and we liked *-on*, which implies "entity." But French members of the SUN Committee raised objection to *meson*. It sounds too much like *maison*, they said, which, although a perfectly proper French word, is a slang term for house of prostitution. This gave us brief pause, but we settled on *meson* anyway.

Before coming to my own focus of interest on muons in the late 1940s, let me mention the "pion industry" that sprang up after the discovery of a second meson. In his extraordinarily prescient 1935 paper, Hideki Yukawa suggested that the force between nucleons (neutrons and protons) might arise from the exchange back and forth between the nucleons of a new (as yet unseen) particle. His theory was built in analogy to the theory that ascribed electromagnetic force to the exchange of photons between charged particles. In such an "exchange theory," there is an inverse relation between the mass of the particle being exchanged and the range of the force—the greater the mass of the exchange particle, the shorter the range of the force. The massless photon leads to an electric force of infinite range: it reaches far outside the atom to macroscopic distances. From the knowledge that the range of the nuclear force was no greater than the size of a nucleus, Yukawa could deduce that the mass of his hypothetical new exchange particle should be several hundred times the mass of the electron.

Since it was published in a Japanese journal that was not widely read (although it was written in English), and was speculative, Yukawa's paper made no immediate splash. But when evidence for a meson of about the right mass (what turned out to be the mu meson, later renamed the muon) became convincing, his paper reverberated through the corridors of theoretical institutes and departments. Once the pi meson (now called the pion) was established as a separate strongly interacting entity (in 1947), enthusiasm for the exchange theory of nuclear forces grew even stronger, and the "pion industry" was running in high gear. It struck me as an example of the herd instinct that every so often grips physicists as they collectively embrace a new vision. I felt in my bones that there had to be a deeper and simpler explanation of the force between nucleons — not that I knew what it was! At any rate, I didn't join this particular movement. It struck me as being akin to postulating an exchange force between atoms if one didn't know that atoms are themselves complex entities and that the forces between them are ultimately simple electromagnetic forces, made to appear complicated by the structure of the individual atoms.

Sharing my distaste for the herd instinct, Edward Teller once gave me his circular definition of an intellectual: "someone who thinks the same things and uses the same words as other intellectuals." Yet I couldn't very well denigrate the pion industry if I didn't have an alternative to propose. To use the slang of the time, I felt that I had to "put up or shut up." Time after time in my career, starting as early as 1934–1935 in Copenhagen, I kept coming back to this problem of the force between nucleons. I was driven by a vision, too — the vision that ordinary electromagnetic forces would ultimately provide the explanation. I did not let this vision become an obsession. Every once in a while, I would pull out a notebook or just sit back and think, reexamining the problem, trying to come up with new ideas. Then I would lay that work aside and go on with other research, hoping that the next time around the right insight would dawn.

There was one fringe benefit of this recurring interest. The paper on polyelectrons that I wrote in 1945 while still at Hanford paved the way for detailed studies of the simplest nonnuclear systems, positronium (consisting of one electron and one positron) and ionized positronium (consisting of two electrons and one positron). These are systems in which, indeed, only electromagnetic forces are at work. I still live for the day when I can see an actual drop of liquid positronium.

As physics has marched forward in the decades since 1950, the pion exchange theory of nuclear force has faded from a leaping flame of theoretical physics to a dying ember, as I hoped that it would. My own vision of electromagnetism as the source of the nuclear force never did catch fire. What we see today are quarks held together by gluons. The gluons, as their name implies, are themselves exchange particles that serve as the nuclear "glue." So, in fairness, one should say that the idea of a pion exchange force has undergone metamorphosis, not extinction. It will be for physicists of the twenty-first century to discover whether quarks and gluons are the last word on nuclear forces.

My interest in muons (I use the modern name for this particle) in the years just following World War II centered on how they play out their lives, on what happens when they cease to be. What is the mechanism of, and what are the products of, the decay of a muon? What happens when a slowed muon is captured by an atom, and cascades through successive energy states as it approaches its lowest, or "ground," state near the nucleus? What happens when a muon in this lowest state ends its life swallowed by the nucleus instead of freely decaying? To what extent can the muon, in the microsecond or so that it lives within the atom, reveal details about the properties of the nucleus at the atom's center? Working out answers to these questions—and comparing theory with experiment—was very satisfying, for it penetrated straight to some of the most basic physics of the submicroscopic world—the domain of physics that was coming to be called elementary-particle physics.

In this work, I had the good fortune to have as a colleague Jayme Tiomno, a graduate student from Brazil, who started working with me and then completed his doctoral work with Wigner after I went off to Paris in 1949. Of medium height and dark-complected, Tiomno spoke always in a low, deep voice, as if sharing a confidence. Back in Brazil, he joined the progressive elements in academia, speaking out against military intervention in the universities and research institutes. He nevertheless hung on to his posts until 1969, when, along with his physicist wife, Elisa, and some two hundred other professors, he was forced into early retirement. In 1973, following a year's leave at Princeton University and the Institute for Advanced Study, he gained a paid position at Pontifical Catholic University (with the Pope's personal permission, he has been told), and later went to Rio's Brazilian Center for Physics Research, where he and Elisa still work actively as emeritus professors. Besides retaining his links to Princeton, he has established an experimental high-energy group that collaborates with Fermilab near Chicago.[4]

I always think of Tiomno as one of the most unappreciated of physicists. His work on muon decay and capture in 1947–1949 was pathbreaking and would still merit recognition by some suitable award.

Immediately following the discovery in 1947 that the "meson" of cosmic

---

[4] Fermilab, named in honor of Enrico Fermi, who didn't live to see its construction, operates what is now the world's highest-energy accelerator. It produces particle collisions at the equivalent of nearly two trillion volts.

radiation is really two particles—the pion and the muon—Tiomno and I set about studying the lighter of the two, the muon. What all of our analysis led to, as reported in papers we submitted the next year, was the likely conclusion that the muon can be looked upon as a "heavy electron." We reached a number of tentative conclusions, all of which were confirmed later as experimental results became more extensive and more accurate. Like the electron, the muon appeared to have one-half quantum unit of spin. Like the electron, it appeared to lack any strong interaction with nucleons. Its decay and its capture by a nucleus both seemed to result from an interaction startlingly like the interaction that governs beta decay (in which a nucleus emits an electron). Just as a neutrino is associated with the electron, a light neutral particle is associated with the muon. We hypothesized that this muon partner might itself be a neutrino, as it has proved to be. In short, we saw intriguing evidence that every property of the electron was mimicked exactly by a property of the muon. This was an exciting conclusion. It was also puzzling. I. I. Rabi later asked, referring to the muon, "Who ordered that?" And it is reported that for years, a corner of Dick Feynman's blackboard in his Caltech office bore the chalked question, "Why the muon?"

The "equivalence" of the muon and electron can be illustrated by a triangle diagram suggested by Tiomno, which we used in one of our papers. It looked like the diagram shown here.

The letters N and P in the upper circle stand for the two nucleons, neutron and proton. The symbols e and ν in the lower right circle stand for the electron and its companion neutrino. The symbols μ and μ_0 in the lower left

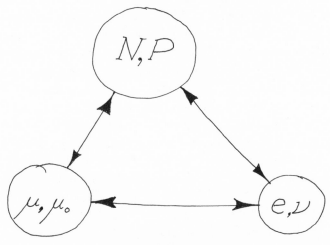

The "Tiomno triangle."
(*Drawing by John Wheeler.*)

circle stand for the muon and its companion light neutral particle. Nowadays we would replace ν in the lower right by ν_e to designate the electron's neutrino, and we would replace μ_0 in the lower left by ν_μ to designate the muon's neutrino.

The three lines connecting the circles designate the weak interactions of identical strength and character that occur among these pairs of particles. The line from the top circle to the one at the lower right designates the beta decay interaction. This interaction drives certain radioactive changes—for instance, the transmutation of a newly manufactured nucleus of neptunium 239 in a reactor into the precious bomb ingredient, plutonium 239. In this spontaneous transmutation, a neutron changes to a proton while emitting an electron with its neutrino, so all four particles in the two circles are involved.

The horizontal line at the bottom designates the muon decay process, in which a muon decays into an electron and two neutrinos, one of each type. (Again, all four particles in the circles are involved.) The third line, connecting the top circle and the one at the lower left, designates a reaction that can occur when a negative muon finds itself in orbit around a nucleus. In this "capture" or "charge-exchange" reaction, the muon and a proton disappear while a neutron and a neutrino appear. (Once again, all four particles are involved.) A neutron might "like" to decay into a proton and a muon, just as it does decay into a proton and an electron, but it doesn't have enough mass to do so. Only the capture reaction can occur. On the other side, an electron capture reaction does sometimes occur, mimicking the muon capture. Notice that for every interaction represented by one of the lines, *no* particle survives unscathed. Every one of the four particles involved in each interaction is either created or destroyed in the process.

This triangle is only a shorthand way of expressing the underlying mathematics, with its identical interaction working in all three directions. We found it to be a very beautiful representation of a pattern in nature. It doesn't answer the question, Why the muon? but it compactly summarizes a lot of particle physics. I always thought that this triangle should be called the "Tiomno triangle." He got there first. But a few months after our paper appeared, Giampietro Puppi published similar ideas in an Italian journal. He, too, saw the great simplicity of a common interaction among nucleons, electrons, muons, and neutrinos. As luck would have it, the Tiomno triangle is now known to everyone as the "Puppi triangle," even though Puppi did not include a diagram in his paper.

While thinking about muons, it occurred to me what a splendid nuclear probe this particle provides. Being some 200 times more massive than the electron, the muon moves in atomic orbits 200 times closer to the nucleus than the orbits of an atomic electron. The muon almost literally climbs inside a heavy nucleus, such as that of lead or uranium. Yet, because it is free of strong interactions, it responds only to the electric charge in the nucleus, not to any other complicating factors. I realized that a close study of the Chang radiation emitted by the muon as it is falling to lower energy states could reveal features of the nuclear charge distribution, including possible deviations from a spherical shape, and nuclear compressibility as evidenced by differences between neighboring nuclei. I wrote up these ideas in the fall of 1949 while I was working in Paris, but was delinquent about publishing them. Finally, in 1953, it came to my attention that L. James Rainwater and Val Fitch at Columbia University were making measurements of just the kind that I had advocated. I swiftly completed a paper under the title, "Mu Meson as Nuclear Probe Particle," and with the cooperation of Fitch and Rainwater, arranged for it to be published in *Physical Review* next to their experimental paper. The paper generated a lot of response, and set in motion a series of measurements of the Chang radiation over many following years, measurements that went beyond even my early imagination in what they revealed about the size, shape, and other properties of nuclei.

Val Fitch is now my Princeton colleague, also emeritus, with an office close to mine. It was by chance that I learned about his and Rainwater's work with muons before they published it. I went to Columbia in 1953, trying to recruit Fitch to come to Princeton, and, of course, we got to talking about our work and interests in physics. I don't know whether our common interest in muonic atoms (as they are now called) played a role, but he did move to Princeton, and has been a valued friend and colleague ever since. We now often meander across campus to join fellow physicists and mathematicians for lunch and conversation at the faculty cafeteria. Sometimes, when I am planning to have lunch at the Institute for Advanced Study, Val offers me a ride in his silver convertible Mazda Miata—he likes to have an excuse to drive it. A bit shy in large gatherings, he is someone that people remember for his direct gaze, unfailing politeness, and attentive listening. Out of a sense of duty to young people and to his profession, he often steps aside from his physics to accept speaking engagements and advisory committee assignments.

As it happened, both Fitch and Rainwater later received Nobel Prizes, but not for this joint work, and not for muons. Fitch's prize was for his work with James Cronin demonstrating that time-reversal invariance is not an absolute law. For most processes in nature, any sequence of events that can occur in one order can occur in exactly the opposite order—that is what we call "time-reversal invariance." But for certain processes, as Fitch and Cronin showed with the radioactive decay of a particle called the kaon, the principle breaks down. A quite amazing consequence of this breakdown of time-reversal invariance is that out of the chaos of the early universe emerged a world contain-

Val Fitch, my fellow Princeton emeritus professor, 1995.
(Photograph by Denise Applewhite,
courtesy of Princeton University.)

ing more particles than antiparticles. As cosmologists like to put it, if Fitch
and Cronin had not found what they found, we would not be here. I will
come to Rainwater's Nobel work later in this chapter.

The muon is now firmly established pretty much as we left it in 1949, itself
a fundamental particle bearing a strong family resemblance to the electron,
each of these particles having its own companion neutrino and each being
immune to the strong interactions. They belong to a family called leptons.
The word *lepton* was coined to mean "light or small," but now we have one
heavy lepton. Besides the electron and the 200-times-heavier muon, there is
a tau particle belonging to the same family. It is nearly 3,500 times as massive as the electron and also has its own neutrino. Most remarkable of all, we
now have good indirect evidence that this is the end of the line: There are
no still-heavier leptons waiting to be discovered.

Establishing the cosmic-ray laboratory in Princeton was, I think, the right thing to have done. Cosmic rays were free, they were available at once, and they reached quite fantastic energies. They were, on the other hand, not copious and not controllable. Inevitably, accelerators would later take center stage in particle research, but cosmic rays provided the right way to get started quickly, and a cosmic-ray laboratory is the right scale of enterprise for a single campus. When, in 1955, Princeton joined forces with the University of Pennsylvania to build and operate the Princeton-Penn Accelerator, I could not work up any enthusiasm for the project. It seemed to me to be too big and expensive to be managed by just two departments. But I held my counsel, and the accelerator was built. It went into operation in 1962 and was decommissioned a decade later. I stick to my view that big accelerators should be operated only by large consortia or national laboratories, so that they can be professionally managed, as the big businesses that they are, to serve all of the physics community.

Accelerators have indeed grown enormously powerful and more useful — and more expensive. Most of what we have learned about particles in recent decades has come through research using accelerators, together with a good deal of ingenious theoretical work. Unfortunately, the U.S. Congress, in 1993, withdrew support for the Superconducting Super Collider (SSC), already under construction in Waxahachie, Texas, where it would have become the world record holder among accelerators, with protons pushed to an energy equivalent to 20 trillion volts.

Cosmic rays have not lost their appeal, for even now their enormous energy surpasses that of every accelerator by a large margin. James Cronin of the University of Chicago is among those planning new experiments to study the great cascades of events triggered when the occasional incoming cosmic-ray particle with an energy 10 million times greater than the energy of a particle in the Fermilab accelerator slams into an atomic nucleus high in the atmosphere.

The years 1948 and 1949 were so busy that, in retrospect, I can't figure out how quite so much happened in my professional life in so short a span of time. My work on action at a distance, muons, elementary particles, and cosmic rays was only a part of what was occupying my mind. As the founder and first director of Princeton's cosmic-ray laboratory, I became a promoter and overseer of experimental research, and had to act as a bit of an administrator and fund-raiser as well. In 1948, the Office of Naval Research (which was one of the first and most far-sighted of the government agencies supporting science after World War II) provided $375,000 of contract research support for our cosmic-ray work.

And I was teaching!—which, after all, is what Princeton was paying me to do. At this time, most of my teaching was at the graduate and advanced undergraduate levels. One of the courses I was assigned was a graduate course on classical mechanics, a subject that was supposed to have been wrapped up and tied with a ribbon in the nineteenth century. I tried to breathe some new life into it, learning as I taught. There were, as the students and I learned, some fascinating ways to treat this subject that tied in with twentieth-century quantum mechanics. A decade later, Ken Ford and I wrote some papers on what is called the semiclassical analysis of scattering, whose ideas went right back to this course.

Teaching is more than giving courses. It is also working with individual students. Over the years, I have guided the work of more than fifty Ph.D. students, as well as a comparable number of undergraduates doing junior or senior theses. That is two-way teaching at its best. Without the help of these students, without the insights and the stimulus that they provided, my own achievements would have been far less. And this doesn't even count the dozens of postdoctoral researchers who have joined my research circle, young people trying their wings as I once tried mine. They, too, have enriched my work.

The head of steam I built up after the war bubbled over into a dozen papers in 1949. One of them was written with a graduate student, John Shanley (also an Army captain), and an undergraduate, Evan Kane (who had graduated by the time the paper was published). Shanley, Kane, and I examined the influence on cosmic rays of "five heavenly bodies" (the Sun, Earth, Mars, Venus, and the Moon). The magnetic fields produced by these bodies deflect the protons and heavier nuclei arriving from outer space and modify their direction, energy, and intensity at the Earth. I have already mentioned another paper I drafted that year with a graduate student, David Hill, with Niels Bohr an intended coauthor.

For his dissertation research, Hill was carrying out some of the first large-scale numerical calculations in physics. With the help of a machine called the SSEC (selective sequence electronic calculator), located at IBM headquarters in New York City, he was using a dynamical form of the liquid-droplet model to trace the sequence of shapes of a fissioning nucleus from the moment it absorbed a neutron until it was ready to divide. IBM, as talented in public relations as in computer design, built models of fissioning nuclei according to Hill's calculated shapes and displayed them in the lobby of its midtown building, next to a panel of lights that flashed as the calculations proceeded. *Time* magazine picked up the story and displayed across a whole page the principal equation that Hill was using. A mystified public was supposed to marvel. His fellow graduate students hooted. It is interesting that Hill, who, even as a young man, could be described as an urbane gentleman, always well dressed and soft-spoken, himself showed a flair for public relations. In later life, after a curtailed career in teaching and research, he sold insurance, promoted mining ventures in South America, and fought for patent rights to "smart" (that is, computerized) cash registers.

Another of my students, John Toll, was starting down a path of bringing modern mathematical methods to bear on problems in relativistic quantum theory. In 1949, he followed me to France; in 1950, to Los Alamos; and, in 1951, back to Princeton. He turned out to be a great asset in pure research, in applied research, and in administration. While in France, working in one of my favorite areas of mathematical physics, the relation between the scattering of particles and their absorption, he obtained elegant new results on what is called the "scattering of light by light"—that is, the collision of one photon with another. Many later researchers built on this work. Toll was a thoughtful person interested in social issues and devoted to service. He had a

John Toll, c. 1955, when he chaired the Physics
Department of the University of Maryland at College Park.
*(Photograph by Al Danegger, courtesy of AIP Emilio Segrè
Visual Archives,* Physics Today *Collection.)*

slight problem with his weight and a slight problem keeping his shirt tucked in, but those attributes only made him more lovable. After working with me, Toll followed a meteoric career in academic administration. He went directly from earning his Princeton Ph.D. to the chairmanship of the Physics Department of the University of Maryland in College Park, building it into one of the best, and at one time the largest, of the physics departments in the country. He moved from there to the presidency of the State University of New York at Stony Brook, where he was again a builder, and returned later to Maryland as president (and then chancellor) of the University of Maryland's statewide system. Retiring from that position didn't slow him down. He is now president of Washington College in Chestertown, Maryland.

I have never been too busy to dream. Dreaming of what might be—of how the world is put together and how its parts interact—provides necessary sustenance for my brain, as nourishing as any calculation. In 1948–1949, in the midst of all the research, teaching, writing, and consulting, I found time to dream—usually at home, usually in a comfortable chair with a pad of paper on my lap, sometimes pacing the room and talking aloud. (Later, inspired by Enrico Fermi's example as well as by rules laid down at Los Alamos, I started keeping my thoughts in bound notebooks. By now, I have filled thirty-nine of them.)

What were some of my dreams at that time? I was beginning to see dimly the outlines of a new approach to nuclear physics. In the 1930s, I had worked on the resonating-group-structure model, which pictured different aggregations of particles constantly forming and dissolving. Then, with Bohr, I had worked on the liquid-droplet model. Beginning in 1948, evidence began to surface that an "independent-particle model" also had validity. Under certain circumstances, the nucleus behaves as if the particles within it moved about almost oblivious to one another. My dream was to develop a coherent way to describe nuclei that took account of all these features. With much valuable stimulus from Niels Bohr, this dream gained substance over the next few year. I called the result the collective model of the nucleus.

These thoughts about nuclei were perhaps too down-to-earth to be called dreaming. They were only a short step from what was already known. But some other things I was pondering at the time were more clearly "far out." Still gripped by my vision that Everything Is Particles, I was dreaming of a world made solely of electrons and positrons. I was even dreaming of applying the idea of action at a distance to Einstein's general relativity theory. To realize this dream would require "sweeping away" the spacetime that connects events, just as Feynman and I had swept away the electric and magnetic fields that connect charged particles.

When, in 1948, I applied for a Guggenheim Fellowship to enable me to go abroad for a year and give uninterrupted attention to new research problems, I chose to emphasize my interest in the physics of electrons and positrons and my ties to Niels Bohr. The nuclear problem was perhaps too much along the track of some of my previous research; I was afraid it might seem mundane to the Guggenheim's officials. In truth, I did not regard it as a problem that required the solitary, concentrated mental effort that a year away could provide. As to the idea of getting rid of space and time, I didn't have the courage to mention it.

The Guggenheim Foundation kindly approved my application for the 1950–1951 academic year, and then, when conditions at Princeton changed to permit me to take leave sooner, approved my request to change to 1949–1950. The plan included the proposition that I spend the year in Paris, with side trips to Copenhagen. Although I wanted to work with Bohr, I did not want to get back fully into the conversational culture of his institute. I wanted time for isolated thinking and calculating, and knew that it would be an easy matter to travel by train from Paris to Copenhagen as often as I wished during the year. The specific choice of Paris was mainly for the family's benefit. Our children, Janette and I felt, would gain most by attending a French school and learning the French language. It would have been fun for them to learn Danish culture and the Danish language, but it would have been less useful to them in their later lives. The children were seven, eleven, and thirteen when they started to school in Paris, and they soaked up French like sponges. (Alison, the youngest, was a little resistant at first to the summer tutoring Janette and I had arranged for them. Once, at lunch, a few weeks after we arrived in France, the two older children entertained us with French songs they had learned. Alison, in a small but determined voice, instead sang—in English—"One Hundred Bottles of Beer on the Wall." She became the most accomplished French speaker in the family.)

We set sail from New York on June 29, 1949, aboard the SS *United States*, our first destination being St. Jean de Luz, a village near Biarritz in the south of France. We settled into the friendly, comfortable *pension* Les Goelands in mid-July for a two-month stay. The children, with the help of a local young woman whom we hired, were to learn French. Janette and I planned to relax and enjoy the surroundings while practicing the language ourselves. And I planned, of course, to pursue my physics in this peaceful setting. John Toll had accepted my invitation to join us for the year. He took a room in another *pension* in the town, and came several times a week to talk physics or dine with us. (Poor John was attacked by fleas in his low-cost *pension*. Until Janette examined his arms and legs and identified the problem, he thought that he was allergic to the rich and plentiful food he was being served. What a relief. It was only fleas. He then methodically trapped fleas and tried out various flea poisons on them. Once he found an effective deterrent, all was well. He could work in his room and in the garden in light summer clothing, and partake of all the good food that his *pension* offered.)

Shortly before leaving for France, I mailed to Bohr a manuscript dealing with certain details of fission, especially the tendency of a uranium nucleus to divide into two quite unequal parts. The authors listed on the draft paper were Niels Bohr, David Hill, and I. Many of the ideas in the paper had come from discussions that I had had with Bohr in previous meetings, so I took the liberty of putting his name on the paper. Hill's contribution came from his dissertation work on the dynamics of fissioning nuclei. We imagined that Bohr would review the paper and make a few editorial suggestions for its improvement, after which we would submit it for publication. I was encouraged by a letter of July 4, 1949, from Bohr (addressed "Dear Wheeler" and signed "Niels Bohr," as was his formal custom). In it, he wrote, "The manuscript that you sent me came as a great surprise but, realizing that it more represents an account of the discussions we through the years have had about the theme rather than some original contribution of which I feel innocent, I do not only agree with the plan, but welcome it as a token of the continuation of our co-operation. [He agrees to be a coauthor.] Since I received it just before leaving Copenhagen for the country, I should like to think a few days whether I might suggest some smaller alterations or additions and shall then write to you again."

Three years later, after Bohr, Hill, and I, taking account of the ever-growing evidence in support of what I was by then calling the collective model, had reworked and rewritten the paper many times, and enlarged its scope beyond fission alone, Bohr decided that his contributions had not been sufficient to justify his coauthorship. He suggested that Hill and I proceed without him. We finished the two-author version in the fall of 1952, and it was published in 1953. The delay was in some ways unfortunate, since some of the ideas in the paper had, in the meantime, been advanced independently by others, hardly surprising at a time when the understanding of nuclear structure was exploding. The bright side of the delay was the chance it afforded to pursue another invigorating, provocative collaboration with Bohr.

My first trip to Copenhagen came in mid-September, just as we were getting ready to move from St. Jean de Luz to Paris. John Toll agreed to accompany the family, assisting with the driving. They made a grand holiday out of it. (Seven-year-old Alison, still the little rebel, declared that she could see all she needed to see of cathedrals from the car window.) Because I had scheduled the Copenhagen trip to mesh with Bohr's schedule, I did not get to join the fami-

ly in Carcasonne and the Loire Valley, although later, before my return to the
United States, Janette and I stole away for a trip to the Riviera and Italy.

In Paris, we stayed at the comfortable and pleasant Pension Domecq, at 70,
rue d'Assas on the Left Bank. It lay next to the Luxembourg Gardens, and
was convenient to l'École Polytechnique [the Polytechnic Institute], where
my host, Louis Leprince-Ringuet, provided office space and a warm welcome
for me. I did not get into any collaborative research with my French colleagues, but I enjoyed conversations and seminars with them. Leprince-
Ringuet was doing high-altitude cosmic-ray work, on mountains and in
balloons, so I was much interested in discussing his results with him. Another stimulating acquaintance was the theorist Bernard d'Espagnat, who shared
my interest in fundamental questions in quantum theory. I loved to walk and
talk with him, around l'École Polytechnique and in the Luxembourg Gardens. I never, for long, forget the 1908 words of Einstein in a letter to his colleague Johann Laub: "This quantum problem is so uncommonly important
and difficult that it should be everyone's concern."

Bohr, it turned out, was more excited around this time by developments
in nuclear physics than by electron-positron pair theory. Like a dog with a
bone, he wanted to gnaw on the nuclear questions until the rough spots got
smoothed on every side. So it was in my private stretches of work in Paris
that I puzzled over a purely electromagnetic world and over a possible world
without even space and time. Much of this work was done sitting at a small
table at Pension Domecq, some of it at the Polytechnic Institute. In October,
I looked again at one of my favorite particles, the muon, and wrote a paper
on how one could use this particle to probe the nucleus. I circulated the paper
privately to some interested physicists, but, as described earlier, didn't get
around to publishing it until 1953, when the experimental results of Val Fitch
and James Rainwater finally got me activated. (The private circulation of
preprints, as they are called, is standard practice in physics, since fellow
researchers often don't want to wait until a paper is published to study it. In
some fast-moving fields, the preprint—now frequently distributed electroni-
cally—is the principal means of communication. The published paper
becomes, so to speak, a record for the file.)

Riding the train in late September 1949 from Copenhagen to my new
"home" in Paris, I found ideas bouncing around in my head on the content of
the paper that I assumed Bohr, Hill, and I would soon be publishing. During our meetings that month, Bohr had emphasized the critical importance of
the new evidence for the independent particle model of the nucleus (according to which, nucleons move freely like gas molecules within the nucleus).
We agreed that our principal challenge was to reconcile this model with the
seemingly quite different model of the nucleus as a liquid droplet. Some-

where in the back of my mind was a piece of information that did not seem obviously connected to these considerations. It was recently published information on the quadrupole moments of nuclei. A quadrupole moment is a measure of how much a nucleus is deformed away from being a sphere (toward either a football or a pancake— shapes technically called prolate and oblate, respectively). The new results showed that many nuclei have a football-type (prolate) deformation larger than could easily be accounted for by either the independent-particle model alone or the liquid-droplet model alone. (To give a more familiar illustration, Earth, being flattened at the Poles by its rotation, possesses a quadrupole moment of the pancake, or oblate, variety. Many nuclei deviate from a spherical shape percentagewise by much more than Earth does.)

Rocking along on the Copenhagen-to-Paris train, it occurred to me suddenly how combining the independent-particle and liquid-droplet models of the nucleus, just as Bohr advocated, could account for the large nuclear deformations. I realized that a single nucleon, caroming around relatively freely within the nucleus in a state of motion that carried it more often to one part of the nuclear interior than another, could push on the walls of its liquid-droplet "cage" and distort those walls quite substantially. The reason was related to the fact that the distortion would provide more elbow room for this particular nucleon to move, which in turn would allow it to sink to lower energy. At the same time, the distortion would increase the energy of the surface tension. But it was easy to see that the decreased energy of this single nucleon (the independent-particle feature) would more than offset the increased energy of the surface (the liquid-droplet feature). Given the chance, a nucleus—or any system—realigns itself to minimize its energy. So this nucleus, hypothetically spherical to begin with, would shift to a quite nonspherical shape.

It wasn't until early December that I communicated these thoughts to Bohr in a letter that covered other work on our joint paper. The principal points that had been occupying my time since my return from Copenhagen (in late September), I said, were "the question of quadrupole moments and the question of justifying the liquid-drop model from the point of view of the individual-particle picture by standard quantum mechanical methods." I went on, "It has turned out to be possible to show very clearly that the quadrupole moment created by deformation of the nucleus by a single particle in an otherwise empty shell exceeds by a factor of approximately 5 the quadrupole moment directly due to the charge distribution of that particle itself."

On December 24, 1949, Bohr wrote in response, "What you tell about your considerations of the quadrupole moment of a nucleus with one particle in an otherwise empty shell seems to us very beautiful and convincing and, as we understand, the point is that the deformation of the nucleus arising from the presence of this particle will imply a comparatively large additional quadru-pole moment of the particles in the closed shells." In the rest of his letter, Bohr dealt with other, more technical questions, and invited me to visit Copenhagen again in January for further work on our paper.

By arriving at this explanation of large nuclear deformations—I sometimes think that some of my best work has been done on trains!— I had uncovered an important new feature of nuclei, one that would prove to be important for the further development of the subject. I gave some thought to publish-ing this result as a separate paper, but in the end decided to let it be incorpo-rated into the larger Bohr-Hill-Wheeler paper. It therefore came as an understandable disappointment when I received, a year later, a preprint from James Rainwater at Columbia setting forth the same idea. For this contribu-tion, published in 1951, Rainwater shared the 1975 Nobel Prize with Aage Bohr (Niels's son) and Ben Mottelson. Aage Bohr had done postdoctoral work with Rainwater at Columbia. Then he and Mottelson, a young American who settled permanently in Copenhagen, pursued the independent-particle–liq-uid-droplet link in great detail, revealing how the particles move in deformed nuclei and how the rotation of the deformed nuclei gives rise to new kinds of nuclear-energy states.

Insights have a way of surfacing at different places at the same time. When that happens, it is usually because ideas bounce around the globe, triggering the same thoughts in different places. Long before the existence of the Inter-net and e-mail, the channels of communication in the international physics community were swift (although measured in days, not milliseconds). In this case, the channels were relatively clear—between me and Niels Bohr, between Niels and Aage Bohr, and between Aage Bohr and Rainwater, among many others. (For instance, Bohr related my quadrupole moment idea promptly in December 1949 to Jens Linhard, a Danish physicist who was back in Copenhagen on Christmas vacation from his research position in Birmingham, England). I have sometimes wondered whether Bohr, despite his scrupulous wish to see researchers get proper credit for their ideas, let fall some remark to his son, which, carried to Columbia, was sufficient to germinate the same idea about nuclear deformation there. It is equally likely that the flow was in the other direction. Perhaps, during my September visit, Bohr made some remark to me, based on his discussions with his son, that was just suffi-cient to set my mind working in the new direction on the train trip to Paris.

At any rate, I learned a lesson. When one discovers something significant, it is best to publish it promptly and not wait to incorporate it into some grander scheme. Waiting to assemble all the pieces might be all right for a philosopher, but it is not wise for a physicist.

On September 23, 1949, the day that Janette, the children, and John Toll left St. Jean de Luz for Paris, and while I was working with Bohr in Copenhagen, President Truman announced that the Soviet Union had exploded an atomic bomb "within recent weeks." (The date was later pinpointed to August 29.) Evidence for the explosion was gathered in the early part of September by the Naval Research Laboratory and the Air Force, whose teams found radioactive fission products in rainwater collected in Kodiak, Alaska, and Washington, D.C., and in a filter paper carried aboard a B-29 flying over the North Pacific.

Two weeks before Truman's announcement, I had taken part in a meeting of the U.S. Reactor Safeguards Committee, meeting jointly with British counterparts at Harwell, in England. My assigned task at the meeting was to report on the chances of a serious reactor accident. I pointed out that sabotage was a risk far greater than the risk of accidental failure of well-designed systems and backup systems and even systems that backed up the backup systems. The saboteur, I said, could well be a trusted colleague, free to move around the control room, knowledgeable about its intricacies. He would be a loner, I suggested, animated by some twisted ideology. Sitting across the table from me as I spoke was Klaus Fuchs, soon to be unmasked as a Soviet spy.

Among my good friends at that meeting were Edward Teller and Dick Feynman—but of course we had no reason yet to speak of the Soviet bomb. When we, and other western physicists, heard the news, we were surprised but not amazed. We had not expected the Soviet bomb quite so soon, but we knew that the talented scientists and engineers in the Soviet Union would achieve it before long. A few months later, we learned that Klaus Fuchs's espionage had accelerated the achievement.

Thirty years later, in 1979, the centenary of Einstein's birth, I undertook to deliver six lectures in Europe on Einstein's legacy, and one of them was in East Berlin. My audience, mostly government officials, included Klaus Fuchs, by then released from prison and working in East Germany. At a coffee break, I asked an attendant to take me to meet him. I approached him with a coffee cup in one hand and a notebook in the other so that I would not have to shake hands.

"What are you doing now?" I asked.

"I am advising our government on nuclear energy for power generation" was his answer. We chatted amicably for a few minutes about physics, a safe subject, and parted.

Harry Smyth's call from Washington reached me in our *pension* one evening that fall. I took the call on the wall phone in the middle of the dinner hour, while a score of French guests put down their knives and forks to watch and listen. Smyth, former chair of my department and author of the famous postwar "Smyth Report" on the atomic bomb, was now an Atomic Energy Commissioner. Would I consider cutting short my leave, he asked, in order to return and contribute to an all-out effort to develop a hydrogen bomb? He didn't use those words. He spoke obliquely in a shorthand code, but I had no trouble understanding his meaning. I had already heard from Edward Teller, who told me that he intended to go back to Los Alamos and renew his work on the "super." Smyth let me know that he shared Teller's conviction that the United States had to react to "Joe 1" (as we had started to call Joseph Stalin's first nuclear bomb) with a renewed effort to design thermonuclear weapons—as well as to improve fission weapons. There was little doubt, he hinted, that the Soviets were already hard at work on an H-bomb. (Despite the indirection of our words, I am sure that a suitably educated spy listening in on this conversation would have divined its meaning as clearly as I did.)

Smyth posed a terribly difficult dilemma for me. I agreed that the Soviet threat was real and that the United States needed to react to it. I felt an obligation to help. But it had been only four years since I had left war work to return to academia. I was enjoying a wonderful, productive year with my family in Paris. There was no end of exciting physics that I wanted to work on. I had students to guide. I had a commitment to the Guggenheim Foundation. Many other physicists to whom the same invitation was issued in 1949 and 1950 simply said no. That was not so easy for me. I shared the fears of my Hungarian friends. I felt a patriotic duty to respond when my country asked for my service. So, in that first conversation with Smyth, I didn't say no. I said maybe. I would think about it. I would consult with family and friends.

I must have received other calls in the *pension*, from Smyth again and from Teller, and I must have been visibly troubled. My children still remember my inner struggle, which I could not hide. I agonized over the decision with Janette. I remember a long conversation on the pros and cons that we had one evening in our bedroom, with the children dropping in now and then. When I went off for a second visit to Copenhagen in mid-January, I had still not decided, although I couldn't help thinking of Europe as a house of cards that would collapse when the first strong wind blew from the east. Over breakfast with Bohr one morning, I told him what was on my mind and how divided I felt. He no doubt sensed which direction I was leaning. "Do you imagine for one moment," he said, "that Europe would now be free of Soviet control if it were not for the Western atomic bomb?" His words stayed with me. When I got back to Paris at the end of January, Janette and I decided that I should go.

9

.

FROM JOE I TO MIKE

SO THE decision was made. I would join the thermonuclear weapons project.
It was early February 1950. I told Harry Smyth and Edward Teller that I would
come to Los Alamos as soon as possible. I notified Henry Allen Moe at the
Guggenheim Foundation. He was understanding and supportive. He readily
agreed to let me postpone the balance of my fellowship to a later date. (My
plan, at the time, was to come back to Europe in the summer of 1951 to work
with Bohr, using up the balance of my fellowship time in that way.) I let my
host, Louis Leprince-Ringuet, know. I let Niels Bohr know. I let John Toll
know — he, like Bohr, was already well aware that I had been wrestling with
the decision. I got in touch with Allen Shenstone, chair of my Princeton
Physics Department. He was a wonderfully straightforward, no-nonsense person of the old school, a Canadian by birth. He let me know exactly what he
thought: I was making a mistake. As I was soon to learn, most physicists in
America shared his view: there existed no Soviet threat, no crisis in world
affairs, sufficient to justify interrupting a career in research and teaching.
But Shenstone put no obstacles in the way. Later that spring, Princeton for-
mally extended my leave of absence to June 30, 1951.

Janette and I agreed at once that we should not interrupt the children's
schooling. She would stay in Paris with them until the end of the school
year. This pleased the children. Now fluent in French, they were enjoying life
where they were. John Toll was also happy to remain in Paris for a while
longer. He was making excellent progress in his work on dispersion rela-

tions—work that would bring him quite a bit of recognition before long—and needed only as much of my guidance as could be handled by mail. I felt good about having him there to help Janette and the children if the need arose. He was already close to being a member of the family anyway.

Facing an extended separation, Janette and I decided on a short vacation in Italy before I departed for America. We climbed into our Renault Quatre Chevaux (four horsepower, four cylinders, four passengers) and headed south.

All of my planning was taking place with such speed that some wheels had not had time to turn. As Janette and I were laying plans for Italy and I was looking into ship passage to New York, I had not yet received an invitation from the director of the Los Alamos lab, Norris Bradbury. As to what sort of compensation I might receive at Los Alamos, and what travel reimbursement, I was still in the dark. Assured by phone that all such matters would be taken care of, I agreed to meet in Europe with a Los Alamos representative to learn about the status of the H-bomb work. Frederic De Hoffmann, a recent Ph.D. in theoretical physics from Harvard and a staff member at Los Alamos, suggested that Janette and I stop off on our way to Italy for a meeting at what he described as his favorite hotel in the south of France, the Negresco in Nice. When I later came to know Freddie (as he was always called), I could see that this approach to a business meeting was typical of him. He was, at the time, twenty-five years old, and already a cosmopolitan world traveler, with refined tastes in wine, food, and lodging. Freddie had emigrated from Vienna to America as a teenager and sailed through his undergraduate and graduate work at Harvard, completing his Ph.D. before his twenty-fourth birthday, even though employed at Los Alamos for two of his school years. When we scheduled our meeting in Nice, he had already been part of a declassification team advising the Atomic Energy Commission on what wartime documents might be made public.

When Janette and I encountered Freddie, we were both impressed. It was impossible not to like this brilliant, talkative young man of the world. When I reached Los Alamos, I found that he had been "adopted" by Edward Teller as a chief scientific assistant and "reality checker." Edward came up with ideas all day long, and Freddie figured out whether they made any sense all night long. Their personalities and modes of work meshed perfectly. Later Freddie followed his restless imagination to California, where he became president of General Atomics and then president of the Salk Institute, in addition to a string of other achievements. He was felled by AIDS in 1989 at the age of sixty-five, having been infected through a blood transfusion in the course of a heart bypass operation.

In February, I sailed for New York. From there, it was natural to stop in Princeton on the way west, to talk with Allen Shenstone and other depart-

Freddie De Hoffmann, when he joined General
Dynamics Corporation, 1955.
*(Photograph from General Dynamics, courtesy of
AIP Emilio Segrè Visual Archives.)*

mental colleagues. In my Princeton department, I found something I had
never experienced before: dissonance with my colleagues. Shenstone was
not the only one to think that I was making a mistake in going to Los Alamos
to work on thermonuclear weapons. Others—both in Princeton and elsewhere—shared his view, notwithstanding President Truman's action the previous month (January 31, 1950) authorizing a "crash program" to design and
build an H-bomb. This designation of national priority by the president did
not sway most of my physicist friends. For many of them, it was just too soon
after World War II. They had done their part. They had returned to their
academic research and teaching. They wanted no more of weapons work.
Some, overcome by the horror of the atomic bomb, could not imagine working on an even more powerful weapon, no matter how they assessed the Soviet threat. Indeed, the advice that the Atomic Energy Commission (AEC)
had received from its General Advisory Committee several months earlier

Myself as a toddler in Glendale, California, c. 1913.

As an NRC Fellow in
Copenhagen, 1934 or 1935.
*(Courtesy of Emilio Segrè Visual
Archives, Wheeler Collection.)*

"Brother Wheeler," a young
faculty member at the
University of North Carolina
in Chapel Hill, 1938.
*(Courtesy of Niels Bohr
Archive, Copenhagen.)*

Walking in Princeton with Albert Einstein and Hideki Yukawa, 1953.
(Photograph by Howard Schrader, courtesy of Princeton University.)

"Every difficulty is an opportunity." Leading a discussion in Princeton in 1971, I
was taking on the doubters, making the case that black holes *will* someday be
observed and that gravity and the quantum *will* someday be joined.
The colleague with the small bald spot is Tullio Regge.
(Photograph by Robert Matthews, courtesy of Princeton University.)

Some members of the General Advisory Committee on Arms Control and Disarmament, meeting with President Richard Nixon and Secretary of State Henry Kissinger, c. 1973. *Left to right:* James Killian, William Casey, Nixon, Kissinger, I. W. Abel, John Wheeler, Cyrus Vance. *(Courtesy of the White House.)*

Next to a drawing of a wormhole, 1973. *(Photograph by Robert Matthews, courtesy of Princeton University.)*

I become a Texan, c. 1976.

I visit Black Hole,
Nova Scotia, in 1981.
*(Courtesy of Emilio
Segrè Visual Archives.)*

This painting by Raymond Everet Kinstler, which hangs in Princeton
University's Jadwin Hall, shows me in a favorite spot in High Island, Maine.
(Photograph by Robert Matthews, courtesy of Princeton University.)

With Janette and our three children (*from left to right*): Alison Lahnston, Jamie Wheeler, and Letitia Ufford, 1985.

Soon after my return to Princeton, 1987.
(*Photograph by Robert Matthews, courtesy of Princeton University.*)

With Janette in High Island, 1980s.
(Photograph by Beverly White Spicer.)

Janette and I, 1991.
(Photographs by Joanne B. Ford.)

With Ken Ford in New York City, 1993.
(Photograph by Cecelia Brescia, American Institute of Physics.)

With my fellow Wolf Prize recipient Yacov Sinai at the Israeli Consulate in New York, 1997. Wolf Prizes in several fields are awarded annually by the Wolf Foundation in Jerusalem. In 1997, the prize in physics came to me; Sinai, originally from Moscow and now my colleague at Princeton University, was one of two recipients in mathematics. *(Photograph by Joanne B. Ford.)*

(in October 1949, and affirmed in December) was against proceeding with the development of thermonuclear weapons. The scientists on that committee, some of the country's most distinguished,[1] wanted to shy away from an H-bomb program for a combination of technical, ethical, and policy reasons. Yet contrary advice, supportive of the development of a superweapon, reached Truman from other members of the AEC; from influential members of Congress; from high-ranking military leaders; and from individual scientists, including Edward Teller, then at Los Alamos, and Ernest Lawrence and Luis Alvarez, both at the University of California, Berkeley.

The antagonism that I felt from some of my colleagues did nothing to shake my conviction that I was doing the right thing, but it did distress me. My past relationships with scientists of every shade of opinion and of every nationality had always been cordial. Open, friendly collegiality was something I had worked at and succeeded at. I determined to keep working at it, and for the most part I was successful. As hostilities surfaced over the next few years among scientists with divergent views on the superweapon, civil defense, nuclear testing, nuclear power, radioactive fallout, the Communist threat, and missile defense, I managed to remain on good terms with most of those who held strong views. Back in my student days, I had found it wise to smile at people I met walking across campus, because my near-sightedness prevented me from knowing whether an approaching person was an acquaintance or not. This practice made me some new friends and helped me keep old ones. Maybe it set the stage for my efforts in later life to look for friendship and common bonds with everyone.

While I was in Princeton that February, Ken Ford, a second-year graduate student, approached me to ask if he might work with me on his dissertation research. I readily agreed, but had to tell him that I was going to be on leave from Princeton for at least a year, to work on weapons. I let him know that I would be glad to have him join me in Los Alamos, but I gave him no sales pitch. I had taken the same approach with John Toll. Having students on hand was always a stimulus to my own work, and having Toll and Ford nearby would be welcome. Despite the urgency of the thermonuclear work, I had no intention of abandoning my work in pure physics. I would also be happy to have these students join me in the Los Alamos laboratory to help with the project. But the decision to leave the campus and follow me west, especially the decision whether to join the weapons work, had to be theirs, made without pressure.

[1] Oliver Buckley, Bell Labs; James Conant, Harvard University; Lee DuBridge, Caltech; Enrico Fermi, University of Chicago; John Manley, Los Alamos (secretary); Robert Oppenheimer, Institute for Advanced Study (chair); I. I. Rabi, Columbia University; Hartley Rowe, Hydrofoil Company; Glenn Seaborg, UC Berkeley; and Cyril Smith, MIT.

Edward Teller felt no such diffidence. He wrote to John Toll in Paris urging him to join the project. On a visit to Princeton that spring, he contacted Ken Ford and made the same pitch in person. Ken (then twenty-three) remembers Edward at his most eloquent as they sat together on the steps of the Institute for Advanced Study. Both John and Ken did conclude that world peace would be more stable if the United States achieved a thermonuclear weapon before the Soviets did. They decided to join me, and it was a happy collaboration, both in the laboratory and in our pure research outside the lab.

I have always loved train travel. It provides structure without distraction. What for some people is monotony is for me the wonderful freedom to think in a relaxed, prearranged block of time. My brain seems to work well on trains. As the Super Chief hastened across the prairie in early March on its way west from Chicago to New Mexico, I was able to reflect on the nonspherical shapes of atomic nuclei, following up on the idea that had occurred to me the previous fall on a train trip from Copenhagen to Paris. In the first of a pair of letters penned to Niels Bohr on Super Chief stationery, I reported my calculation that a typical nucleus could exist in either of two semistable shapes, prolate (football) or oblate (pancake), with an energy barrier making it difficult for a nucleus, once in one shape, to change to the other. In the second letter, dated two days after the first, I recanted—perhaps the trip over Raton Pass had cleared my head—noting that the nucleus could easily sneak *around* the barrier, changing smoothly from one shape to the other through a succession of intermediate shapes more complicated than either footballs or pancakes. This means that a nucleus will easily find its way to whatever shape has the lowest energy. Before long, experimental physicists discovered that most nuclei are footballs. It took longer for theorists—especially Aage Bohr, Ben Mottelson, and Sven Nilsson in Copenhagen—to figure out why nuclei would rather be footballs than pancakes.

Following the same procedure as the scientists who had come to Los Alamos in its earliest days, I got off the Super Chief at Lamy[2] and was driven to Santa Fe, where I checked in with Dorothy McKibbin at her inconspicuous office at 109 East Palace Avenue, just off the Plaza, before proceeding on to the still fenced-in city of Los Alamos. During the war, Los Alamos and everyone in it had a single address: P.O. Box 1663, Santa Fe, New Mexico. This box was the conduit for things. The conduit for people was Dorothy McKibbin. A young widow raising a teen-age son, she was warm and efficient, loved

---

[2] Physicists and mathematicians are the kind to be tickled by the information that the main line of the Atchison, Topeka, & Santa Fe Railroad passes through neither Atchison nor Santa Fe.

by everyone on "the hill." She processed them all in her office labeled "U. S. Eng-rs." Whether it was Mr. Farmer (Fermi), Mr. Baker (Bohr), or a technician who had never before been west of New Jersey, every new arrival was made equally welcome and sent onward feeling that things were not so strange after all in this strange place. Dorothy McKibbin's office was not closed until 1963, when she retired after twenty years as the lab's gatekeeper.

Back in 1942, when General Groves had selected Los Alamos as the site for Project Y (the wartime code name for the secret bomb laboratory), it was not because of the breathtaking beauty of the place. Although the pellucid air and striking views of the Jemez and Sangre de Cristo mountains may have influenced Oppenheimer when he recommended the site, Groves picked it for its remoteness and the ready availability of a great deal of nearby land, much of it already government-owned, where experiments with radioactivity and explosives could be conducted. Exercising the rights of the federal government, Groves took over whatever private property he needed, including the Los Alamos Ranch School for boys, whose few buildings provided an initial base of operations.

By the time I arrived in 1950, Project Y had become Los Alamos Scientific Laboratory (and would later become Los Alamos National Laboratory). Pending the family's arrival, I was put up in Fuller Lodge, a delightful log structure that had been the principal building of the Ranch School. Edward Teller welcomed me on the afternoon of my arrival, and told me that in response to President Truman's order to accelerate weapons work, including work on an H-bomb, the lab was now on a six-day week. This set me back a little, since I expected to find time, as always, for my "Princeton work." When I met Edward for breakfast the next morning, I was able to tell him that by chance I had opened the Gideon Bible next to my bed to just the right page for the occasion, finding there the words, "Six days shalt thou labor." (Female lab employees had their own form of mild protest. Many of them wore jeans to work on Saturdays.)

That morning I walked to the lab—still downtown—and plunged into the task of learning about past work on thermonuclear weapons, some of it going back to 1942. The ideas then current made it look very difficult, if not impossible, to design an H-bomb. It is not surprising that when the AEC's General Advisory Committee (GAC) met the previous fall to consider how the United States nuclear weapons program should react to Joe 1, some members of the committee doubted the technical feasibility of the superweapon. Teller and I saw the difficulties clearly enough, but had no reason to give up without a lot more exploration. I return later in the chapter to our scientific efforts.

Like my path of arrival through Lamy and Dorothy McKibbin's office, starting to work without any contract or financial arrangements was reminiscent of the wartime period. My wish to continue on Princeton's payroll, with Los Alamos reimbursing Princeton, couldn't be worked out. Both Allen Shenstone and Norris Bradbury found things to object to in the proposed arrangement, partly related to my continuing consulting work for Du Pont. Finally, more than a month after I had started working in Los Alamos, Norris Bradbury informed me that I would be paid $45 per day as a consultant to the lab (counting only days actually worked) plus $90 per month for subsistence. This was perfectly satisfactory. Bradbury also generously authorized payment of full travel costs for the whole family from Paris to Los Alamos and back to Paris. As it worked out, only Janette and I went back to Paris the next summer. The children returned to Princeton.

When Janette and the children arrived in late April, we were provided, at a very modest rent, what was probably the finest house in Los Alamos at the time, 1300 20th Street, the house on "Bathtub Row" where Oppenheimer had lived during the war. It was one of the few original houses from the Ranch School, equipped—unlike all the wartime housing—with bathtubs, not just showers. It was a large, comfortable house, big enough for the parties that Oppenheimer liked to stage, and big enough for my family to settle into. In the large backyard was a small excavated Indian ruin. In a spare room with windows on three sides, John Toll, Ken Ford, and I set up desks for evening and weekend work.

Los Alamos housing was "socialized," and everyone loved it. The benevolent Zia Company provided furniture, fixed leaks, replaced lightbulbs, and cleared balky plumbing. Getting three desks was another matter. Behind the benevolence lay a bureaucracy not easily cracked. Desks were not on any approved furniture list. For a few days in July, after my students had arrived, I spent as much time prying loose desks as I did working on weapons. But, in both cases, persistence paid off, the desks more quickly than the weapons.

Janette and the children took at once to the beauty, the wonderful climate, and the convenient living in the city on the mesa. Our next-door neighbors on Bathtub Row were the Ulams, Stan and Françoise, Polish and French respectively, and their daughter Claire, quite American. Claire and our children became playmates, and Janette and I grew very fond of Stan and Françoise. Stan was a brilliant mathematician who, like John von Neumann, was curious about everything and possessed a deep intuition about physics. It was about a year after we first met that Ulam came up with the breakthrough idea (quickly pounced upon, improved, and elaborated by Teller) that turned the H-bomb from being possible to near-certain. Stan liked to say, as he wandered from the lab around 4:00 P.M., "I don't know how you physicists work so hard.

Stan Ulam in his Los Alamos study, late 1940s. He has a picture of John von Neumann on his mantelpiece.
(*Courtesy of AIP Emilio Segrè Visual Archives, Ulam Collection.*)

I can't work more than six hours per day." Then he went home to do pure
math in an office next to his bedroom, sometimes into the night. Françoise's
housekeeping was an eye-opener for our children. They learned that in her
scale of values, reading a book or chatting with friends came before dishwashing and ironing. In their affection for good food and culture, the Ulams
were soul mates of Freddie De Hoffmann. One of Stan's "entrepreneurial"
initiatives later in Los Alamos was to found a European-style coffeehouse.
Unfortunately, those who shared his cultural values were not numerous
enough in Los Alamos for the financial success of the venture.

The Tellers we counted already as friends. I had traveled with Edward and
Mici from Copenhagen to London in 1934, just after their marriage. Janette
and I had come to know them well during the time that we were in North
Carolina and they were in Washington, D.C. Edward was persuasive, engaging, and moody. He could be endlessly optimistic and imaginative in the
face of technical obstacles. He made strong friends and strong enemies. When
people drove him to distraction—which was often—the piano was his outlet.

Edward Teller at the piano.
(Photograph by Fred Rothwart, courtesy of AIP Emilio Segrè Visual Archives.)

He could, and often did, use me as a sounding board when the piano was not enough and he needed to write or talk. Mici was a devoted wife, yet independent and spirited. She, too, was Edward's sounding board, but she remained sunny when he turned into a black cloud. The Tellers' presence in Los Alamos made our year there more agreeable than it might otherwise have been. In such a "company town," a newcomer does not so easily penetrate the social structure.

Globe-hopping makes for complex automobile logistics. When Janette set out to sell the Quatre Chevaux in Paris, she found she couldn't because it was in my name. A properly executed power of attorney, submitted through the French consul in New Orleans, was required. I prepared the needed document with the help of Ralph Carlisle Smith, the senior legal officer at the Los Alamos lab. "Don't sign over all your rights to Janette," he said. "She might divorce you and take everything." I told him I'd run that risk. In April 1950, with the car safely sold, Janette and the children set sail from Le Havre for the ocean crossing to New York. I was supposed to be on the dock, but when they disembarked, they saw no sign of me. *Is my husband so buried in physics somewhere that he has forgotten the time, or worse, the date?* Janette wondered. Fortunately, I was only a little late, and arrived in time to cope with the baggage and dry the tears.

Without tarrying in the East, we piled our luggage, our children, and ourselves onto the Broadway Limited for the overnight trip to Chicago. There we picked up a used car, purchased for us by Jack Hutchings, the brother of my brother Joe's widow. It was a Studebaker, with the unsettling feature that it looked about the same from the front and the back. Janette called it the "Hollywood Hearse." But it handled the deserts and mountains of New Mexico, made a side trip to California, and got us back to Princeton the next year.

As December 1950 approached, Edward Teller, John Toll, Ken Ford, and I decided to present papers at the meeting of the American Physical Society to be held at UCLA in the week between Christmas and New Year's. This was the major West Coast meeting of physicists, held every winter on the Pacific coast, alternately north and south. (It is hard for present-day physicists to imagine that at that time, all of the participants at a major Physical Society meeting could fit comfortably in one lecture hall, with no simultaneous sessions required.) Janette and Mici were ready for a break from Los Alamos and wanted to check out southern California. It was agreed that they would be accompanied by one pair of men westbound and another pair eastbound. Edward and I boarded the Super Chief in Albuquerque for the ride to Los Angeles, while Janette and Mici and John Toll and Ken Ford took off in the Studebaker, loaded with luggage and four sleeping bags. According to their report, a roaring fire in a Grand Canyon lodge was the most beautiful sight they had ever seen after camping out in zero-degree weather on the rim of the canyon. On the way back, Ken and John climbed on the train in Pasadena on New Year's day, after we had all marveled at the Rose Bowl Parade. Returning by car, Edward and Mici and Janette and I opted to sleep in a motel, not outdoors in sleeping bags—although, truth to tell, Janette and Mici might have been ready for the adventure of sleeping under the stars again.

The disapproval of some of my colleagues over my choice to join the H-bomb effort at Los Alamos was deeply troubling. "Why did you let the AEC persuade you to do this sort of thing?" asked one. "I see Teller has hypnotized you," said another. In my mind, I was answering a call to national service. It was not scientific curiosity that made me interrupt one line of work to take up another—uprooting my family in the process. Nor was it the pursuit of financial reward. Nor was it weakness of will in the face of someone else's persuasive powers. Although thermonuclear problems offered intriguing scientific challenges, the reason to work on them was entirely practical. I considered it urgent that the United States react to Joe 1 with a priority program to develop a thermonuclear weapon before the Soviets did. When President Truman made this a national priority, I felt obligated to accept the call to help. Never did I feel any pangs of self-doubt. It was a great disappointment to me that so few of my colleagues shared my view that a national scientific mobilization was called for. Fortunately, as it turned out, a small group of us got the job done, and done rather quickly. It took the cooperative effort of some senior consultants, a cadre of talented staff members already at work at Los Alamos, and a score or so of young scientists who were recruited to the effort.

I was no sooner in Los Alamos than I contacted Harry Smyth in Washington to see if he could arrange to establish a patriotic base of support for my choice to join the project. I asked him to seek from President Truman a letter, to be directed to Harold Dodds, president of Princeton, and to Henry Allen Moe of the Guggenheim Foundation, requesting my service in the national interest. Smyth resisted the suggestion, arguing that since both Dodds and Moe supported my decision, they needed no high-level arm twisting. Let's save the president, Smyth said, for some case in which a scientist's institution is opposed to his joining the project. I wasn't convinced. I had been so hurt by the unexpected and first-ever hostility of some of my colleagues that I still wanted to establish that I was responding to an inescapable national obligation, not making a free choice for personal reasons. Smyth continued to resist my request, and the matter was dropped. I can see in retrospect that he acted properly, but I was guided at the time by intense feelings.

According to a secondhand report, Robert Oppenheimer had said, "Let Teller and Wheeler go ahead. Let them fall on their faces." His attitude at the time (which he modified later in the face of technical evidence) seemed to be this: the hydrogen bomb can't be done, or if it can be done it will take too long, or if it can be done and doesn't take too long, it will require too large a fraction of the nation's scientific manpower, or if it doesn't require too large a fraction of the nation's labor force, it will be too massive to deliver, or if it is deliverable, we oughtn't to make it.

Some others were nearly as negative as Oppenheimer. Perhaps, had I been in the thick of politics all my life, I could have shrugged this off. But the international collegiality and harmony of the world of physics had not prepared me for such divisive forces. I was nevertheless at peace with myself and made peace with my fellow scientists of all persuasions. As other issues came along in the future—civil defense, missile defense, nuclear power, weapons tests—I and my friends often had to agree to disagree. Some—Wigner, Teller, von Neumann, and Alvarez, for example—were on my wavelength. Some—such as Bethe, Christy, Goldberger, Oppenheimer, and Schwinger—were not. It is a great happiness to me that I maintained cordial relations with all of these people. Our mutual respect and our common commitment to probing basic questions of nature overrode differences on policy issues.

Teller's temperament is different than mine, and I regret that he has made his share of implacable foes. During the war years, when he wanted to leap

Robert Oppenheimer at the Institute for Advanced Study,
unknown date, 1950s or 1960s.
(Photograph by Alan Richards, courtesy of Robert Matthews.)

beyond the yet-to-be-designed atomic bomb and investigate an H-bomb, he alienated Los Alamos laboratory leaders, notably Oppenheimer and Bethe. While I was at Los Alamos, his impatience with laboratory procedures and his perception of wrongly assigned priorities brought him into conflict with lab director Norris Bradbury. He lost more friends when he campaigned for a new weapons laboratory. By the time he testified against Robert Oppenheimer at Oppenheimer's security hearing in 1954, it was hard to find a physicist anywhere who admired Teller. He is a man of conscience who has continued to campaign relentlessly and by whatever means are at hand for what he believes in, and it has cost him dearly. His impact on scientific and military policy is sure to be debated by historians for generations to come. I have disagreed with his tactics but never with his goals.

On April 27, 1954, the day before Teller's testimony at the Oppenheimer hearing, I happened to be staying in Washington. After dinner, Teller joined me in my hotel room for discussion and agonized aloud over what he should say in his testimony. He asked my advice. "Edward," I said, "tell the story as you see it." It was no surprise for me to read later in the printed proceedings his actual words, "I would like to see the vital interests of this country in hands which I understand better, and therefore trust more."

Teller's testimony in the Oppenheimer matter was the last straw for many of his colleagues and former friends, some of whom refused to be civil or to shake hands with him when they met. Delivering a eulogy in 1984 for our mutual friend Montgomery Johnson, Teller, with his usual theatrical flair, reportedly said, "As a result of those hearings, I lost every friend I had but one. This week I lost that friend."

Given the reaction I experienced from my colleagues within weeks of making the decision to go to Los Alamos, I should not have been surprised by our very limited success in attracting other senior scientists to the project. Teller was the principal recruiter. Of all those in senior ranks whom he contacted, only one, Lothar Nordheim, from Duke University, accepted and joined us for a full year. A still-junior but very brilliant scientist from California, Marshall Rosenbluth, also answered the call. Others—some of very great distinction—came for shorter periods. Von Neumann and Bethe gave the work at least a few days per month. Fermi spent two full summers at Los Alamos, and came for brief visits during the academic year. Emil Konopinski from Indiana University gave some time to the project.[3] People like this can make invaluable contributions even on a part-time basis. Von Neumann and Fermi, in particular, were enormously helpful. Both had the most stunning ability to listen to a recital of current problems for only an hour or two and then provide comments or calculations that would show the way to overcoming the problems. They also enriched the life of the lab by giving colloquium talks on almost every visit. On summer days, when Fermi was speaking on classified problems in his powerful voice, we joked that we should close the windows to prevent leakage of secret information across the fences a hundred feet away.

The Los Alamos lab was divided into fourteen divisions. T Division (T for theoretical) housed the theoretical physicists, mathematicians, a battery of young women who carried out calculations on electric calculators (the "computresses"), and the first primitive electronic calculators—predecessors of computers. P Division (P for physics) housed experimental physicists and their equipment. There were, among others, a CMR Division (chemistry and metallurgy research), a J Division (weapons testing), a D Division (documents and library), a W Division (weapons), and a GMX Division (explosives). H-bomb design work was centered initially in T Division. P Division became involved quickly because we needed to know more about the characteristics and probabilities of reactions involving deuterium and tritium—the

---

[3] While at Los Alamos in World War II, Konopinski had demonstrated convincingly through calculation that an atomic explosion could not ignite thermonuclear reactions in the atmosphere or in the earth.

John von Neumann, 1950s.
(Courtesy of AIP Emilio Segrè Visual Archives.)

heavy isotopes of hydrogen. As designs progressed, CMR and other divisions became involved. By mid-1951, with ideas getting translated into hardware, the center of gravity of the thermonuclear work shifted toward W Division, headed by the extremely competent, no-nonsense engineer, Marshall Holloway. We theorists could draw our spheres and cylinders on the blackboard. Holloway had to make them real. (Holloway was one of the people driven nearly to distraction by Teller. Teller, in turn, had such antipathy to Holloway that he resigned from the lab when, in 1951, Holloway was given general responsibility for the H-bomb work.)

During the war years, Hans Bethe had headed T Division. Among its many strong members were Richard Feynman, Robert Serber, Teller, and the head of the British mission, Rudolf Peierls.[4] Now, in 1950, the cast of characters was largely new. Carson Mark, a wise Canadian, had become the division head. He had around him some remarkably talented young physicists, including Conrad Longmire, Ted Taylor, and, beginning that summer, Marshall Rosenbluth. Stan Ulam could have taken his pick of academic jobs but, for reasons not everyone could fathom, found Los Alamos a congenial place to work.

[4] Peierls, originally German, was knighted by Queen Elizabeth in 1968.

Only much later in his life did he leave Los Alamos for the University of Colorado. Cornelius Everett, a shy, retiring mathematician, was a frequent collaborator with Ulam and a leader in numerical computation. When Everett walked down a corridor, he kept close to one wall, tapping it as he went, so that he could concentrate on his thoughts, not on his path. Teller had rejoined the division from Chicago, and, with De Hoffmann, formed what was in effect a two-person group. I headed another small group. John Toll, Ken Ford, Burton Freeman, and several other young physicists were members of it. Lothar Nordheim was a group of one. Actually, Teller, Nordheim, and I, and the people with whom we worked directly, were not officially in any group. We were assigned to TDO (the T Division Office), reporting directly to Carson Mark.

Mark did not have the stature in the outside world that comes with achievement reflected in publications, but he was an astonishingly astute and knowledgeable physicist (trained originally as a mathematician). He was a perfect research manager. He understood deeply all of the science he was overseeing. He could ask just the right questions to ferret out weak spots and get at ways to fix them. He was infinitely patient, and he was respected by everyone. He was tall and ruggedly handsome, with a gift for clear diction. Like Niels Bohr, he spoke slowly, often sucking on a pipe and seeming to ruminate. Yet he was the opposite of delphic. What he had to say was lucid and to the point.

Around the time of my arrival in Los Alamos in early 1950, Bradbury formed an interdivisional committee, the Family Committee, with Teller as its chair. "Family" referred not to the twenty-five members brought together from half a dozen lab divisions, but to the new family of weapons that was foreseen. Teller, although the nominal chair of this committee, was too volatile to effectively coordinate the work of this diverse group. Moreover, he left the lab in a fit of pique in September 1951, to return to Chicago. While Teller was still at Los Alamos as well as after he left, Carson Mark served more often than not as the de facto chair of the Family Committee. He was as ideal a choice for this coordinating role as he was for managing the theoretical division. He and the practical Marshall Holloway got along fine. Mark is the unsung hero who made all of the parts come together for the successful test in the fall of 1952. When only twenty months elapse between a clear design idea and the successful large-scale test of an entirely new kind of weapon, it is because a dozen different developments charge down parallel tracks to meet at the end. It takes a master orchestrator, a lot of communication, and a lot of mutual trust.

People like Conrad Longmire are also rare gems in an applied laboratory like Los Alamos. Conrad, the always cheerful, unflappable father of seven children, was a powerful theoretical physicist who also had a keen sense of all the practical engineering difficulties. Some theorists bog down if they try

Carson Mark, 1951.
*(Courtesy of University of California
Los Alamos National Laboratory.)*

to visualize or calculate what happens in anything other than a simple ideal-ized shape. Some engineers can't discern the fundamental processes going on in their real-world devices. Longmire could bridge the gap. He was a strong right hand to Carson Mark.

Ted Taylor, who later turned away from weapons work to focus on problems of energy and the environment, was, at that time, the most imaginative design-er of new fission weapons at Los Alamos. In 1945, there were two designs, Fat Man (used in Alamogordo, New Mexico, and Nagasaki, Japan) and Lit-tle Boy (Hiroshima). Both were near the limit in size of what a B-29 could carry—and near the weight limit, too, with full fuel on board. A decade later, there were bombs of all sizes and weights and intended modes of delivery. More than any other person, Taylor was responsible for the richness of the proliferation. He was an intuitive person, a little dreamy. He could somehow visualize accurately what some complex maze of steel and uranium and plu-tonium and wires and bolts and high explosives would do. Since an H-bomb requires an A-bomb trigger, Taylor's input was critical to the thermonuclear program.

It isn't possible to describe all of the other people who contributed to the success of the thermonuclear program. What is most remarkable, perhaps, is how small was their number. A few dozen physicists, a handful of mathe-maticians, a score of engineers and metallurgists. The active talent that made it happen numbered no more than a hundred—backed up, of course, by a larger number of technicians, assistants, and "computers" (as people who calculated were then called).

The first order of business when I reached Los Alamos was to learn what had gone before. The "super," as then conceived, had fatal flaws. There was question whether even an atomic bomb could provide a high enough temperature to assure ignition. And there was question whether, once ignited, even the mighty furnace of nuclear fusion could provide enough energy to keep the thermonuclear flame alive. A thermonuclear fusion reaction is literally combustion, or burning. High temperature causes the deuterium or other fuel to react. As with a log in a fireplace, the part of the fuel that is burning must generate enough heat to ignite nearby parts. Otherwise the flame will die out. And there has to be sufficient temperature in the first place to get the whole thing started. This is quite unlike chain-reacting fission, which does not require high temperature and does not spread locally from point to point.

The temperature required to keep a thermonuclear flame going is at least 60 million degrees Celsius, even hotter than at the center of the Sun.[5] At this temperature, a great deal of the energy is contained not in the material (such as deuterium) but in electromagnetic radiation. If this radiation is allowed to escape, not enough energy is left behind to heat the material, just as a coal fire that is not banked will die out because too much of its energy radiates away. Moreover, fusion produces high-energy neutrons that fly far from the combustion zone before they give up their energy, so they don't add to the heating of the nearby material. Calculations showed that even if the radiation were contained within a massive shell, the temperature of the material would not remain high enough to assure that the deuterium fuel, once lit, would stay lit.

What to do? The nature of the problem—or, to put it differently, the deficiencies of the design ideas—had been known for eight years. There was good reason for pessimism, but no reason for despair. We set about doing two things. First, for the existing design ideas, we calculated the progress of combustion in more detail than had been done before, nudging the parameters this way and that to see if we could alter the conditions enough to make a successful thermonuclear explosion. Simultaneously, we tried to come up with as many new design ideas as possible, crazy or not, and put them to the test of calculation. We considered and discarded designs based on fission and fusion fuel blended into a solid cylinder ("yule log") and fusion fuel occu-

---

[5] The Sun's central temperature of 16 million degrees Celsius is sufficient to sustain thermonuclear fusion reactions there because gravity compacts the material to a density 150 times that of water. Even so, the rate of reaction in the Sun is exceedingly slow—the Sun will burn for 10 billion years!

pying scattered pockets in fission fuel ("Swiss cheese"). Teller's "alarm clock" idea, dating from 1946, we resuscitated and studied. Consisting of interleaved spherical layers of fusion and fission fuel, it held promise for yields of hundreds of kilotons, but not megatons. Independently invented by Andrei Sakharov and named "layer cake" by the Soviets, this design was the basis of the first Soviet thermonuclear explosion in 1953.

During the summer of 1950, my small group was totally caught up in this effort, as were Teller and De Hoffmann. Others in T Division, especially Ulam, Longmire, and Taylor, were involved in the brainstorming. Ulam and Everett were doing more calculations. There weren't many of us, and we interchanged ideas daily. Carson Mark kept himself fully informed and always made good suggestions. For a time, I shared a single large office just off the seminar and coffee room with four of my young associates. Since we could reach our office—and others could reach us—only by going through this public room, it was hard not to keep in touch with others. The morning gathering for coffee and pastry was a time when we freely shared ideas that we had been puzzling over or putting to numerical test the previous day. Compartmentalization of information did not apply to this group, nor did pride of ownership get in the way of the free exchange of ideas. Teller would drop into the coffee room or our office like a dark-haired bushy-browed prophet Isaiah to foretell a new approach. Ulam seemingly spent more time in the coffee room, the corridors, and other people's offices than he spent in his own office. He was the wandering minstrel, as offhand as Teller was intense. But when Ulam tossed off one of his disarming questions or diffidently suggested a new approach, we listened.

One of the early priorities for Toll, Ford, Freeman, and me, along with Teller and De Hoffmann, was to prepare a comprehensive report on all that was then known about thermonuclear reactions and to present and assess the various design ideas for a weapon. This report was intended for a September 10, 1950, meeting of the AEC's General Advisory Committee. As the date approached, the report's thickness began to resemble that of the Manhattan telephone directory—so, not surprisingly, we started calling it the telephone book. When Teller and I presented the report to the GAC, we could not say, "We know that an H-bomb will work, and this is how." We could only say, "There has been a lot of progress on a lot of fronts, and some of the ideas hold hope." Bradbury, the lab director, took care to identify our report as our personal assessment, not an official lab statement. Despite the reluctance of some of its members, the GAC concluded that research should continue. Oppenheimer, trying to be objective despite his personal doubts, expressed the "frustrated gratitude" of the GAC for our efforts. For the time being, there was to be no scale-up and no scale-back.

Oppenheimer's word "frustrated" was a good one. Not that we lacked enthusiasm, or even optimism. Not that we weren't eager to push forward. But we lacked data and we lacked computing power. Without experimental data on thermonuclear processes under conditions that might exist in a bomb, we had to insert a lot of guesses ("informed estimates") in the calculations. Without computing power, we had to make do with crude approximations.

A decision had already been made to include a test of certain features of thermonuclear fusion reactions in the series of tests code-named "Greenhouse," scheduled for late April 1951 in Eniwetok. My group helped to prepare for Greenhouse, but not to the exclusion of our continued work aimed at actually achieving a thermonuclear weapon. It fell to me at a late October 1950 meeting of the GAC to report on our detailed plans for Greenhouse.

Prior to World War II, the use of calculating machines in the service of science was limited principally to spectroscopists, who measured and calculated wavelengths of light to great precision, and to astronomers, who, beside being concerned with spectra, were beginning to understand the evolution of stars and to calculate the complex behavior of matter within stars, including the thermonuclear reactions that give them light. As a byproduct of their work on stars, astronomers contributed some of the mathematical tools used to solve equations numerically.

As scientists set about designing an atomic bomb at Los Alamos during the war, they needed some of the astronomers' techniques. Within a bomb, just as within a star, nuclei react under extreme conditions of temperature and pressure. Many of the calculations needed in wartime Los Alamos were carried out in the manner of the prewar astronomers, using electrically driven (but not electronic) calculators. The resulting rows and columns of handwritten numbers, penciled onto large sheets by the young women with their Marchant calculators, looked like medieval versions of spreadsheets. The calculators, each about the size of a breadbox, were full of small and not-so-quiet gears. Their muted clackity-clacks sounded like distant freight trains.

Toward the end of the war, the first crude electronic calculators came into use. These were modified versions of IBM accounting machines. In 1950, Los Alamos had, if I remember correctly, three of these monsters. Each one was several times larger and heavier than a refrigerator. Called card-programmed calculators (CPCs), they read punched cards at the rate of a little more than one per second, and they were noisy. We put many of our ideas to the test on these machines, but continued to use young women at their Marchants as well.

The first "modern" computer—one that stored instructions internally—was the ENIAC, completed at the University of Pennsylvania in 1946 and delivered the next year to the Army's Aberdeen Proving Grounds in Maryland.

Weighing in at more than 30 tons and consuming 174 kilowatts of electric power, it was the granddaddy of all computers to follow. By the time of our major computing on the H-bomb, 1950–1952, the ENIAC was already outdated.

Von Neumann had the vision to see the great future potential of computers, and he set about designing one for the Institute for Advanced Study in Princeton. Working from the von Neumann design, Nick Metropolis undertook to build the MANIAC (mathematical analyzer, numerical integrator, and computer) in Los Alamos. Urbane, well-dressed, and reckoned an eligible bachelor by every matchmaker in town, Metropolis was later the cofounder with Ulam of the European-style coffeehouse in Los Alamos. The MANIAC, being still under development and testing in 1950 and 1951, was just short of being ready to assist in our work. In those days, a standard method of checking the MANIAC's performance was to walk around its innards, hitting the steel framework that held its electronics with a rubber mallet, and then repeating a calculation to see if the MANIAC gave the same answer. The MANIAC never achieved routine reliability, but it provided a vital testing ground for early ideas of computing and computer design. In 1957, it was honorably retired, replaced by new IBM machines designed to meet the laboratory's needs.

In the late summer of 1950, Fermi and Ulam, still reluctant to use the slow and balky computers, set up an informal competition to see who could first get answers to certain questions about thermonuclear burning, with one using a calculator, the other a pair of dice. Fermi had the assistance of a dazzling young beauty, Miriam Planck (unrelated to Max Planck), who carried out calculations on her faithful Marchant calculator. Ulam, with Everett, applied the relatively new "Monte Carlo technique" by literally throwing dice. How far an atomic nucleus or a photon of light moves through a material, and what happens when two particles collide, are matters of chance, and the dice are used to decide which among the possibilities will be chosen before proceeding to the next step. Fermi loved competition, and he was bent on winning this one. That summer, we noticed, he found it necessary to spend a great deal of time in his office with Miriam, reviewing her calculations and laying out new ones. Ulam pretended it was all in fun, but was no doubt bent on winning too. In the end it was declared a draw when both sets of calculations yielded about the same answers at about the same time—answers that made the early thermonuclear designs even less promising. Shortly after her work with Fermi, Miriam married the physicist David Caldwell and went with him to Santa Barbara, later earning a Ph.D. in sociology. Most of the computresses who stayed at Los Alamos soon became computer programmers.

In late February 1951, Stan Ulam came up with an idea that he called "a bomb in a box." He pictured an atomic bomb set off within a large container such that energy from the bomb could make its way to a far corner of the container and there compress some nuclear fuel to hitherto unattainable density, causing this secondary nuclear fuel to explode, releasing more energy. Until then, implosion had been achieved only with conventional high explosives. Ulam visualized an atomic bomb being used as the source of implosion for another bomb. Since implosion with an atomic bomb source might be expected to produce far higher density than conventional explosives could produce, Ulam reasoned that the efficiency of energy production in the secondary bomb might be quite high.

Carson Mark relates that when Ulam came into his office unannounced to tell him about his new idea, he (Mark) was annoyed because he was hard at work on details of planned atomic tests in Nevada. He didn't want to set aside that urgent work to listen to some new, "far out" idea. But Ulam persisted and Mark listened. The next day, Ulam took his idea to Teller. Whereas Ulam had visualized the secondary bomb as another atomic bomb, Teller immediately saw the possibility of having the secondary consist of thermonuclear fuel. Thus was born the "two-stage idea," using energy from an atomic bomb to compress and ignite an H-bomb. Moreover, out of their discussion came another key idea, that energy should be transferred from the atomic trigger to the thermonuclear fuel not by the brute force of shock waves but by the flow of radiation. The secondary would be illuminated by a great pulse of light, causing the secondary to implode and become extremely hot. Mark attributes this idea of radiation implosion to Teller, and I have no reason to think otherwise. Teller had a mind prepared. He had been "thinking thermonuclear" for years, and needed only the stimulus of one new idea to conceive a new design. There was always a bit of tension between Ulam and Teller, and later it became difficult to sort out from their recollections who thought of what. But at that time, they were in a relatively cooperative spirit, and promptly wrote a joint report on what came to be known as the Teller-Ulam or Ulam-Teller idea.

In a letter to von Neumann describing the new idea, Ulam said, referring to Teller: "Edward is full of enthusiasm about these possibilities; this is perhaps an indication they will not work." Both Ulam and Teller were, in fact, so full of ideas that inevitably more of them were wrong than were right. But this one was right. What had been standing in the way of success in H-bomb design was the enormous appetite of empty space for energy. When a container of deuterium or lithium deuteride (possible thermonuclear fuels) at ordinary density is heated to such high temperatures that it will burn, a great part of the energy used for heating is soaked up by the space in which the material resides. In other words, the input energy goes only partly into heating material substance. Much of it goes into heating space. (Inside your living room, there is also radiant energy occupying the space, but at the modest temperature of your room, the energy in the molecules of air exceeds the energy in the space. If the room were heated to tens of millions of degrees, the reverse would be true. Most of the energy in the room would be radiant energy in space, not material energy in the molecules.)

The Teller-Ulam idea, because it called for enormous compression of the thermonuclear fuel, reduced the amount of space occupied by a given amount of material and thereby prevented that space from soaking up so much of the available energy. More energy was left to heat the fuel, preventing the thermonuclear flame from dying out.

Extensive calculations using the new idea would be necessary before we could be certain, but suddenly the odds seemed in our favor. The reaction of all of us was, "Of course! Why didn't we see it sooner?" But in science, ideas do not come out of the air. They are built on top of prior exploration and discovery. In this case, the myriad of ideas and calculations that led to Oppenheimer's "frustrated gratitude" were probably necessary before the new idea could germinate.

Even before the Ulam-Teller approach could be proved out definitively through calculation, it gave a new impetus to thermonuclear work at the lab. Above all, it made a full-scale test appear feasible, and we began to talk of scheduling such a test for as early as the fall of 1952. A reinvigorated Family Committee met more often, and work along parallel tracks in different divisions got under way—in chemistry, metallurgy, engineering, cryogenics (creating and managing very low temperatures), nuclear measurements, and test logistics. By the fall of 1951, plans were on track for an "Ivy" series of tests, to be conducted at Eniwetok in the Pacific a year hence. I don't recall when our particular thermonuclear device got a name, but it came to be called "Mike." It is recorded that President Truman was presented with a model of Mike in June 1952, some four months before its test.

The least expensive thermonuclear fuel is deuterium, or heavy hydrogen. It is abundant in ordinary water, and can be extracted at modest cost. Each nucleus of deuterium contains one proton (which makes it hydrogen) and one neutron (which makes it "heavy"). At sufficiently high temperatures, these deuterium nuclei, or deuterons, can react and fuse to form helium nuclei, with the release of a large quantity of energy (even more per unit mass than in nuclear fission). In the Sun, "light" (ordinary) hydrogen participates in the thermonuclear burning, but the process involving light hydrogen is too slow for use in a weapon or even for use in controlled thermonuclear burning for power production.

A more expensive but even more effective thermonuclear fuel is a mixture of deuterium and tritium. Tritium is still heavier hydrogen. Its nucleus contains one proton and two neutrons. Almost none of it is found naturally on Earth, because it is radioactive with a half-life of about 12 years. Small amounts made by cosmic rays quickly decay. Nuclear physicists knew by the 1940s that the DT (deuterium-tritium) reaction takes place more readily than the DD (deuterium-deuterium) reaction under comparable conditions. This means that the chances of success in gaining a thermonuclear explosion should be greater with DT fuel than with pure deuterium fuel. On the other hand, tritium can be produced in quantity only at great cost—and, once produced, it lasts only about twelve years. Yet when the outlook for success in the H-bomb project seemed gloomy in 1950, the members of the AEC decided that it would be prudent to create the means of making large quantities of tritium, in case it should be needed. In mid-1950, Du Pont was asked to undertake the massive project, just as it had been asked to build and operate the reactors at Hanford. Again, reactors would be required. Eventually, five large reactors went into operation at the Savannah River Plant, near Aiken, South Carolina, capable of making great quantities of tritium. These reactors differed from the earlier ones mainly in having more neutrons to spare (that is, more neutrons beyond those required to keep the chain reaction going). These extra neutrons could stimulate nuclear reactions in the element lithium that would produce tritium. In case the need for tritium proved to be modest, the reactors could easily be modified to produce plutonium or other materials instead of tritium. So, one way or another, their purpose was to produce materials for atomic energy. One can think of a large reactor as being analogous to a power plant. The power plant produces electricity that can be used for a great many purposes. The reactor produces neutrons, which likewise can be put to use in many ways.

During my year at Los Alamos, we explored the idea of using a particular form of the compound lithium hydride as a thermonuclear fuel. (This was not a new idea. Teller had advanced it several years earlier—and, as we learned later, the Soviets had thought of it too.) Lithium is the third element in the periodic table (after hydrogen and helium). Its common isotopes are lithium 6 and lithium 7 (whose nuclei contain three and four neutrons, respectively). The molecular combination useful as thermonuclear fuel consists of one atom of lithium 6 and one atom of deuterium (heavy hydrogen), so it is sometimes called lithium-6 deuteride. This compound has two great advantages over pure deuterium. First, it is a solid at ordinary temperature, whereas deuterium is a gas. If used as a fuel, deuterium must be liquefied by cooling it to 20 degrees above absolute zero, and it must then be kept under pressure. Moreover, it is highly flammable—witness the 1937 *Hindenburg* disaster.

(There was a joke in the lab that a large container of liquid deuterium really *is* a hydrogen bomb.)

Lithium-6 deuteride has a second advantage: It "manufactures" tritium during the burning process. It is as if you poured low-grade petroleum in your gas tank and it automatically created high-octane gasoline as it burned.

So far as I know, all hydrogen bombs now have lithium-6 deuteride as their fuel. In 1950–1951, however, it was a relatively new idea, and there was a limit to how many new ideas we wanted to try out all at once. So we continued to explore the liquefied isotopes of hydrogen as a thermonuclear fuel at the same time we were developing the superior new lithium-based fuel.

I could never forget for long while I was in Los Alamos that my stay was to be temporary. I had to think about when I would leave, where I would go, how I could continue to mix pure physics and teaching with government service, and what would be best for my family—Janette's life and the children's schooling. I felt sure that the thermonuclear effort would require more than one year. Janette and I talked about the possibility of sending the two older children to the International School in Geneva in 1951–1952, while she and I and Alison, the youngest, stayed a second year in Los Alamos. But dividing the family in this way was not appealing, and Janette longed for a stable life somewhere else. Although she had a few friends in Los Alamos, she was not happy with the social structure in this one-company town, where friendships outside the lab tended to match the hierarchy of positions inside it.

Throughout my career, I have received regular invitations to consider relocating to other institutions. In general, I have responded with a polite no. Princeton afforded a good working environment and was an agreeable community. But at this time I was a little more receptive to alternatives. I wondered whether my commitment to serve the nation by working on weapons had driven a wedge between me and a few of my Princeton colleagues, who had already registered their disapproval. I wondered, too, whether Princeton would be a place where it would be easy to divide my time between academic pursuits and applied projects. So when an invitation came to visit UC Berkeley to explore the possibility of a position there, I expressed interest, and spent a few days there in December 1950. Ernest Lawrence and his colleagues proposed that I become a professor of theoretical physics in the department and devote some time to working at the nearby Radiation Laboratory. The Rad Lab, as it was called (later christened Lawrence Berkeley Laboratory), was supported by the AEC and housed work in nuclear physics and high-energy physics as well as weapons-related research.

My friend Edward Teller, ever vigilant, ever sensitive, ever suspicious, briefed me thoroughly on Berkeley before my visit. He had been there not long before me. He was entertaining an offer from UCLA at the time and liked the idea of our being at least in the same state. He was quite concerned, however, about the rift in the physics faculty (and in other departments) produced by a loyalty oath recently imposed by the Regents of the University of California, on top of an oath required by the state legislature. Anti-Communist hysteria was being fanned nationwide in the early stages of what became known as the McCarthy era, and loyalty oaths were springing up everywhere. Although I was deeply concerned about the Soviet threat and about subversion sponsored by the Soviets, I considered the loyalty oaths useless and the anti-Communist witch hunting reprehensible. (One of my ancestors, Mahitable Towne, had been burned as a witch in Salem.)

Los Alamos staff members, being employees of the University of California, were subjected to the California Regents oath requirement. After reading the oath carefully, I signed it even though I saw no value in it. Nearly everyone else at Los Alamos signed it, too, and went back to work, after at most grumbling about how silly it was.[6] Not so at Berkeley and some other California campuses. Some faculty members objected loudly in public, some resigned in protest, and some thought that the requirement was entirely reasonable. Debate on the issue was not always friendly. It was getting in the way of normal collegiality and probably interfering with research as well. What I found at Berkeley was a department divided. Gian Carlo Wick, an excellent theorist, was one of those who had resigned. Lawrence was one who had no patience with the protesters. (Teller, I should add, was anti-oath. He declined the invitation from UCLA, partly for this reason. He didn't see enough evidence that the Regents would reverse their position.) So Berkeley did not look to me like the right place at that time. Since Janette had no great enthusiasm for California, we laid that idea to rest.

In Washington, Harry Smyth was pleased by this outcome. He had become upset by what he interpreted as the Rad Lab and Los Alamos competing for my services using the same (AEC) money. He wasn't shy about trying to call off Lawrence from any effort to lure me from one AEC facility to another. (In my mind, of course, it wasn't Berkeley versus Los Alamos. It was Berkeley versus Princeton.) Some years earlier, Smyth had stepped in forcefully when he learned that Arthur Compton had approached me at the Metallurgical Lab

---

[6] Among some 2,300 Los Alamos staff members, it is my recollection that only one, John Manley, a senior experimental physicist who had remained at the lab after the war, did not sign the oath. Even for Manley, not signing was at best a mild form of protest, for he had decided to leave the lab anyway on account of his opposition to the H-bomb program. He moved to the University of Washington, returning to Los Alamos in 1957 after the University of California rescinded the oath requirement. My student Ken Ford at first refused to sign, but relented as the deadline for his dismissal approached.

about coming to Chicago after the war. Compton did not renew the discussion.

Even before the Berkeley visit, I had begun to think about setting up a separate weapons research group in Princeton, so that I could return to the working environment that I liked so well and to a town where my family was happy, and where I could yet continue to serve Los Alamos. My confidant Teller liked the idea. My wife liked the idea. Norris Bradbury was understandably cool to it when I brought it up with him in the fall of 1950. He had already been irritated by the beginning of Teller's campaign for a second weapons laboratory. (By the time Teller got his way and Livermore was designated as the site of a second competing lab eighteen months later, Bradbury had been rubbed raw.)

Bradbury was a reasonable man, however. I was more than happy to accede to his suggestion that the Princeton group act as an arm of Los Alamos, complementing, not competing with, activity at Los Alamos. With that understanding, we came to a meeting of the minds. I had more reason to wonder whether the Princeton authorities would raise objections, because of the secrecy requirements. But when I contacted Allen Shenstone and Harold Dodds, I was agreeably surprised to encounter no resistance. Shenstone and Dodds recognized that universities had an obligation of national service, and they were willing to accommodate my wishes in order to bring me back to Princeton. As I learned later, they had spoken to Oppenheimer, by then director of the Institute for Advanced Study, and received his blessing.

The particular thing that made the proposal palatable to the university authorities was the chance to move the project a few miles away from the main campus. Princeton had just acquired from the Rockefeller Institute for Medical Research its large plot of land—823 acres—just off Route 1 three miles from the campus. (The Rockefeller Institute was relocating to New York City to become Rockefeller University.) Princeton University renamed the plot the Forrestal Research Center (to honor James Forrestal, a Princeton graduate and America's first Secretary of Defense, who committed suicide in 1949). Now it houses the Princeton Plasma Physics Laboratory, among other enterprises. In the early 1980s—before cutbacks in funding for plasma physics—Forrestal was home to some 1,500 workers. Our little project was its first occupant.

During my year in Los Alamos, anti-Communist fervor was building in the United States—the California loyalty oath was one manifestation of it—and my colleague David Bohm back in Princeton got caught in its crossfire. Called before the House [of Representatives] Un-American Activities Committee, he refused to answer the committee's questions about his alleged Communist ties dating from the time when he was a student in Berkeley working with Oppenheimer. He was cited for contempt and, in December 1950, arrested. (To make bail, he had to borrow money from then graduate student Silvan Schweber.) The university, reacting swiftly, banned him from the campus, with the result that his graduate students had to go to his residence to meet with him. On May 31, 1951, after a trial, Bohm was acquitted on all counts. Nevertheless, the university did not renew his appointment, and he had no luck finding other suitable work in the United States. In October, he left the country for a post in Brazil, later relocating to Birkbeck College in London, where he became famous for his writing and teaching, much of it a strange mixture of mysticism and fundamental physics.

Since the Bohm affair—which understandably polarized the campus—occurred while I was away, I played no part in it. Had I been there, I'm not sure I would have been outspoken in Bohm's defense. Even though I am descended on both my mother's and father's sides from long lines of dissenters and activists, and am a passionate believer in freedom of expression, I found it hard to accept Bohm's decision to shield those who adhered to Communist ideology at a time when the Soviet Union was suppressing its own people and threatening world peace. The university was gauche in its manner of dealing with Bohm, yet I could sympathize with its goal, to preserve its reputation as a center of unbiased scholarly inquiry, not the home of blind loyalty to one ideology or another. Bohm chose to proclaim neither his independence of thought nor his commitment to Soviet doctrine. Admirably stubborn, he proclaimed only his abhorrence of the Communist witch hunt that was under way.

I was largely responsible for bringing Bohm to Princeton in the first place. Shortly after World War II, I had visited Berkeley and, at my department's request, interviewed Bohm. Upon my favorable recommendation, Princeton offered him a temporary appointment, and he joined the department in 1947. I was interested in Bohm because he had given great thought to the interpretation of quantum theory, and had also advanced some ideas on critical conditions in plasmas. As I became better acquainted with his work, I came to realize that it was based more on intuition than the rock-solid foundation I had at first imagined. This makes it harder for me, in retrospect, to say whether my failure to step forward vigorously in his defense—even from a distance—rested more on my assessment of his behavior or his physics. In his later career, he did make some provocative contributions to debates about the meaning of quantum theory and the relation of consciousness to reality. When he died in 1992, he was revered by many.

Some time in late winter, probably in February or March of 1951, Lyman Spitzer, an athletic patrician and head of the astrophysics program at Prince-

Lyman Spitzer, late 1960s.
(Courtesy of AIP Emilio Segrè Visual Archives.)

ton, stopped by in Los Alamos following a ski holiday in Colorado. He had agreed to work with me on the bomb project in Princeton starting later that year and was in Los Alamos to be briefed. While skiing, an idea had come to him—an idea about to deflect him into a new path—and he was eager to share it. He had thought of a configuration, a doughnut twisted into a figure eight, in which he thought hot hydrogen plasma might be contained by a magnetic field and stabilized long enough for thermonuclear reactions to occur and generate energy—not explosively, but at a controlled rate. A hydrogen plasma is a gas so hot that all the electrons have "boiled off" the atoms. What is left is a soup of charged particles—electrons and bare nuclei, in this case deuterons or tritons. If a magnetic field threads through the twisted torus (the figure-eight doughnut), the charged particles execute tiny circles, while also drifting both along and across the magnetic field lines. Spitzer's calculations indicated that his geometry would deter the drifting across the field lines, and thus keep the hot plasma particles from hitting the walls. He wanted to get the reaction of laboratory scientists to the idea. From the laboratory leaders, he wanted a blessing (and funding!) so that he could expand the Princeton research program to pursue this idea.

The intersection of ideas was obvious. I was interested in explosive ther-

monuclear reactions. Spitzer had just become interested in controlled thermonuclear reactions. We both wanted to work in Princeton. His scientific
ideas got a good reception, and so did his plan for an expanded Princeton
research project. (Money was not quite so hard to come by in those days.)
Within a few months everything was arranged for us both to set up shop at
the new Forrestal Research Center.

Roy Woodrow, Princeton's chief financial officer, reminded us in a meeting
that spring that our combined project needed a name. Spitzer had a suggestion. "How about Project Matterhorn?" he said. He loved mountaineering
and to him, *Matterhorn* meant challenge. Moreover, he had polished his idea
while on a ski holiday in the mountains. "Fine," I said. "Now your device
needs a name. How about Stellarator?" He liked it, and the name stuck.
(Loosely, *Stellarator* means "star machine." Its purpose was to harness the
same kind of fusion energy that powers the Sun.) Five years later, when for the
first time Soviet and American researchers on controlled fusion came together and exchanged ideas at an Atoms for Peace conference in Geneva, the
participants were startled by how many of the same problems had been recognized on both sides and how many of the same ideas had been generated
independently on both sides. The Stellarator was an exception. It had not yet
been devised on the Soviet side. But the Soviets had come up with a configuration called a Tokamak. As it has turned out, the Tokamak is an even more
robust design and finally displaced the Stellarator as the principal configuration used by fusion researchers in many countries.

In April 1951, I went back to Princeton to work out some of the practical
details of setting up our new enterprises. Projects Matterhorn B (for bomb)
and Matterhorn S (for Stellarator) were assigned space in the same galvanized
iron building, one that had served to house experimental animals for the
Rockefeller Institute. John Toll was my strong right arm in setting up the security and office procedures and arranging for the proper lines of communication to Los Alamos. He and Ken Ford came back in June to recruit a secretary,
guards, and computing assistants. I set about seeking scientists.

It had been my hope that Princeton, being a major academic center, might
attract senior scientists who were reluctant to move to Los Alamos. I had
even advanced this as an argument for setting up the Princeton group. But I
misjudged. Among the 120 senior colleagues I solicited by letters, telephone
calls, telegrams, and personal visits, only one, Louis Henyey, a professor of
astrophysics at UC Berkeley, accepted my invitation. Few, I think, judged
our work to be immoral. Rather, most simply doubted the urgency of the
project and saw no reason to interrupt their own work. One New England colleague reportedly ridiculed my letter openly in his department. So I turned
to bright, interested younger people. In the end, the team consisted mostly

Most of the Matterhorn B team in 1952. *Front row, left to right:* Margaret Fellows, Peggy Murray, Dorothea Reiffel, Audrey Ojala, Christine Shack, Roberta Casey. *Second row:* Walter Aron (with mountain-climbing rope), William Clendenin, Solomon Bochner, John Toll, John Wheeler, Ken Ford. *Third and fourth rows:* David Layzer, Lawrence Wilets, David Carter, Edward Frieman, Jay Berger, John McIntosh, Ralph Pennington, unidentified, Robert Goerss. *(Photograph by Howard Schrader, courtesy of Lawrence Wilets.)*

of graduate students (such as Toll and Ford) and fresh Ph.D.s. Fortunately, the mental muscle of a theoretical physicist can reach peak performance early. In this mostly under-thirty group we had a lot of talent, enough to get the job done. We had, in total, about twenty scientists and half a dozen support staff.

For its most intense fifteen-month period, summer 1951 to fall 1952, the Matterhorn B team concentrated on refining the design and calculating the expected results of the "Mike" shot, which took place on October 31, 1952 (November 1 in Eniwetok). It just so happened that the design of the device lent itself to a division of labor between Los Alamos and Matterhorn. Crudely speaking, there were three aspects to the design (though in reality there were many more): the fission-bomb trigger, the thermonuclear combustion, and the engineering and metallurgy of the device. People at Los Alamos concentrated on the fission-bomb trigger and on the flow of energy from this trigger to the thermonuclear part of the device. We in Princeton took it from there, working on the ignition of the thermonuclear fuel, its burning, and the explosive expansion of the entire device. (All hardware matters were handled in Los Alamos.)

Many in the Matterhorn group traveled back and forth to Los Alamos every month or two, and we kept in close touch by phone and letter with the Los Alamos group, but because of the nature of the design and the split responsibilities, there was no need for daily communication. After the successful Mike shot, my team pursued follow-up research and refined our calculations, looking toward actual weapons (as opposed to test devices). The center of gravity of the work shifted gradually back to Los Alamos, and our Princeton efforts gradually wound down until Matterhorn B officially ended its work late in 1953.

Matterhorn S, under Spitzer's leadership, made great progress on the Stellarator design and the theory of controlled fusion, with a group of no more than a dozen people, also almost all quite young. In the language of physics, the barrier between Matterhorn S and Matterhorn B was quite permeable. In other words, there were fruitful interactions between people in the two groups, and some joint seminars. The fundamental problem, that of nuclear fusion at high temperature, was the same for both groups. Each group could look to the other to analyze its own ideas and to get new ideas.

In 1991, Matterhorn S celebrated forty years of productive work. It had long since changed its name to the Princeton Plasma Physics Laboratory. At its peak, before cutbacks in plasma funding, it had an annual budget of $65 million and employed more than 500 people. The metal shack in which it all started is still there, empty and abandoned. Hidden behind larger modern buildings, it isn't easy to find if you don't know what you're looking for.

When Lyman Spitzer showed up in Los Alamos, bubbling with enthusiasm as he sketched out his twisted doughnut and explained how it might hold superhot plasma together, he predicted that it might take five years to prove the principle. Years later, as the target kept receding, a joke among physicists went: "Controlled fusion energy is five years away and always will be." In fact, no one says five years anymore, but no one says it won't be achieved. As a potential long-term power source for humanity, using the nearly limitless supply of deuterium in the ocean as fuel, it is too important not to pursue.

Back in June 1951, after the Ulam-Teller approach to the H-bomb had given those of us most directly involved a clear sense of optimism that success was quite likely, we still had to convey that optimism convincingly to the jury of senior advisors, the AEC's General Advisory Committee. The ability to compute the consequences of the idea was still a great challenge. Part of the chal-

The metal shack in which Project Matterhorn B did its work
in the early 1950s and where the Princeton Plasma Physics Laboratory
got its start as Project Matterhorn S. This photo dates from 1960.
(Photograph by John Peoples, courtesy of Princeton University.)

lenge was reducing the equations governing the process to an approximate form simple enough to feed into the primitive computers then available without making them so simple that they completely lost their validity. Toll, Ford, and I put our heads together on this, and Ford wrote the computer code for the IBM CPCs. He made a few test runs on CPCs on the night shift at Sandia Laboratories in Albuquerque, then, in early June, shifted to a CPC located in an IBM office in New York City. We arranged for the use of the machine from about 8:00 P.M. to 8:00 A.M. every night. We figured that with more than one hundred hours of computer time available to us in two weeks, we could make sure that the computer code was correct and that the results were sensible—and then go on to make meaningful computer runs, feeding in different assumptions about the device. Ford took the train from Princeton to New York in the evening and returned by train to Princeton in the morning, bringing the results of the night's work.

The GAC's June 1951 meeting was held at the Institute for Advanced Study in Princeton on the 16th and 17th of the month. Among the members present were Oppenheimer, Fermi, and Rabi. Bradbury and his Los Alamos team made their presentations on the first day. My turn came on Sunday morning, June 17. (The GAC traditionally met on weekends for the convenience of its academic members.) Ford's calculation during the previous week had been encouraging, and we agreed that he would make a final run with the most realistic assumptions on Saturday night. He brought the data back on an early train and went to the Matterhorn building, where he made graphs of the principal results on sheets of paper about two feet by three feet in size. Then, with only minutes to spare before my scheduled presentation, he rolled up the graphs and sped to the Institute. He approached a window of the first-floor conference room where we were meeting, tapped on the window, and caught my eye. I had just started my presentation. I interrupted to go to the window, open it, and accept the roll. There was time only for him to say, "Looks good." I unrolled the graphs, taped them to the blackboard, and explained their meaning. The reaction was electric. Here, for the first time, was convincing evidence that a thermonuclear flame, spreading in a device based on the Ulam-Teller principle, would propagate, sustain itself, and burn up a significant fraction of the fuel. I can't remember whether it was at this meeting or later that Oppenheimer called the design "technically sweet." But the clear sense of the scientists present was that we had a sure route to success, and that there was no longer any technical reason to delay the program.

There was still plenty to do. Computing power was a bottleneck. We bought time on just about every available computer anywhere. The Matterhorn team ran marathon sessions for weeks and months on the SEAC (Standards Eastern Automatic Computer) at the Bureau of Standards in Washington, D.C. (again the graveyard shift) and on Eckert and Mauchly's new Univac in Philadelphia. The U.S. Weather Service, scheduled to take delivery of that Univac, kindly agreed to delay delivery for our benefit. Foster and Cerda Evans, a husband-and-wife team from Los Alamos, ran more calculations on the ENIAC at Aberdeen. Ulam and Everett used the Los Alamos MANIAC.

When we made our regular trips to Los Alamos to coordinate efforts, we usually went by train. The two-day trip gave us a chance to get away from the urgency of programming and calculating, and spend some relaxed time talking over the general principles, reviewing the equations, and wondering about the possible effects of design changes. The Pennsylvania Railroad management was kind enough on several occasions to have the Broadway Limited[7] make an unscheduled stop at Princeton Junction to pick us up. Once a conductor asked if we were the Princeton basketball team. He couldn't imagine any other reason for a special stop. I didn't look much like a coach, but the tall, young, healthy physicists with me could very well have passed for basketball players.

---

[7] As a railroad buff, I mourned when the Broadway Limited made its last run in 1995.

Keeping my own physics alive during the Los Alamos–Matterhorn period was not easy, but I tried to think and write a little almost every day. Having my students John Toll and Ken Ford with me in Los Alamos was a stimulus. We spent time at our desks in the Bathtub Row house on many evenings and during some weekend hours. John was completing his dissertation work, and Ken was getting started on his. My efforts were devoted to the nuclear physics work with Niels Bohr and David Hill that was still percolating, and to keeping up with recent developments such as the new insights in quantum electrodynamics provided by Schwinger, Feynman, and Tomonaga. Although quantum electrodynamics (or pair theory, as we called it in its early days) had been and remained a love of mine, I was reluctant to jump on the Schwinger-Feynman-Tomonaga bandwagon that was attracting literally hundreds of theorists throughout the world. When I see a herd running one way, I like to march another way. I was speculating on new states of nuclear matter and wondering whether nuclei could exist as cylinders and doughnuts as well as the standard spheres and ellipsoids.

In April 1950, soon after settling into Fuller Lodge in Los Alamos and diving into the thermonuclear work at the lab, I headed east to spend a few days working with Bohr and Hill in Princeton before meeting the family in New York. We were still assuming that our long-gestating paper on the collective model of the nucleus would have triple authorship. At the same time, I made preliminary plans to spend the summer of 1951 with Bohr in Copenhagen. The patient and understanding Henry Allen Moe at the Guggenheim Foundation was agreeable to using the remaining portion of my fellowship for that purpose. As it worked out, the pressures of getting Matterhorn started compressed the Copenhagen visit to less than a month, but it was revitalizing, as always, to spend time with Bohr and with other visitors to his institute.

While in Copenhagen in the summer of 1951, I was able to participate in a meeting of the International Union of Pure and Applied Physics, an especially significant body in the early post–World War II years when ties among scientists in all countries needed to be reestablished. The meeting included animated discussion of the possible creation of an international laboratory for high-energy physics in Europe, an idea first proposed, as far as I know, not by a European but by the American I. I. Rabi. What was still at the discussion stage in 1951 eventually came into being as the European Center for Nuclear Research (CERN—the acronym is based on the name in French). Now a leading center for particle research, CERN will soon house the world's highest-energy accelerator.[8]

---

[8] In CERN's present accelerator, particles cross an international border thousands of times each second as they circle under France and Switzerland.

Back in Princeton with the responsibilities of getting Matterhorn up and running and managing a bigger staff than I had ever directly managed before, I needed self-discipline to find niches of time for my own research. During the very first weeks back in Princeton, before Janette and the children had arrived and before I could get back into our house on Battle Road, Toll, Ford, and I set up cots in a corner of the Rockefeller Institute's cavernous power house. While Toll and Ford slept until some reasonable hour like 7:00 or 7:30, I got up quietly around 5:00, showered, and went to my desk to work on physics before the day started. Back in my own house, I kept up this early-rising routine for some time. After a demanding day at Matterhorn, I was ready for bed right after dinner. I set the alarm for 3:00 A.M., and used a few quiet hours to think about matters other than thermonuclear burning and computer codes.

After Bohr bowed out of the collective-model paper, Hill and I set to work to finally finish it off. I recall Hill's visit to Princeton in the summer of 1952 to work with me on drafting the text and preparing what turned out to be fifty-two illustrations with lengthy captions. The paper, embryonic so long, had grown into a tome. I could hardly ask Hill to rise at 3:00 A.M., so I altered my own schedule to work with him in the evening. But by 9:00 P.M. on the first few evenings, I was fighting drowsiness. Then I discovered that if I took a break for a hot bath, I came to life again and could work into the night.

Finally, in mid-October, just before I left for the Pacific to witness the Mike test, David and I submitted the collective-model paper to *Physical Review*. Sam Goudsmit, *Physical Review's* editor, told us that the paper was far too long, but that he would accept it if all the figures could be collected as an appendix at the back instead of being scattered throughout the text. So that's the way it was published—eighteen double-spaced pages of text and twenty-four pages of figures and captions. When it went to the printer, Goudsmit suggested that I send a box of chocolates to the editor who had handled it.

The military plane that carried me to Eniwetok in late October landed in a rainstorm heavier than any I had ever seen. The next day was beautiful. I was given a helicopter ride around the atoll with its dozen or so small islands and clear lagoons. Technicians and scientists assembled the Mike device on one of those small islands, Elugelab. After the test, Elugelab no longer existed.

We were told that during World War II, the Japanese had used Eniwetok as a military training ground. As part of their fitness training, Japanese soldiers ran across an island, swam to the next one, ran across it, and so on. Given the number of sharks that we saw on our helicopter ride, it sounded pretty risky to me.

On the day of the test, November 1 in the Pacific, I was on board the SS *Curtis*, a cruiser stationed about thirty-five miles from Elugelab. Before the moment of the explosion, I could see nothing through the dark glasses I had been issued. Then came what appeared to be a dull spot on the horizon, only a little less dark than the general background. It quickly grew in size and brilliance, as if the sun had suddenly made its appearance. Then came a mass of churning clouds, mottling the face of the explosion. As the brilliance gradually dimmed, the boiling display climbed high in the sky. All of this in absolute stillness. The rolling thunder and burst of warm wind hit us some three minutes after the first light. It sounds trite to say that the experience was engraved in my memory, but it is the plain truth. There are some sights and sounds that one does not forget.

My first reaction was simply relief. It worked. And apparently it worked as planned. I'm glad I was there. I carried into every future discussion of nuclear war that vivid image, so much more powerful than equations and graphs and numbers alone.

The preliminary measurement of the yield was 10 megatons, 800 times the energy of the Hiroshima bomb. A final figure, according to the official history of the AEC,[9] is 10.4 megatons. The prediction of the Matterhorn team, won from Ford's long nights nursing the SEAC in Washington, was 7 megatons. We had erred on the conservative side. Yet, given the uncertainties of knowledge about physical processes occurring at conditions of density, temperature, and pressure never before encountered, and given the primitive computing power available to us,[10] we had no reason to be red-faced over a prediction that missed by 30 percent.

On November 4, 1952, the day that Dwight Eisenhower defeated Adlai Stevenson to become President of the United States, I wrote a letter to Niels Bohr from Hawaii, where I had stopped on my way back to Princeton. "The election going on today," I said to Bohr, "however it may turn out, is obviously not the only factor that creates a new situation favorable to talking once again about conditions necessary for peace in the world." Bohr, I felt sure, could surmise why I was writing on stationery of the Royal Hawaiian, and my reference to "not the only factor" was my way of letting him know that the test had gone well.

Since 1944, Bohr had advocated, first privately, then publicly, that nuclear weapons come under some kind of international control. With both Truman

[9] Richard G. Hewlett and Francis Duncan, *Atomic Shield, 1947/1952* (University Park, Pa.: Pennsylvania State University Press, 1969).

[10] The cheapest personal computer that you can buy today has orders of magnitude greater speed, storage capacity, and reliability than the room-sized SEAC of 1952. Yet the SEAC was state-of-the-art in its time, already vastly superior to hand calculation. As I recall, we paid $20,000 per month to use it; it was worth it.

and Churchill, his words had fallen on deaf ears. They could not discern the practical realist behind the facade that Bohr presented—an apparently vague, disorganized professor out of touch with reality (the man "with his hair all over his head," wrote Churchill in a September 1944 memorandum to his science advisor, Lord Cherwell). One is reminded of Churchill's encounter nearly thirty years earlier with Henry James. According to James's biographer, Churchill "was impatient, irritable; he could not wait for the end of such long and intricate sentences."[11] Bohr had established better rapport with Franklin Roosevelt, with whom he had a long meeting in August 1944. In the text of a speech the following spring, one he did not live to deliver, Roosevelt wrote, "[I]f civilization is to survive, we must cultivate the science of human relationships—the ability of all peoples, of all kinds, to live together and work together in the same world, at peace." I like to think that Bohr had an influence on that thought and those words.

Following his August 26, 1944, meeting with Roosevelt (perhaps a few days later), I met Bohr and his son Aage on the street in Washington, and the three of us went off to a restaurant to talk. Bohr had spent more than an hour with the president arguing (in his low-key but determined way, I'm sure) that a postwar nuclear arms race could be averted only by open sharing of information, and he was suddenly feeling humble. "How could such a man as I," he asked rhetorically, "speak to the greatest man in the greatest nation on Earth in the midst of the greatest war in history? But I put it to him man to man."

I lacked the personal passion to push for Bohr's concept of an open world, and did not join in promoting it. I could not convince myself that it was workable in practice. Yet my respect for Bohr was so enormous that I wanted him to get a hearing at the highest levels. I tried through my contacts in Washington—especially Senator Henry ("Scoop") Jackson and his assistant Dorothy Fosdick—to get something started that would lead to Bohr being invited officially for consultation. I told Bohr this, but pointed out that my efforts might lead nowhere, so I suggested that he try to initiate something himself through the Danish ambassador to the United States. I hoped that a new president might be more receptive to Bohr's ideas, particularly in the context of the new realities of multimegaton weapons. In fact, Bohr was never consulted officially by high U.S. officials.

The very temporary lead of the United States in thermonuclear weapons may have preserved peace and stability at a sensitive time in a nervous world. Soon the Soviets had such weapons, too. Andrei Sakharov, whom I came to know and admire later for his heroic defense of liberty and openness, was the leader of the Soviet design team. He independently conceived the "Ulam-

---

[11] Leon Edel, *Henry James, the Master: 1901–1916* (London: R. Hart-Davis, 1972), p. 526.

Teller idea" (without benefit of espionage), a reminder that ideas can't be locked away in vaults. Just as Teller and I feared in 1950, the Soviets pushed for a hydrogen weapon after the success of Joe 1, and succeeded in getting it. Had our small band in America not pushed as hard as we did, the Soviets would have won this race.

The Mike shot came only nine months before the first Soviet H-bomb test. What might have been the history of Europe and the world, I have asked myself, if we had begun a serious effort to build an H-bomb only after the Soviets had achieved one? Fortunately we got there first, if only by a little. I am sometimes asked to name the most important peacetime use of nuclear energy. My answer is simple: a nuclear device to keep the peace.

As soon as I could after the Mike shot, I sent a telegram to the Matterhorn team to let them know of the success of the test. Or at least that was my intention. It was, of course, an open, unclassified message. I was so clever in my choice of words—too clever by half, as the British would say—that my young colleagues in Princeton were only perplexed, not enlightened, by my message. (Unfortunately, my exact words have been lost.) As it happened, however, they learned through a different channel that the test had succeeded. Teller stationed himself next to a seismograph in Berkeley at the expected time—allowing for the travel time of the seismic wave—and saw the needle react with large swings right on schedule. He could infer roughly the energy of the explosion, and concluded that it was consistent with the magnitude we had predicted. After waiting a while to see if any more signals would arrive (they didn't), he sent a telegram to the physicist Elizabeth Graves at Los Alamos saying, "It's a boy." His choice of words was more successful than mine. Elizabeth correctly interpreted the message. It was the first news of the test to reach Los Alamos. Elizabeth's husband Alvin Graves, head of J Division (the test division), was at the Pacific test site. Direct word from there came later. She or one of her associates called the Matterhorn team in Princeton, giving them the first sure news of success.

Matterhorn's main business was concluded. In the months that followed, we worked on analysis of Mike and worked with Los Alamos colleagues on modified design ideas. Most important, perhaps, we put together a summary report of all that we had done and learned. John von Neumann was a big help to us as we tried to organize that knowledge and bring in the insights from our Los Alamos colleagues. The resulting report, "PMB-38" (PMB for Project Matterhorn B) was used, I have been told, as a standard reference work by H-bomb designers at Los Alamos and Livermore for some ten years afterward.

10

THE FORCE OF GRAVITY

ON MAY 6, 1952, I came home after a busy day at Matterhorn (the Mike test was less than six months away), took a new bound notebook from my shelf, gave it the name "Relativity I," and wrote on page 1: "5^{55} pm. Learned from Shenstone ½ hour ago the great news that I can teach relativity next year. I wish to give the best possible course. To make the most of the opportunity, would be good to plan for a book on the subject." Such was my first step into a territory that would grip my imagination and command my research attention for the rest of my life.

I wanted to teach relativity for the simple reason that I wanted to learn the subject. That fall, fifteen graduate students enrolled in my course—it was the first time that a relativity course had been offered at Princeton—and together we worked our way through the subject, trying to get behind the mathematical formalism that had dominated the theory for decades, looking for real, tangible physics. By May 1953, a year after my first note to myself, my "Relativity I" notebook was full of class notes and thoughts on relativity—as well as records of student performance.

It was actually nuclear physics and quantum theory that drew me into relativity. In January 1952, I had taken some time to study two classic 1939 papers, one by Robert Oppenheimer and George Volkoff, the other by Oppenheimer and Hartland Snyder. Both were concerned with gravitational collapse, the predicted fate of a massive star after its thermonuclear fuel was burned up. Oppenheimer and his colleagues had shown that the collapse to a "singularity"—a geometric point of infinite density—is a real consequence of relativity theory, not some meaningless mathematical artifact. Perhaps it must be taken seriously as the actual fate of a massive dead star.

Eventually, I embraced gravitational collapse, picked the name "black hole" for its result, and, with my students, studied properties of black holes. But in 1952, I was still troubled by the idea of a singularity. In the notes I entered in one of my bound notebooks on the Oppenheimer-Volkoff-Snyder papers, I was looking for a way out. Something new should happen at the tiniest dimensions, I felt, that would prevent the total collapse. Aristotle said that nature abhors a vacuum. I was convinced that nature abhorred a singularity. I carried this conviction with me for many years, and even now hold out the hope that some new physics will reveal the center of a black hole to have some structure.

Here is an example of an apparent singularity that is not a true singularity. Near a proton, the electric field becomes stronger and stronger the closer you get to the particle. If the proton were a true point, the field would become infinitely strong when you reach the proton. But in fact that doesn't happen. The proton has a finite size. It has structure. To be sure, everywhere outside the proton, the electric field has the same strength *as if* the proton were a point. But when you penetrate inside the proton, you find that the electric field is merely very large, not infinite. The proton's size and structure put a lid on the strength of the field. Something of this sort should happen, I thought, when the whole mass of a star collapsed to a very small size, even if it were the size of a pinpoint or smaller. Wouldn't quantum mechanics enter the picture, I reasoned, and change the prediction of general relativity, preventing total collapse? Or perhaps a collapsing star finds a way to radiate away its mass and energy until it becomes a cinder too puny for further collapse.

But if I was going to fight against the idea of a true singularity, I had to understand the "enemy." I had to add general relativity to my arsenal of knowledge so that I could explore that new domain of supercondensed matter where both relativity and quantum mechanics might share the stage. That's why I wanted to teach the course, and why I was so delighted when Allen Shenstone approved my request. He had already, in April, approved my going back on the university payroll quarter-time that fall, leaving three-quarter time to devote to Matterhorn and Los Alamos in 1952–1953. The pieces of a physicist's time, of course, do not add up to 100 percent, but to more than that. The course took more than 25 percent of a normal work schedule, Matterhorn took more than 75 percent, and other activities—such as a review article I prepared on nuclear fission—took time, too. Somehow I found time for the family and for reading—silently to myself and aloud to Janette. I don't hunt, fish, bike, climb mountains, or play golf. Reading is my one avocation.

Einstein's special theory of relativity, advanced in 1905, shook the foundations of classical physics. From space and time the theory created spacetime, showing that we live not in a three-dimensional world but in a four-dimensional world. It united energy and momentum, knit electricity and magnetism more tightly together, showed that time is relative, provided a speed limit in nature, and, most dramatically, revealed the equivalence of mass and energy, giving the twentieth century its most famous formula, $E = mc^2$.

Yet in many ways, Einstein's general theory of relativity, completed a decade later, in 1915–1916, was more revolutionary. It showed that spacetime can be "curved" or "warped" and that gravitation, the force that rules the universe, is but a manifestation of spacetime curvature. Einstein gained world fame in 1919 when his prediction that light could be deflected by gravity was confirmed experimentally. Scientific expeditions from the Greenwich and Cambridge Observatories in England headed for equatorial destinations where that year's eclipse of the Sun was total. Arthur Eddington led the Cambridge expedition to the island of Príncipe, off West Africa. The Greenwich expedition went to Sobral, Brazil. At the moment when the Moon fully blocked the Sun's light at its location, each team photographed the field of stars near the Sun. Back in England, they carefully compared their African and South American eclipse photographs with photographs of the same part of the sky taken at night at a different time of year. Tiny differences in positions of stars in the two sets of photographs showed that starlight passing near the sun was being deflected, and through an angle consistent with Einstein's prediction. Einstein got the good news by cable, and showed it to one of his students. She asked him what he would have said if the results had not confirmed his theory. "Then I'd have to feel sorry for the dear Lord," he answered, "because the theory is correct."

My parents probably heard about this striking result because it was reported in all the newspapers. I was eight years old at the time, and don't recollect hearing about it, but I would be surprised if my parents had not mentioned it at home.

Why did this result so grip the public imagination? Largely, I think, because of its implication of warped space. If there is one thing we all think we have a firm grip on—something that is itself simple, reliable, knowable—it is space. Space is *there*. It is the stage on which we play out our lives and observe all that goes on. It is not—or so ordinary experience makes us want to believe— a *participant* in the action. Einstein's theory changed that. It removed a prop from under the way we look at nature. If space is malleable, if it is part of the action and not just the stage, what else might turn out to be other than it seems? One may speculate that the romance of an expedition to Africa also played a role in the way the news of the deflection of light was received. Certainly, this way of doing science was far from the stereotype of white-coated scientists with heavy glasses and unkempt hair moving from one gurgling liquid to another in a laboratory. And one should not underestimate the very appeal of the mysterious. Almost no one, after all, really knew what warped space meant, or in what sense time was a fourth dimension. Einstein became a mythic figure because he *did* understand these things.

Not all the consequences of general relativity were far removed from everyday observation. The theory offered, at last, a deep explanation for one of the most familiar of everyday facts, that all objects acted on by gravity fall with the same acceleration. (Galileo supposedly demonstrated this fact in the early seventeenth century by dropping wooden and metal balls from the Leaning Tower of Pisa. John Philoponus of Alexandria reported the same finding in 517 A.D.) Once spacetime curvature is taken into account, a freely falling object is not being pushed or pulled by an outside force. It is "floating," responding not to a pull from Earth (or any other body) but to the spacetime geometry in its immediate vicinity. The question Why do a heavy object and a light object, which have different weights—different gravitational pulls— nevertheless fall side by side with the same acceleration? is a question that no longer need be asked. The question should be, rather, Why do two adjacent objects near a gravitating body remain close together as they fall (or orbit)? The answer is that nothing acts to separate them. They are floating side by side in the same part of spacetime.

In the decades since I taught my first relativity course, numerous quite remarkable predictions of the theory have come to light. But at that time, several stunning consequences of general relativity were already known. One was that our universe is dynamic, expanding, evolving. Another was the prediction, alluded to earlier, that matter can collapse to a singular point of infinite density.

The "force" in this chapter title refers not just to the gravitational force that bodies exert on one another. It is also the force that the theory of gravitation exerted on me. I have been caught in its spell since that first course at Princeton. What I learned in teaching the course was that the riches of Einstein's theory had been far from fully mined. Hidden beneath the equations— simple in appearance, complex in application—was a lode waiting to be brought to the surface and exploited.

Workers in relativity in the 1930s and 1940s had been mainly concerned with mathematical niceties in the theory. Few saw the possibilities of confronting the theory with new experiments and observations, beyond the classic three, which, by the 1920s, had convinced the scientific community of the correctness of general relativity. The first of the three tests was the precession of the orbit of Mercury: a line drawn between the points of Mercury's elliptical orbit that are farthest and nearest to the Sun—its aphelion and perihelion—rotates more than could be explained by the effect of other planets. The second test was the deflection of starlight by the Sun (what Eddington's team observed during the solar eclipse in 1919). The third was the "gravitational red shift"—the slight slowing in the pulse of electromagnetic frequencies radiated from the surface of a massive star.

Oppenheimer, through his 1939 papers, provided an exception to the outlook of the time that general relativity, or gravitation theory (I use the terms synonymously), was simply a playground for mathematicians. He gave attention to a new phenomenon, gravitational collapse. Yet even Oppenheimer did not pursue the subject, and perhaps did not really take it seriously. He may have considered it an intriguing (or, in the lexicon of physicists, "amusing") oddity of the theory that would have nothing to do with what we actually observe in the universe around us.

With relativity, my approach was different, as it had been with quantum theory and nuclear physics. It is precisely in the strange limiting cases of theories, I believed, where new insights lurked, waiting to leap forth. Moreover, it was my philosophical conviction that nature would avail itself of all the opportunities offered by the equations of valid theories. If nuclei could exist in doughnut shapes, I felt, then some of them would exist in such shapes. If heavy matter could be made from electrons and positrons only, then some heavy matter would be made of these particles. If matter could collapse to infinitesimal or even zero size, then some matter would collapse. We physicists should think about where such extreme behavior might occur, and look for it.

By pushing a theory to its extremes, we also find out where the cracks in its structure may be hiding. Early in the twentieth century, for example, when Newton's theory of motion, flawless in the large-scale world, was applied to the small-scale world within a single atom, it failed. Some day, some limit will be found where general relativity comes up short. That limit will show itself only after we have pushed the theory in every extreme direction we can think of. In doing so, we will very likely find some strange new ways in which the theory *is* valid before we find out where it is *not* valid.

The next best way to learn, after teaching, is writing. In that first entry in my "Relativity I" notebook, half an hour after being informed that I could teach a relativity course, I wrote that it "would be good to plan for a book on the subject." The book did get written. It was called *Gravitation*, ran to 1,279 pages, and appeared in 1973, twenty years after I dove into the subject. It got written because of the energy and dedication of two skilled coauthors, Charles Misner and Kip Thorne. They had been my students, and both have pursued distinguished careers in relativity theory. Misner is now a professor and master educator at the University of Maryland in College Park. Thorne is the Feynman Professor at Caltech, a leader in the project to construct a gravitational-wave observatory, and a brilliant expositor for the general public.[1]

Actually, *Gravitation* wasn't my first book on relativity. As a kind of "warm up," I worked with another younger collaborator, Edwin Taylor (no relation to Ted Taylor), on an undergraduate textbook on special relativity. In September 1962, Taylor arrived in Princeton on a sabbatical from Wesleyan University. We got to talking about a new approach I was about to try in an honors physics course for freshmen—to *begin* the course with special relativity. Taylor agreed to sit in on the course and take notes. I didn't know much about him at the time. I was taking a chance, but I like to do that when somebody shows an interest and is willing to work. As I learned later, Edwin Taylor is not just a good teacher; he is one of those rare individuals who is a gifted teacher, and he is a graceful writer as well. His contributions to physics have been in curriculum and pedagogy, no less important to the whole enterprise than the research that moves the edges of knowledge forward.

By the year following my freshman course on special relativity, Taylor and I were ready to try out the notes on students and submit them to colleagues for evaluation. Then we set about transforming the notes into a book. Taylor, full of initiative and good ideas, kept *my* nose to the grindstone. In 1966, the results of our labors appeared as *Spacetime Physics*. Thanks to his dogged later efforts from his base at MIT, we managed a second edition in 1991, and a later book, *Scouting Black Holes*, in 1996.

The scope of general relativity made it seem more suited to book-length presentations than subjects I had worked on earlier. In 1962, I pulled together a set of related reprints of research papers in a book called *Geometrodynamics*. The title word is one I coined as a companion to the old familiar *electrodynamics*, a word used to describe the theory of electricity and magnetism—how electric and magnetic fields are created, propagate, and get absorbed. I wanted to emphasize the analogous content of relativity: a theory, in its essentials, of spacetime geometry—dynamic, changing geometry, influenced by mass, capable of propagating, and in turn influencing mass.

In 1964, three of my students—Kent Harrison, Kip Thorne, and Masami Wakano—and I put together another book, *Gravitation Theory and Gravitational Collapse* (published the following year). As I tell at the beginning of Chapter 14, this book resulted from two of my failings: a tendency to write at length and a tendency to procrastinate. The book appeared before I had intro-

[1] Kip S. Thorne, *Black Holes and Time Warps: Einstein's Outrageous Legacy* (New York: W. W. Norton & Company, 1994).

duced the term "black hole," but not before I had begun to ponder the implications of that central feature of relativity theory, gravitational collapse. In
1965, responding to an invitation from German colleagues to help celebrate
fifty years of general relativity, I prepared a long paper entitled "Einstein's
Vision." The vision that animated Einstein throughout his later years—a
vision I had many an occasion to discuss with him—was a vision of a totally
geometric world, a world in which everything was composed ultimately only
of spacetime. I will try to make clear below what "composed of spacetime"
means. This paper, with additions, made its appearance several years later as
a short book in German, *Einsteins Vision*. (Although I had somehow won a
German prize in high school and used the language regularly for reading
papers, I needed expert assistance to prepare this book in German. I relied
on several German-speaking colleagues, especially Helmut Krikava, a visitor
to Princeton from Vienna.)

These several books served as stepping stones to the Misner-Thorne-Wheel-
er monograph *Gravitation*. They got done mostly because I had students to
inspire me, teach me, and spur me on. When, years later, I accepted an invitation from Peter Renz of W. H. Freeman to prepare a book for the nonspe-
cialist, I faced a much harder writing task than ever before. I was on my own!
Oh, how easy it was, with no coauthors to badger me, to lay the book aside
in favor of other projects. In the summer of 1989, after work on this book
had dragged on for years, I was under the gun to get it done. Every morning
that summer (or so it seemed) I got a call at my High Island, Maine, retreat
from my ever-cheerful editor, Susan Moran. "What are you planning to do on
the book today?" she asked. Every evening (or so it seemed) she called back to
ask, with the same bounce in her voice, "What did you do on the book today?"
It worked. *A Journey into Gravity and Spacetime* got done, and made its
appearance in 1990. I have the pleasure now of seeing its versions in four
languages lined up on my bookshelf. An editor like Susan Moran is a precious
jewel.

One of the pillars of physics is Newton's third law. A common way to state it
is that for every action there is an equal and opposite reaction. Another way
to say it is that forces come in balanced pairs. The force of the Sun on Earth
and the force of Earth on the Sun are equal in magnitude and opposite in
direction. A person pulls Earth up as strongly as Earth pulls the person down.
The law applies also to a property of motion called momentum. A proton,
rebounding from a nucleus that it has struck, undergoes a change of momentum. The nucleus, in recoiling from the blow, undergoes an exactly opposite
change of momentum.

In its modern, general form, Newton's third law applies to all interactions of one thing with another. It tells us that anything that affects something else must, in turn, be affected *by* that something else. Paul Hewitt, a gifted teacher and textbook author, has expressed the idea in engaging human terms: "You can't touch without being touched." To supply a theme for my book *A Journey into Gravity and Spacetime,* I wanted something that would express the essential content of Newton's third law applied to gravity and curved spacetime. Adapting from the earlier book *Gravitation,* I wrote, "Spacetime tells matter how to move; matter tells spacetime how to curve." In other words, a bit of matter (or mass, or energy) moves in accordance with the dictates of the curved spacetime where it is located. (A ball falling to Earth is responding to the curved spacetime in its vicinity.) At the same time, that bit of mass or energy is itself contributing to the curvature of spacetime everywhere. (The falling ball is affecting spacetime and therefore affecting the motion of other bodies elsewhere.)

Newton's third law played a role in my first paper on general relativity. One of the three great confirmations of Einstein's theory had been the deflection of light by the Sun. While teaching the relativity course in the spring of 1953, I began to think about the fact that if light is influenced by gravity, then gravity must be influenced by light. To put it differently, light not only responds to gravity; it creates gravity. This was not itself a new idea. Einstein had showed that *all* energy, not just the energy locked up in mass, is a source of gravity. Since light has energy, it can be the source of gravitational force. What I did was to push the thought to an extreme. How much light would it take, I asked myself, to create so much gravity that the light would hold itself together? Starlight passing the Sun undergoes a tiny deflection. It is not hard

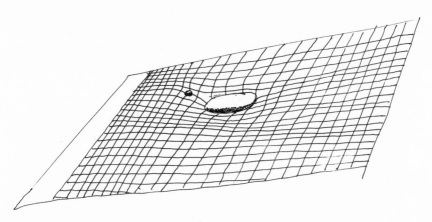

Matter (the large stone) tells spacetime how to curve;
spacetime tells matter (the pebble) how to move.
(Drawing by John Wheeler.)

to calculate that if the Sun were replaced by something vastly more massive (but no larger), light could be caused to bend so much that it would circle around the object in the same way that a satellite circles Earth. Then, I reasoned, what if we got rid of that central supermassive object and cranked up the light intensity to such a level that the light itself created an equally strong gravitational field? The light would continue to circle, held in its orbit by its own gravity. It would need no central attracting body.

This hypothetical entity, a gravitating body made entirely of electromagnetic fields, I called a geon (g for "gravity," e for "electromagnetism," and *on* as the word root for "particle"). There is no evidence for geons in nature, and later I was able to show that they are unstable—they would quickly self-destruct if they were ever to form. Nevertheless, it is tempting to think that nature has a way of exercising all the possibilities open to it. Perhaps geons had a transitory existence early in the history of the universe. Perhaps they are formed and quickly dissipate in today's universe. Perhaps (as some students and I speculated much more recently), they provide an intermediate stage in the creation of black holes.

Apart from the general pleasure that I feel in pushing a theory to its limits, I had a special reason for being attracted to geons. They give "mass without mass." Let me try to explain this cryptic phrase. All of physics, both classical and quantum, faces a conceptual problem in dealing with point particles. We think of electrons and neutrinos and quarks as existing at mathematical points. We think of photons being created and being absorbed at mathematical points. Yet we cannot really deal with the infinite density of mass or the infinite density of charge implied by point particles and point interactions. These points are annoying pinpricks in the body of physics. We endure them because we have to, while hoping that someday we will identify and understand an inner structure in what today seem to be points. A beauty of the geon, in my eyes, is that it exposes just such an inner structure—not, to be sure, in an infinitesimal entity, but in a body that from a great distance looks and acts like a point source of gravity.

In the academic year 1953–1954, I was back to my full-time position of teaching and research at Princeton for the first time in four years. While attending to my classes and research students and continuing to work in nuclear physics, I was also finding time to think about geons. In the 1930s and 1940s, I had been intrigued with building the universe (or at least a model universe) from particles without fields. My papers with Richard Feynman on action at a distance described such a world. Now I was gripped by an opposite vision: a world without particles (more exactly, a world in which the things we call particles are accounted for by an inner spacetime structure).

In the summer of 1954, our children went off with their separate friends to "do" Europe, while Janette and I settled for a month into a lovely apartment in Geneva, Switzerland, provided by Janette's "aunt" (actually her mother's cousin) Madeline Doty. We looked out over the falls where Lake Geneva empties into the Rhone River. It was a peaceful, restful spot for writing. After a year of ruminating on geons, I was ready to put something down on paper. Late in the summer, with the geon paper drafted, the children joined us and we drove together to Rome.

The following summer I found myself back in Geneva, this time for the first International Conference on the Peaceful Uses of Atomic Energy. I was excited to meet certain Soviet colleagues for the first time. We managed to sneak away from some of the sessions to walk and talk about physics outdoors. Very little of this talk was on fusion, for the walls of secrecy around thermonuclear research still stood. Not until 1958, at the fourth of these "Atoms for Peace" conferences, did the two sides fully share fusion information. In his autobiography, the Russian plasma physicist and space scientist Roald Sagdeev[2] gives an entertaining account of the 1958 conference, from the perspective of a twenty-six-year-old on his first trip outside the Soviet Union. He looked in vain for the oppressed masses he expected to see on the streets of Geneva. (Sagdeev could scarcely have imagined then that Susan Eisenhower, granddaughter of the president whose initiative led to these conferences, would someday be his wife.)

In the geon paper written to the muted strains of the Rhone waterfall in 1954, I concluded that the smallest "purely classical" geon (a geon for which quantum effects could be ignored) was a doughnut the size of the Sun with a mass of about a million Suns. (This mass, of course, is the equivalent mass of the electromagnetic energy coursing around the doughnut racetrack. It is "mass without mass" in the sense that it relies on no material particles.) Larger geons were in principle possible, I found, up to the size of the universe. The enormity of even the smallest classical geon was not a reason to stop thinking about these entities. For one thing, I felt that pursuing the beautiful theory of general relativity down whatever lane or into whatever thicket it might lead was an obligation — not to mention a thrill. For another thing, it was conceivable that once quantum effects were taken into account, much smaller geons might be possible. How could I not be lured by the prospect of a miniature quantum geon as small as a single elementary particle?

When I got back to Princeton, I sent a copy of the paper to Einstein (who was then seventy-five), asking him for his comments and asking him also to show the paper to John von Neumann. In a brief written response, after admit-

---

[2] Roald Z. Sagdeev, *The Making of a Soviet Scientist* (New York: John Wiley & Sons, 1994).

ting to "a pretty bad conscience" for not responding right away, Einstein said, "The reason for not approaching you earlier was, of course, that I have not a well founded conviction and that I know that we are seeking in somewhat different directions. It is so much easier to understand the other man if he tells you his reasons orally." In late October, we did talk (only by phone, not in person, to judge by my notes on the conversation). He said that he had himself considered geon-like compressed energy, but of smaller size. He hadn't pursued the idea because he thought it was "unnatural" (that is, he didn't see any link with anything known in nature).

With his usual astonishing intuition, Einstein said in this conversation that he was prepared to admit that his equations of relativity allowed for geon solutions of the kind I was exploring, but he doubted the stability of a geon. A few years later, in the course of preparing a lecture for New York University, I came to understand that geons are indeed unstable. A geon is like a pencil standing on its point. Under perfectly ideal conditions, the pencil will stand there, but in practice, the slightest disturbance will cause it to tip over. Similarly, a geon, under perfectly ideal conditions, will persist as a compressed entity holding together its vast energy. In practice, the slightest disturbance will cause a geon to collapse or dissolve, radiating away its energy to the cosmos.

The paper I sent to Einstein for his review was devoted almost entirely to classical (nonquantum) general relativity. But it did contain a few remarks about quantum physics—for instance, how quantum phenomena might change the nature of a geon if it is small, and how a geon might radiate away some of its energy in electron-positron pairs. This was all the opening Einstein needed. He took the occasion of our conversation to tell me again that he did not like the probabilistic nature of quantum theory. Here was an echo of his famous debates with Niels Bohr, going back some twenty-five years. Einstein, the most brilliant, creative physicist of the twentieth century, clung tenaciously to the view that nature at its core is deterministic, not probabilistic. "God is subtle, but He's not malicious," as he put it in the famous statement. Einstein's philosophical stance will no doubt reverberate through physics for as long as there are unanswered questions about the laws that govern the very small and the very energetic.

In my geon paper, published in 1955, I discussed spherical as well as doughnut-shaped geons, and geons built of neutrinos as well as those built of light. Later, I explored an even "purer" geon, made up of gravitational energy alone, around which gravitational waves of fantastic intensity swirl, creating enough gravity to hold themselves together. Such a geon, it seemed to me, might offer a transitional state between gravity waves and a black hole.

The waves in a stormy ocean sometimes come together to give a momentary wave of exceptional height—172 feet is the highest of which I have found any record. Likewise, gravitational waves arising from the battles of matter and energy in the early universe will occasionally coalesce. The resulting geon-like concentration of gravitational wave energy, if conditions are right, will not mimic the towering ocean wave but will instead collapse to a black hole. The result again is "mass without mass." Strong gravity as a *source*, not mere-ly the *result*, of black holes, is a topic I pursued in 1993 with Daniel Holz, a Princeton senior; Warner Miller, a Los Alamos staff member and my former graduate student; and Masami Wakano, a frequent collaborator, then teach-ing physics in Japan.

More than forty years have passed since I entered the field of gravitation physics. It has been a delight over all those years to see my students, and the students of my students, building on ideas that I explored in that first paper on geons. Now they are the leaders. It is they who inspire me to keep tugging at the curtains that block our view of the mechanism behind existence. What a pleasure it is when one of them invites me to collaborate on a piece of research.

My discussion of geons with Einstein occurred only a few months before his death in April 1955. He had been unfailingly kind to me since I first met him in 1933. In the last two years of his life, he invited my relativity classes to his house for tea, and he and I had several stimulating conversations on the subject that had gripped his imagination for forty years and had just taken hold of mine.

"Mass without mass"—pure energy holding itself together—was not the only theme of the 1955 geon paper. Another idea surfaced there: "charge without charge." Ordinarily, we say that the lines of electric field emerge from, or begin at, a positive particle such as a proton and converge on, or end at, a negative particle such as an electron. Yet if we want to develop a theo-ry—the geon theory, for instance—in which we sweep away point particles and their strange attributes of infinite density and infinite field strength, it seems at first that we must also sweep away charge. Can there be electric charge if there is no place for lines of electric field to begin or end? Cer-tainly not in a Newtonian world of "flat" space. But in an Einsteinian world of "curved" space, geometry can save the day. Think of a washtub with a perfectly flat bottom—the kind in which cowboys take baths in old movies. That corresponds to flat space. The water in it (corresponding to an electric field in space) has no place to go. By contrast, a modern bathtub has a curved bottom, and indeed a bottom so strongly curved in one place that it connects to a drainpipe. The water can "disappear" from the bathtub by flowing down the drainpipe. The water is not lost to the universe, but it is lost to the small

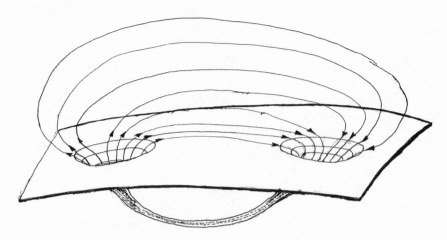

The idea of "charge without charge": Electric field lines that seem to begin
at one place and end at another may be connected, thanks to
a wormhole in "multiply connected" space.
(*Drawing by John Wheeler.*)

part of the universe occupied by the bather. Electric field lines in strongly
curved space behave like water, which can disappear from one place and
reappear somewhere else far distant.

Space with properties like bathtubs and drainpipes is called "multiply connected." Lines of field can *seem* to disappear at one place and *seem* to appear
at another place without ever ending or beginning. Hermann Weyl, the great
mathematician who so inspired me in Princeton, was the first to visualize
electric field lines trapped in multiply connected space, back in 1924.

In the diagram shown here, the two-dimensional sheet represents three-
dimensional space. Into the right hole, lines of field disappear; the right hole
acts like a negative charge. From the left hole, lines of field emerge; the left
hole acts like a positive charge. The "handle" that joins the holes beneath
the plane provides a continuous path for the field lines. They do not begin
or end. This handle also makes the space multiply connected. From a great
distance, no experimenter could tell the difference between these two holes,
respectively ingesting and disgorging field lines, and a pair of particles at
which field lines end and begin. The same idea applies in three dimensions.
The theory admits the possibility that field lines disappear from view—go off
the map, so to speak—at a place that looks like a negative electric charge
and that somewhere else—and it could be trillions of miles distant—these
field lines emerge from what looks like a positive electric charge.

I later gave this handle the name "wormhole." It is the path through which
field lines can sneak from one place to another in multiply connected space.

The size of the mouth of the wormhole could be imagined to be arbitrarily small, so that the mouth could approximate as closely as you like a point particle. What a vision swam across my view! A space riddled with billions upon billions of wormholes, their myriad mouths indistinguishable from electrically charged particles. Truly charge without charge.

Beautiful or not, the wormhole theory of charge has difficulties. It suggests a perfect balance in nature between positive and negative particles of each type. Yet we see in our universe more electrons than positrons, and more protons than antiprotons. And what of the stability of wormholes? "Won't a wormhole collapse?" Niels Bohr asked me in 1962, not long before his death. It was a question that my student Robert Fuller and I were just then trying to answer. Yes, we found, Bohr's concern is justified. The diameter of a wormhole can shrink so fast that not even a photon can shoot through from one of its mouths to the other before it pinches off. Yet, in the strangely distorted world served up by relativity, this pinch-off takes an infinitely long time according to someone looking on it from outside. And in some circumstances, the very existence of the lines of field plunging through the wormhole can prevent the collapse. So the wormhole as a theoretical entity and the visions it conjures up remain very much alive.

Whatever can be, is. (Or, more strongly put, *Whatever can be, must be.*)[3] This article of faith—and it is only faith—makes me want to believe that nature finds a way to exploit every feature of every valid theory. Every faith, including this one, must have some boundaries. The boundary of this one is supplied by the finiteness of the universe. There is a finite number of particles in the universe, a finite amount of mass, a finite duration of it all—the universe began and will likely end. Therefore, not every one of the infinite number of predictions that any theory can make can be realized. Yet I cling to the faith that every general feature will be realized. If relativity is correct, and if it allows for wormholes, then somewhere, somehow, wormholes must exist— or so I want to believe.

Quantum mechanics may provide the compelling logic for wormholes. One thing that the quantum theory taught us in the late 1920s and early 1930s is that the submicroscopic world, viewed through an imaginary microscope of high enough power, is a chaotic place of random fluctuations, in which particles are constantly being created and annihilated. It is a place where even that pillar of physics, the law of energy conservation, is subject to transitory violation. Moreover, the smaller the domain in space and in time, the more violent the fluctuations—the greater the deviations from "normality." There is

---

[3] Some physicists, tongue in cheek, have speculated that nature is the perfect tyrant, decreeing that whatever is not forbidden is obligatory.

ample evidence for these fluctuations, because they alter the measurable properties of particles and atoms.

So an electron sitting quiescent in empty space is not quiescent at all. As we zoom in on it with hypothetical microscopes of higher and higher power, we see a more and more lively neighborhood around the electron. Other electrons and positrons are coming into existence and vanishing. Photons are being born and are dying. Heavier particles join the dance of incessant creation and annihilation. And the closer we get, the more violent the activity becomes. The "isolated" electron is the nub of a seething volcano. Just about everything that can happen in the particle domain is happening right there in this tiny microcosm of the whole universe.

Then, as we recede again to great distance, looking through microscopes of ever lower power, all becomes again simple and orderly. There, viewed from afar, sits a lone electron, with its single unit of negative charge and its characteristic mass and spin, seemingly in splendid isolation. Yet if we set about measuring its magnetic moment (the strength of the magnetism generated because it is a spinning charge), we find not the value predicted by the original Dirac theory of the isolated electron, but a value $1.001\,159\,652$ times as large. The swarm of virtual particles that make up the retinue of the electron, although invisible, lets us know they are there. Even far from the electron, they imprint their undeniable trace on the electron's magnetism, making it a bit more than one part in a thousand larger than it would otherwise be.

Experimental and theoretical advances related to quantum fluctuations had converged dramatically at a conference on Foundations of Quantum Physics, held on Shelter Island, New York, in 1947. I want to say a little about this remarkable conference because it signaled the maturing of American physics after World War II.[4] Planning for it had begun in late 1945, almost immediately after the war. It was conceived by Duncan MacInnes, a talented and energetic physical chemist at Rockefeller University in New York. Through his friend Frank Jewett, president of the National Academy of Sciences, he secured a commitment of sponsorship by the NAS. MacInnes turned also to Karl K. Darrow, the flamboyant, opinionated, and highly visible secretary of the American Physical Society, for assistance in planning. The conference they visualized was to be small—no more than thirty participants. It was to be informal—set perhaps in a rural inn. Its participants were to be mainly

[4] An excellent, detailed account of the Shelter Island conference is given by Silvan Schweber in QED and the Men Who Made It: Dyson, Feynman, Schwinger, and Tomonaga (Princeton, N.J.: Princeton University Press, 1994). Schweber's careful historical research is enhanced by his own deep knowledge of physics.

Some of the participants at the Shelter Island conference, 1947. *Standing at the left*: Willis Lamb. *Seated, left to right*: Abraham Pais, Richard Feynman, Herman Feshbach, Julian Schwinger. I am standing at the rear.
(Courtesy of AIP Emilio Segrè Visual Archives.)

young, and mainly American. They were to gather to discuss basic problems in quantum physics.

The focus on American scientists was partly necessitated by limited financial resources. Given the size of America, this limitation produced, in fact, a focus on eastern American scientists. (MacInnes originally estimated the cost of the conference to be $3,120. The NAS provided a grant of $1,500, and of that, less than $1,000 was spent! The only European scientist in attendance—the Dutch physicist Hendrik Kramers—had to travel only from the nearby Institute for Advanced Study in Princeton, New Jersey. Two of the twenty-five participants—Robert Oppenheimer and Robert Serber—came from California; one—Enrico Fermi—came from Chicago; all of the others came from the East Coast.) But it was not only for practical financial reasons that the conference was dominated by Americans. The time had come, following the wartime successes with radar and the atomic bomb, to let American physicists flex their muscles. American physics had come of age.

Nevertheless, the planners turned first to Europeans (in America) for advice. Léon Brillouin, a French theorist, then in New York, offered information on the highly successful Solvay conferences held in Brussels in the

1920s and 1930s, at which so many notable exchanges had taken place among the (mainly European) leaders in physics. Wolfgang Pauli, a Swiss physicist who, after receiving his Nobel Prize in 1945, was spending some time in Princeton, recommended a relatively large conference that would include most of the established names from Europe as well as America. This advice did not sit well with MacInnes. He wanted a small number of "younger men," the ones who might be expected to make the future breakthroughs. Since he, Darrow, and Pauli categorized me as such a "younger man" (I was then thirty-four) and since I was nearby, they turned to me for advice on an invitation list and a list of topics to discuss. I was honored by the request and happy to comply. The final conference, as it turned out, included most of the people I had proposed, and included topics that I considered important. Among people I suggested were my former student Dick Feynman, then at Cornell, and Julian Schwinger, already a tenured faculty member at Harvard. Feynman and Schwinger, both still in their twenties, would later share, with Sin-itiro Tomonaga of Japan, the Nobel Prize in Physics for work that was inspired by experiments described at Shelter Island.

For just two full days, on June 2–3, 1947, twenty-five physicists (all men, typical for those times) took over the Ram's Head Inn on Shelter Island near the eastern tip of Long Island. We ate together, and talked, and listened, and talked some more. Two topics dominated the animated conversations: new measurements by Willis Lamb at Columbia University showing subtle deviations of the hydrogen atom's energy levels from what the Dirac quantum theory predicted they should be, and the puzzling properties of mesons in the cosmic radiation.

Addressing the second topic, Robert Marshak from the University of Rochester proposed that oddities in the observations would all be resolved if two different mesons existed, a strongly interacting one created in nuclear collisions high in the atmosphere and a lighter, weakly interacting one created as a secondary particle and prominent in the cosmic rays that reach Earth's surface. Within a year, his hypothesis was confirmed by results obtained at the University of Bristol in England by Cecil Powell and his group, who studied tracks left by cosmic-ray particles in special photographic emulsions. Suddenly, as Marshak had forecast, all did become clear. The pion lives out its short life high in the atmosphere. The longer-lived muon can reach Earth's surface, and accounts for most of the clicks that you hear if you place a Geiger counter far from any radioactive source.

Causing even more excitement than mesons at the conference were the new results reported by Lamb. With his student Robert Retherford, Lamb had discovered that the energy of an electron in the hydrogen atom is not exactly what one would calculate using the well-accepted Dirac quantum theory.

Lamb and Retherford had used experimental techniques derived from wartime radar developments. Most of the physicists assembled in the Ram's Head Inn, including me, were convinced that the theoretical explanation of the experimental finding lay with quantum fluctuations. That didn't mean we knew how to calculate the effect. But one among us, Hans Bethe of Cornell University, figured out how to do that on his way home. Riding the train from New York City to Albany on the day after the conference, he carried out an approximate but convincing calculation that produced nearly the correct magnitude of the energy shift that Lamb had reported. This set off a wave of activity at various universities. Within little more than a year after the Shelter Island conference, physicists understood in detail the origin of the Lamb-Retherford result. It was a phenomenal year, in which nearly all of the numerous participants regularly and fully communicated with each other, in the best tradition of open science.

One way to say what Lamb and Retherford had demonstrated is that the electron, as it zooms around the proton at the center of the hydrogen atom, feels more than the force that "reaches out" from the central proton. In addition, it reacts to the quantum fluctuations that exist everywhere in space. It is as if Earth, in its trip around the Sun, were to feel not only the force of attraction of the distant Sun, but also millions of tiny forces arising from tiny lumps of clay that get in its path as it cruises through space.

Not long after the Shelter Island conference, Oppenheimer wrote, "For most of us, [Shelter Island] was the most successful conference we had ever attended." Twenty years later, Feynman said, "There have been many conferences in the world since, but I've never felt any to be as important as this."[5] All of us felt the same way. It was just the right combination of people in the right setting at the right time.

In one sense the Shelter Island conference was a uniquely American event. Twenty-four of its twenty-five participants were citizens or future citizens of the United States. Yet it was, at the same time, an international conference. Nearly half of the participants were born in Europe. We owe to Fascism in Europe both the great influx of talent and the wartime achievements that led directly to American ascendancy in science.

[5] Oppenheimer's assessment was in a letter to A. N. Richards, Feynman's in an oral-history interview with Charles Weiner. Both are quoted in Schweber, *QED and the Men Who Made It* (see footnote 4 on page 242): Oppenheimer on p. 175 and Feynman on p. 156.

11

·

QUANTUM FOAM

EVEN AS I was learning general relativity for the first time by teaching it in the spring of 1953, I was pondering its link to quantum theory. If quantum theory controls the electric field, the magnetic field, and the neutrino field, should it not just as surely control the gravity field, which is spacetime itself? I mentioned this idea in my 1955 paper on geons, but it took several more years for a "Technicolor" vision of quantum gravity to take shape in my head—a vision greatly stimulated by conversations with my student Charles Misner. The vision of quantum gravity is a vision of turbulence—turbulent space, turbulent time, turbulent spacetime. And even more than turbulence. If I was right, spacetime in small-enough regions should be not merely "bumpy," not merely erratic in its curvature; it should fractionate into ever-changing, multiply connected geometries. For the very small and the very quick, wormholes should be as much a part of the landscape as those dancing virtual particles that give to the electron its slightly altered energy and magnetism.

As we get out our hypothetical microscopes again and zoom downward to the vicinity of some material particle, say a proton, keeping an eye on space-time itself, what do we see? When we look at a bit of space 1 ten-millionth of a billionth (10^{-16}) of a meter, which is within the dimensions of the single proton, and a speck of time 1 millionth of a billionth of a billionth (10^{-24}) of a second, less time than it takes light to go from one side of the proton to the other, we see the expected frenzied dance of particles, the quantum fluctuations that give such exuberant vigor to the world of the very small. But of such effects on space and time we see nothing. Spacetime remains glassy smooth.

To see "quantum foam," spacetime stirred into the writhing turbulence of myriad multiply connected domains, we must go deeper—much deeper.

Max Planck, discoverer of the quantum, first noticed that certain basic physical constants can be combined to produce a quantity with the dimension of length. He did not know what this length might represent, but he argued that it is a "natural" length, more significant than any length based on the sizes of objects in our everyday world (the meter, for instance, was originally chosen to be 1 ten-millionth of the distance from the Equator to a Pole). The constants that Planck combined were the gravitational constant G from gravitational theory, the speed of light c from electromagnetic theory, and his quantum constant h. (Nowadays we divide Plank's constant h by 2π to get $\hbar$, or h-bar, the fundamental constant of quantum theory.)

What emerged from my discussions with Misner was the realization that it is this "Planck length" that sets the scale of quantum fluctuations in spacetime. It is quite incredibly small. How small? Imagine a line of spheres, each with a diameter equal to the Planck length. The number of these spheres that would have to be lined up to span the diameter of one proton would be the same as the number of protons that would have to be lined up to reach across New Jersey. It takes 100,000 protons in a row to equal the size of one atom, and 1 million atoms in a row to go from one side to the other of the period at the end of this sentence. You don't need to be told that it would take a lot of these periods to stretch across New Jersey. Relative to the Planck length, even the minuscule entity we call an elementary particle is a vast piece of real estate. To change the analogy from distance to money, a penny relative to the U.S. annual budget is a million times larger than the Planck length relative to the size of a proton.

Just as there is a Planck length, there is also a "Planck time." It is the time needed for light to move through one Planck length.[1] If a clock ticked once in each Planck unit of time, its number of ticks in one second would be greater by many billionfold than the number of vibrations of the quartz crystal in your watch during the lifetime of the universe. The Planck time is simply too short to visualize—but not too short to think about! If interesting things happen within the dimension of a Planck length, they also happen within the duration of the Planck time.

On our imaginary downward voyage to ever smaller domains, after reach-

---

[1] The Planck length is given by $L = \sqrt{\hbar G/c^3}$, the square root of Planck's constant times the gravitational constant divided by the speed of light cubed. In scientific notation, its magnitude is 1.6×10^{-35} meters. The corresponding Planck time is 5×10^{-44} seconds.

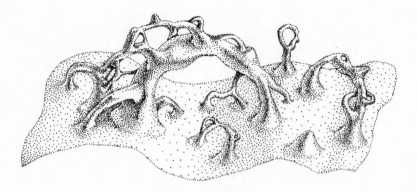

Quantum foam.
(*Reprinted with permission from Fig. 13.7c in Kip Thorne,* Black Holes
and Time Warps: Einstein's Outrageous Legacy [*New York:*
W. W. Norton, 1994], *p. 478.*)

ing the size of a single proton, we would have to go twenty powers of 10 further to reach the Planck length. Only then would the glassy smooth spacetime of the atomic and particle worlds give way to the roiling chaos of weird spacetime geometries. The wormhole would be but one simple manifestation of the distortions that could occur. So great would be the fluctuations that there would literally be no left and right, no before and no after. Ordinary ideas of length would disappear. Ordinary ideas of time would evaporate. I can think of no better name than quantum foam for this state of affairs.

The kinds of ideas outlined above took on more substantial form during a wonderful eight months—January to September—that I spent in Leiden, Netherlands, in 1956. There I had the stimulus of discussion with students and professors at the University of Leiden, as well as with three young Americans who joined me: Charles Misner and Peter Putnam, who were graduate students at Princeton, and Joseph Weber, a professor of electrical engineering on leave from the University of Maryland who joined our little group because he wanted to get into relativity research. Cornelius Gorter, director of the Kamerlingh Onnes Laboratory in Leiden, had invited me to fill the H. A. Lorentz Visiting Professorship for the spring semester. I was delighted to accept (with, of course, the blessing of my colleagues and my department chair in Princeton). The course of lectures that I gave covered quantum theory and relativity and the links between these two great theories—lectures that enabled me to sharpen my own thinking. It was one of the most productive and pleasant periods of my life. Janette and I look back with pleasure on those Leiden days.

Martin Schwarzschild, a Princeton astrophysicist, and his wife, Barbara, were, as it turned out, leaving by ship from Le Havre, France, to return to America shortly before Janette and our daughter Alison and I were scheduled to land there in late January. So we arranged to buy their small Renault Quatre Chevaux, sight unseen. Sure enough, there it was, ready to take us on our way without any red tape. We climbed in and headed northeast across France and Belgium, into the Netherlands and on to Leiden in snow and bitter cold (or bitter it seemed, given the skimpy capability of the car's heater). There, on the frozen Dutch canals, were skaters, just as we had learned to picture them from books like *Hans Brinker Or, The Silver Skates*.

In Leiden we put up at Het Gulden Vlies [The Golden Fleece], across a cobblestoned street from the town hall. Sixteenth-century history lay literally at our doorstep. In the middle of the street, flush with the cobblestones, sat a hexagonal blue stone. It marked the place where the burghers who were unwilling to take up arms against the Spanish forces in the siege of 1573–1574 were hanged. Under the leadership of William I (William the Silent), Count of Nassau and Prince of Orange, the Dutch prevailed over the Spanish in 1574 by opening dikes and sailing their warships to Leiden, driving out the Spanish. (Leiden lies now some four feet below sea level). In gratitude for the successful defense, William established the University of Leiden in 1575, the year after the victory. Nine years later, with a price put on his head by Philip II, King of Spain, William I was assassinated. Part of William's legacy was the religious tolerance for which he fought. It inspired the Pilgrims, including some of my ancestors, to come to Holland before they set sail for America. About forty Pilgrims lived in Leiden for the better part of a year.

How could I not be interested in this history? The oldest building on my Princeton campus—which served briefly in 1783 as the meeting place of the Continental Congress and headquarters of the U.S. government—is Nassau Hall, its name rooted in twelfth-century Europe. The prince for whom the town of Princeton (and, in turn, the university) was named is William III, King of England and, like his great-grandfather William I, ruler (*Stadholder*) of the Netherlands and Prince of Orange. Needless to say, one of Princeton University's official colors is orange. At my neighboring state university, Rutgers, in New Brunswick, stands a statue of William I bearing a plaque with the words, "When he died, the children cried." Dutch settlers were among the first Europeans in New Jersey and have left their mark.

When we sailed for Europe and the Lorentz Professorship, we left our two older children behind in Cambridge, Massachusetts. Nineteen-year-old Letitia and seventeen-year-old Jamie were freshmen, she at Radcliffe, he at Harvard. They joined us in the summer. Alison, then thirteen, came with us and enrolled in the International School in The Hague. Every morning, she boarded a tram for the railroad station and then rode a train to The Hague, returning again by train in the afternoon. Her classes were in English, so we were surprised on one Dutch holiday to hear her reciting an oath of allegiance to Queen Juliana in Dutch.

The Dutch are famous for their full breakfasts. Janette and I always made the Golden Fleece breakfast see us through two meals. The cheese and meat and bread and boiled eggs that we couldn't eat for breakfast we wrapped up to provide us with a very nice lunch. Alison's school lunch, as well, consisted of "leftover" breakfast.

Being well aware of the religious tolerance for which Holland is justly famous, and knowing of the astonishing courage displayed by so many Dutch citizens in shielding Jews in World War II, we were surprised to find in Leiden more overt evidence of religious differences than we were used to. There were stores identified as Catholic and stores identified as Protestant. Those in uniform could visit a Catholic military tearoom or a Protestant military tearoom (so identified, not so named) for refreshment and relaxation—Dutch versions of the USO canteens that had served American soldiers and sailors.

When I filled out my papers at the University for the stipend that I was to receive, I was asked to select a religion because a small portion of my pay would be deducted to help support a church or synagogue. Not finding "Unitarian" among the choices offered, I picked what looked like the broadest umbrella, "Christian." Only later did I learn that I had unknowingly designated a particular denomination, the Christian Church. Eventually, Janette and I did find a compatible worship group in Leiden, the Waldensians, tottering along on the verge of extinction. Two pastors conducted Sunday morning services, one in French and one in Dutch. Our presence swelled the congregation by a significant percentage.

The car that we had acquired from the Schwarzschilds was, we discovered, a great social asset. At this time, only a little more than a decade after the end of World War II, automobile ownership in Holland (as in other European countries) was relatively rare. Few of our Dutch friends had cars. Janette's women friends were more than happy to accompany her on outings in our little car.

As I became more and more happily involved with gravitation and general relativity in the 1950s, I kept asking myself, Why not big effects? For forty years, the focus of most scientists working in this area had been on small effects. Because of general relativity, for instance, the perihelion of Mercury's orbit around the Sun advances (precesses) by only an extra 40 seconds of arc per century. This means that the relativistic effect on the perihelion is to rotate it once around the Sun in 3.2 million years. In this time, Mercury has completed 1.3 billion trips around the Sun. The perihelion's advance is indeed a small effect. Similarly, the deflection of starlight passing close to the Sun, whose measurement propelled Einstein to world attention in 1919, amounts to only 1.7 seconds of arc (about one two-thousandth of a degree), again a small effect. To pick another example, atomic clocks in orbit used as part of the Global Positioning System (GPS) must be corrected for the effect of general relativity on their rates. Time runs faster high above Earth than at Earth's surface. At the 20,000-km altitude where the GPS satellites orbit, an atomic clock outpaces its sibling clocks on Earth by a little less than 25 millionths of a second in each trip around Earth. (Additional effects of special relativity must also be considered.)

It is a characteristic of the physics of the last hundred or more years that new theories have, paradoxically, overturned past theories without harming them in the slightest. Let me explain what I mean. Back in the 1860s, James Clerk Maxwell predicted that electric and magnetic fields could propagate as waves over great distances. Electromagnetic radiation is commonplace now but was a hard concept for Maxwell's contemporaries to accept. Maxwell's revolutionary theory offered a new phenomenon—radiation—without undermining any of the familiar electric and magnetic phenomena that had already been well studied in the laboratory.

Discoveries of the twentieth century were even more dramatic. Einstein's theory of special relativity predicted a whole range of new phenomena—mass that could turn into energy, moving meter sticks measured to be less than a meter long, a space-traveling twin who ages more slowly than a stay-at-home twin. Yet it altered not one whit the familiar and well-tested mechanics of Newton—"classical" mechanics, as we now call it, valid, as it turned out, whenever speeds are much less than the speed of light and energies are much less than the equivalent energies of rest mass. We say that the new theory "reduces to" the old theory in a suitable domain, the domain where the old theory is well tested and valid. For many years after Einstein proposed his special theory, its experimental tests were few in number and not very accurate. This is because the tests were conducted in ranges of speed and energy where deviations from the old theory were small. (When Einstein received the Nobel Prize in 1921 for his work on the quantum theory of light, the Nobel Committee went out of its way to let him know that the prize was *not* awarded for his work on relativity.) Only later came particle accelerators that routinely pushed matter to speeds near the speed of light and nuclear reactions that revealed the link between mass and energy. These, finally, were the big effects, but they followed the theory and its small effects by some twenty-five years.

Quantum mechanics is similar in having only small effects in already familiar domains, reserving its big effects for domains outside of normal human experience. In our classical (nonquantum) world, a particle is a particle and a wave is a wave and they are different. We can measure speeds and positions of objects to high precision. We see causal chains linking one event to another. In the quantum world, the world of the very small, waves and particles are different faces of the same thing, position and speed cannot be simultaneously measured to great precision, and simple causal chains are replaced by probabilistic links. Yet in the large-scale world, quantum physics reduces to classical Newtonian physics. As it turned out historically, there was no great time lag between the discovery of quantum physics and the identification of its "big" effects. This is because the behavior of even so tiny an entity as a single atom, which is ruled by quantum mechanics, easily makes itself felt in the large-scale world (through its radiation of light, for instance).

General relativity is likewise revolutionary without being destructive of anything that came before. It dramatically altered our view of space, time, and gravity, yet in areas of familiar physics, it had practically no perceptible effect. It "reduces to" Newton's inverse-square law of gravitational attraction in regions where spacetime curvature is small. Within our own solar system, only the planet Mercury could reveal, and then only barely, a difference between the new theory and the old. On Earth the subtle effect of gravity on time remained beyond the reach of experiment until, in 1960, Robert Pound and Glen Rebka of Harvard University found a way to use a nuclear gamma ray itself as a clock, measuring the two parts in a million billion that its "ticking" (that is, its frequency) changed as it ascended or descended a distance of about seventy feet. Later, in the 1970s, several teams, including one led by Carroll Alley of the University of Maryland, compared the timekeeping of atomic clocks carried aloft in aircraft and identical atomic clocks fixed on Earth. They found the expected tiny difference attributable to gravity (as well as a difference predicted by the special theory of relativity arising from the relative motion of the clocks). But minuscule effects can have big consequences, not just for our overall understanding of nature but at a very practical level. The astonishing precision with which the pilot, boater, driver, or hiker can pinpoint location is possible only because the atomic clocks in GPS satellites are corrected for relativity.

Not surprisingly, most scientific efforts on the theory of general relativity for several decades following its introduction in 1915 were focused on the small effects, the tiny deviations of its predictions from those of classical physics in the weak gravitational fields that seemed to pervade not only our own solar system but galaxies and the whole universe. There was no evidence that gravitation was strong enough anywhere in the universe to reveal the potential big effects of general relativity.

I was not content with the direction of research that focused mainly on small effects nor with the assumption that gravity was not truly strong anywhere. My first paper on general relativity, which introduced the geon, hypothesized energy densities of electromagnetic radiation so enormous that the radiation propagated not from "here" to "there" (like the beam of a flashlight), but right around in a circle, holding itself in a confined space through the gravitational fields created by its own energy. (If light from a flashlight behaved like light in a geon, the flashlight beam would bend around to illuminate the back of your head.) Was the geon a mathematical exercise designed merely to show what was contained within the theory of relativity, or was it a real entity that might be found somewhere in nature? I wasn't prepared to say. On the one hand, I considered it proper to push the mathematics of the theory to its limits, whether or not the predictions were actually realized. On the other hand, I couldn't help believing that nature would somehow, somewhere, avail itself of the possibilities contained within the theory (if the theory were valid).

Such thinking led me, during my stay in Leiden, to speculate in my course of lectures on a whole range of phenomena where gravity might be enormously strong and spacetime might be drastically curved. Thanks to an outstanding group of young colleagues who had decided to work with me on such problems, I could also go beyond speculation and, with them, work out the details of some of these ideas. Joseph Weber, Charles Misner, and Peter Putnam were at hand. Several others, working with me just before and just after my time in Leiden, made important contributions too.

As my own enthusiasm for gravitation physics grew, my willingness to involve students in the enterprise also grew. In earlier years, I had feared that a student trained in general relativity might have trouble finding a job where his or her training would be an asset and where further work in the field would be encouraged. That is why I had shied away from working in relativity myself in my younger days. I was afraid that students drawn into the work might not be well served. And I could not picture serious work in a field by myself without students.

I had seen too many "one-legged" young physicists working with Einstein at the Institute for Advanced Study. I didn't want to promote any similar imbalance at the university. But after teaching that first course in relativity at Princeton, my outlook changed. I began to see a wealth of possible projects, and to see that we might confront theory with experiment as we learned about the universe and its history. So, by the time of my stay in Leiden, I was willing to accept students who wanted to pursue spacetime physics. As it turned out, many students began to get excited about the field and were not deterred by worries over job prospects. Fortunately, relativity research has since burgeoned at many centers, and those of my students who wanted to continue working in this field have been able to do so.

I was fortunate to have Misner with me in Leiden because I learned so much mathematics from him. He had been a mathematics major at Notre Dame University, and quickly soaked up the mathematics of relativity theory. Fortunately, he was as interested in the physics as the mathematics, and we had great fun working together. It was a happy collaboration, as agreeable and productive as my work in the 1940s with Dick Feynman. Misner was able to provide the solid mathematical underpinnings for wormholes, and explored the general way in which electromagnetic fields can be "read" from the curvature of spacetime. This led to a long paper, "Classical Physics as Geometry," which we published the next year.

Peter Putnam, my other student in Leiden, possessed a brilliant, searching, troubled mind. He was as much philosopher as physicist. Indeed, his undergraduate senior thesis, "Considerations Concerning Nature," completed under my guidance in 1948, was such a confusing mix of philosophy, mysticism, and physics that neither I nor the faculty colleague who shared the marking responsibility with me knew how to assign it a grade. We could only dimly perceive meaning in his arguments for the proposition that the structure of nature must correspond to the structure of thought and the structure of the mind. In the end, we adopted an "algorithm," figuring where Peter's course grades fell relative to the course grades of his fellow seniors, then assigning his senior thesis a grade that put it in the same position relative to the thesis grades of other seniors.

Between that senior thesis and his Ph.D. in physics a dozen years later, Peter followed a circuitous route—and an even more convoluted route later. First, following his mother's wishes, he enrolled in the Yale Law School. His only brother had been killed in action in World War II, and his father had died soon after. His mother, Mildred, active in business in Cleveland, Ohio, was wealthy and strong-willed. But Peter's heart wasn't in the law. He dropped out of law school and took a part-time job with an electronics firm in New Hampshire, leaving time for him to read physics and philosophy. Through letters and visits, Peter kept in touch with me. Under the influence of Sir Arthur Eddington's *The Nature of the Physical World*, he had come to believe that all the laws of nature can be deduced by pure reasoning. Try as I might, I couldn't seem to disabuse him of this belief.

Finally, Peter followed my suggestion that he say good-bye to Eddington and get back to something timely and tractable in the world of physics. He enrolled as a graduate student in physics at Princeton, and asked to accompany me to Leiden. He audited my lectures and contributed some beautiful large drawings to illustrate ideas I was covering, but didn't get seriously into research until he got back to Princeton. Then he finished a Ph.D. dissertation on the distribution of mass and energy in a star that is radiating at a prodigious rate.

In Leiden, we learned to our surprise of Peter's homosexuality. Mildred Putnam, on a visit to Leiden, learned the same thing to her horror. Bad enough, she thought, that he renounced the family riches. Now this. Nevertheless, she and he remained close. Although Peter would not accept money from his mother, he encouraged her to give to Princeton. Eventually she made a generous gift for outdoor sculpture. Peter had decided that sculpture is the most vibrant genre of contemporary art and had himself started a fund for sculpture at Princeton by donating shares of stock in his New Hampshire electronics company. Because of my link to Peter, I became, in effect, an arm of Princeton's fund-raising department, communicating often in letters and sometimes in person with Mildred Putnam before her gift was realized. A fringe benefit of all this for me was the chance to visit the noted sculptor Henry Moore at his studio in England and later to welcome him to Princeton, where his sculpture *Oval with Points* now sits, thanks to the Putnam gifts.

During a postdoctoral teaching stint at Columbia University, Peter offered a physics course so full of philosophy that it attracted students from the nearby Union Theological Seminary. Before long, he obtained a teaching post at Union, where he was, as one fellow teacher there told me, the only person who could out-argue the great theologian Reinhold Niebuhr.

Whatever mechanism in Peter's head propelled him through this world, it produced a jagged path. When, around 1971, his appointment at Union The-

Peter Putnam in Leiden, 1956.
(Photograph by Alison Lahnston.)

ological Seminary was not renewed—largely, I suspect, for lack of publications—he decided to cast his lot with the civil rights movement and moved
to Houma, a town in the bayou country of Louisiana, where he offered legal
services to blacks for little or no fee. To provide simple food and rent on a
tiny house that he shared with a companion, he worked as a night janitor in
a church. When Janette and I, on our way to Texas in 1976, stopped to visit
Peter, one look told us that he was truly impoverished. His mother visited
more than once but was unsuccessful in getting him to leave Houma or to
accept money. One night in 1987, cycling between his residence and his
janitorial job, Peter was struck by a drunk driver and killed. He left behind
reams of notes and manuscripts, none of which, as far as I know, has found
its way into print, although a former seminary student who admired Peter
has been at work organizing them for possible publication.

Peter was not one of my better students, and made no lasting contributions to physics. His talents did not flower in publications. He was perhaps a
bit mad. Yet he deeply affected a few people, me among them. In our long
correspondence over many years and in our occasional long conversations, he
always had a way of raising questions and challenging accepted explanations
that helped me sharpen my thinking about physics and about the way we
humans describe and understand the world around us.

Einstein's equations of relativity tell us that spacetime can not only be
"curved," it can also wiggle, or vibrate. Vibrating spacetime can transmit energy from one place to another. In short, gravitational waves can (according to
the theory) exist.

The stage of a theater provides an analogy. (Let's leave washtubs and bathtubs behind.) If the stage were absolutely rigid, it would correspond to the
flat space of our ordinary experience and of Newtonian physics. It is just
"there," the stage where action occurs but not itself part of the action. If the
stage had some give to it, it would sag a little where an actor was standing
and sag more where a heavy part of the scenery was resting. It would then
correspond to the curved space of general relativity, where the greatest mass
produces the greatest curvature. If the actor then jumps up and down or a
weight is accidentally dropped onto the stage, the flexible floor will vibrate
and the vibrations will propagate from one part of the stage to another. The
stage then corresponds to the dynamic spacetime of relativity that admits
waves as well as curvature. The idea of a flexible stage ("warped" space) was
already unsettling enough in people's minds. The dynamic stage (gravity
waves) strains our powers of visualization even more. Yet we must follow
where gravitation theory leads. Spacetime is an important part of the action,
not just the place where action occurs.

In some of my research, as I tried to push gravitation physics to its limit, I even asked the question Can spacetime be everything—both what *is* and what *happens*? Whatever the final answer to that question may be, there is no doubt that gravity waves can roll across the cosmos, carrying energy from one place to another.

Since the earliest days of general relativity theory, some scientists have been moved to question the reality of its implications. It predicts an expanding universe. But surely, said some—including, for a time, Einstein himself—the universe cannot be expanding.[2] Yet, as Edwin Hubble proved with his telescope in the 1920s, it surely is. General relativity also predicts the collapse of matter into black holes. An oddity of the theory with no relation to the actual world, said some. Yet every year the evidence for the reality of black holes becomes more convincing. Their existence can hardly be doubted. In addition, the theory predicts geons and wormholes. Are they real? We don't yet know. I would not want to bet against them. And general relativity predicts gravitational waves. When my postdoctoral colleague Joe Weber and I got interested in this aspect of gravity physics during our term in Leiden in 1956, we had first to confront the arguments of other scientists that such waves do not in fact exist. Of course, we could not produce evidence for their reality. (The first such evidence was reported in 1978.) All we could do was try to show that such waves were entirely consistent with everything else in the theory, that there was no theoretical bar to believing in them. Our paper making this case was published the next year. Other work around the same time by Hermann Bondi, Felix Pirani, and others helped nail down the case, and gravitational waves have been generally accepted since the late 1950s.

What could cause a gravitational wave? Some gravity waves could still be bouncing around the universe today as a result of the violence of the Big Bang at the beginning of time. Supernovas exploding and matter collapsing into black holes inevitably produce such waves. Whenever two neutron stars coalesce or two black holes collide or whirl in orbit around each other, they must send gravitational waves careening across the cosmos. Whatever the source of a gravitational wave, it is likely to be cataclysmic by our ordinary standards. Yet all calculations indicated that the intensity of gravitational waves reaching Earth, regardless of the drama attending their birth, must be extremely weak. Detecting them poses an enormous technical challenge.

Joe Weber was the first to accept that challenge. Following our work together in Leiden, he embraced gravitational waves with religious fervor and has pursued them for the rest of his professional career. I sometimes ask myself

[2] Einstein, in a later conversation with George Gamow, called his renunciation of the expanding universe the greatest blunder of his life.

Joe Weber, c. 1952.
(Courtesy of AIP Emilio Segrè Visual Archives.)

whether I imbued in Weber too great an enthusiasm for such a monumentally difficult task. Whether, in the end, he is the first to detect gravitational waves or whether someone else, or some other group, does it, hardly matters. In fact, he will deserve the credit for leading the way. No one else had the courage to look for gravitational waves until Weber showed that it was within the realm of the possible.

What does a gravitational wave do when it encounters matter? Its effect is "tidal." This means that it stretches matter in one direction and squeezes it in another, much as the Moon stretches Earth's oceans on the near and far sides of Earth (making high tides) and sucks the oceans downward on the in-between sides of Earth (making low tides). A gravitational-wave detector is designed to respond to such tidal forces. To get an idea of how it works, imagine that you have welded four metal rods together to make a square. You lay the square flat on the ground, and label the four rods N, E, S, and W to match the points of the compass. Now a gravitational wave arrives from outer space, propagating downward from the sky. At one instant, rods N and S will be pulled a little closer together and rods E and W will be pushed a little further apart. An instant later, N and S will be pushed apart while E and W are pulled together. How often does the pushing and pulling occur, and how big is the effect? The frequency of arriving gravitational waves will depend on their source, but in any case is expected to be extremely low relative to frequencies of light or even radio waves. Around 1,000 hertz (vibrations per second) might be typical. As to the size of the effect, it is almost unimaginably small. As a gravitational wave from some cosmic source rolls by your metal square, the arms of the square are expected to be displaced by far less than the diameter of a single atomic nucleus.

Joe Weber designed a "bar detector," a giant cylinder of aluminum with extraordinarily sensitive piezoelectric detectors affixed to it. These can pick up even the tiniest vibration, provided the incoming wave jiggles the bar at the bar's resonant frequency (the frequency at which it would naturally vibrate if lightly tapped). In 1969 and the early 1970s, Weber reported some measured vibrations that *might* have signaled the passage of gravitational waves. Most of the physicists interested in gravitational waves doubt that these were real signals. Nevertheless, it is Weber's leadership and inspiration that have led to another generation of even-more-sensitive detectors.

Detectors of the new class use interference between beams of laser light to detect the slight relative motion of suspended masses. Instead of arranging rods in a square as described above, imagine three weights suspended from a ceiling at three of the four corners of a square. The three weights define an $\llcorner$ (two sides of a square). We can label one of those sides W (the west side) and the other one S (the south side). When a gravitational wave arrives from above (or below), the weights at the end of side W will separate slightly (very slightly!) as the weights at the end of the side S will move slightly closer together. These tiny motions can be detected by a beam of laser light split into two beams that travel along sides W and S and then reflect back to recombine into a single beam.

A major experiment of this kind is being organized by a Caltech-MIT group, with Kip Thorne as one of its founders. They plan to have three detectors, two located in the state of Washington and one in Louisiana. One in each place will have arms (the sides of the $\llcorner$) that are 4 kilometers long. This grand undertaking, with the acronym LIGO (Laser Interferometer Gravitational-Wave Observatory) will usher in a new age of gravitational-wave astronomy. Multiple detectors are needed to provide assurance that the extremely weak signals truly result from gravitational waves. Each detector vouches for the others. With the help of a similar Italian-French detector called VIRGO, scientists will use the slightly different arrival times of signals at the different detectors to find out from what direction the waves are coming and to extract their full information content.

Those of us studying gravitational waves feel as confident of their reality as Einstein felt about the deflection of starlight by the Sun in 1919. Nevertheless, it is nice to have in hand solid indirect evidence for gravitational waves. In 1993, my Princeton colleagues Joseph Taylor and Russell Hulse were recognized with the Nobel Prize for their measurement of the rate at which a pair of neutron stars, one of them a pulsar, are "running down." These objects, which rotate around each other, are not maintaining their separation (as Earth maintains its separation from the Sun), but instead are gradually spiraling toward one another—diminishing their separation at the rate of 2.7 parts per billion per year. This inward spiraling means that they are losing energy (even though, in a seeming paradox, they speed up as they get closer together). No mechanism other than gravitational radiation has been proposed for such loss of energy. Indeed, the rate of inward spiraling is just what one would expect if it is gravitational radiation that is draining their energy.

While in Leiden, I met the widow of Paul Ehrenfest, the great Dutch theoretical physicist, friend of Einstein and teacher of many of Europe's leading physicists. In 1933, under the awful spell of mental illness, Ehrenfest had shot and killed his mentally retarded son, then himself. Peter Putnam, who made it his business to meet the notable people in Leiden, brought me and Mrs. Ehrenfest together. I spoke to her of the importance of preserving her husband's papers. She did not respond at the time, but months later, going through my daily mail back in Princeton, I laid aside a large brown envelope from Leiden in order to deal first with the smaller envelopes likely to contain correspondence. When at last I opened the big envelope, out poured a stack of postcards, written by Einstein to Ehrenfest over the years. These two had enjoyed each other immensely.

Not for a minute did I think of keeping this collection for myself. It was obviously a resource for historians and a wider public to enjoy. So I called Helen Dukas, Einstein's longtime secretary, who was still living in the Mercer Street house and was now an executor of his estate. She suggested that I get in touch with Otto Nathan, another executor, in New York City. Nathan, as I soon learned, was a short man with a short fuse whose self-importance rested on his control over Einstein's papers. When he learned of the postcards, he fired off a ferocious letter to me, then showed up in my Princeton office, prepared to remove the cards by force. Had he known me, he would have known that such a threat was the worst way to get what he wanted. Instead of yielding to his demand, I contacted Herbert Bailey, head of the Princeton University Press. At his suggestion, I turned over the postcards to John Stachel of Boston

Paul Ehrenfest, c. 1924.
(Courtesy of Niels Bohr Archive, Copenhagen.)

University, the editor-in-chief of the Einstein papers for the press. Now the cards are appearing in their right chronological places in the published papers.

That was not the end of my contact with Otto Nathan. In 1981, I testified before an arbitrator in New York in a related dispute between Nathan and the Princeton University Press. Nathan wanted to have Stachel removed as editor-in-chief of the papers, but the press prevailed and Stachel remained (although he later stepped down after years of distinguished work on the project).

During our first months in Leiden, Janette did a lot of touring in our little car, and devoted her spare time to Henry James. Alison adapted to her daily train ride to The Hague, improved her French and German, learned a little Dutch, and became even more a citizen of the world. When June rolled around and the Harvard-Radcliffe semester ended, our children Letitia and Jamie boarded the SS *United States* and sailed across the Atlantic together to join us for a while in Holland before they went off to other pursuits in Europe. To keep them with us for a week, we chartered a boat and made our way through the canals and lakes of Friesland—on our own, without benefit of a captain or guide. We had a delightful time, even crossing a water bridge over a dam across the Zuider Zee. We could see why so many of our Dutch friends look on a boat as a ticket to privacy. Without leaving the country, one is suddenly free of the crowds and congestion that are everywhere in this heavily populated nation.

I have to admit that I never stop thinking about physics. What kind of physics I think about and how my mind works depend on where I am and what else I am doing. If the setting is a family outing in which I am thoroughly engaged with the people around me and thoroughly relaxed, as on our boat trip in Holland, my scientific thinking is likely to be almost subliminal. It circles around deep questions of principle rather than some particular calculation or some particular paper in preparation. I have never been able to let go of questions like: How come existence? How come the quantum? What is my relation to the universe and its laws? Can spacetime be all that there is? Is there an end to time?

These questions, I should add, have nothing to do with my religious convictions, which center on guides to living, guides to civilized intercourse among humans. The deep questions that I wrestle with belong to science, as I define it, not religion.

I can remember no more stimulating, productive—and agreeable—eight months in my life than the period I spent as Lorentz Professor in Leiden. Forty years had passed since Einstein had advanced his general theory of relativity. The theory was ripe for exploitation, and soon the world of physics would be ripe for experimental tests of it. Fifteen years after my stay in Leiden, Charlie Misner, Kip Thorne, and I wrote in our book *Gravitation*: "For the first half-century of its life, general relativity was a theorist's paradise, but an experimentalist's hell. No theory was thought more beautiful, and none was more difficult to test. The situation has changed. . . . A half-century late, the march of technology has finally caught up with Einstein's genius." I had the great good fortune to start my work on relativity in what was still a theorist's paradise, and I must say that it has been a theorist's paradise ever since—not least because of the new opportunities that arose for experimental tests. Theory is doubly exciting when it predicts new experimental results and when it can confront measurement.

Indeed, a great part of being a theoretical physicist is to develop concepts and equations that tie experience together. That is the standard definition of the craft. The theorist may be staring at a chalkboard or a piece of paper or a computer screen, or talking to a colleague, but the central topic when he or she is working in this mode is the relation of theory to experiment, either yesterday's experiment or tomorrow's. Yet that is not the whole of being a theoretical physicist. There is a place, too, for free invention. And a place for theory guided by aesthetic considerations: Is the equation beautiful? Do the concepts hang together like the elements of a master artist's composition?

For me, there is still more. I have not been able to stop puzzling over the riddle of existence. From the calculations and experiments that we call the nitty-gritty of our science to the most encompassing questions of philosophy, there is one unbroken chain of connection. There is no definable point along this chain where the truly curious physicist can say, "I go only this far and no farther."

12
·
NATURE AND NATION

IN 1955, at a conference in Rochester, New York, I met a bright and bubbly young red-headed Italian graduate student named Tullio Regge. I was introduced to him by Robert Marshak, a talented theorist and a good administrator who, at that time, chaired the University of Rochester Physics Department. Marshak had organized the first conference on high-energy physics in Rochester in 1950, and continued to direct what rapidly became an essential annual pilgrimage to that city for physicists interested in elementary particles. (Elementary-particle physics is often called high-energy physics because high-energy accelerators—or high-energy cosmic rays—are needed to create new particles.) Starting in 1958, the conference became a road show, meeting in various centers in the United States and Europe. But it remained affectionately known as the "Rochester conference," whether it was meeting in Berkeley, Geneva, Tbilisi, or Tokyo. (In a similar way, the January meeting of the American Physical Society continued to be known as the "New York meeting" long after it left New York to meet in various cities around the United States. The New York meeting finally expired, done in by meetings focused on narrower fields of physics.[1] The Rochester meeting, a specialty meeting from the first, has survived and prospered, although it doesn't lack for

[1] More properly, the January meeting has undergone metamorphosis. No longer a meeting of the American Physical Society, it has gained new vigor as a meeting of the American Association of Physics Teachers.

competition. Known now as the International Conference on High Energy Physics, it meets biennially in cities around the world.)

Marshak described Regge to me as a brilliant mathematical physicist with some interest in general relativity. My conversation with him confirmed Marshak's assessment. It happened, at the time, that I was trying to come to grips with the problem of the stability of what we then called a "Schwarzschild singularity," named after Karl Schwarzschild (Martin's father). Karl Schwarzschild had been director of the Potsdam Observatory just outside Berlin before World War I and came to know Einstein. In 1916, just after Einstein published his theory of general relativity, Schwarzschild worked out the geometry of spacetime around a hypothetical mass concentrated at a point—an achievement all the more remarkable because he was at the time in active service in the German army, having volunteered soon after the war began. His solution can now be found in every textbook on relativity.

This "Schwarzschild solution" was Karl's last major piece of work. Felled by an infection contracted in the war, he died in the same year the work was published, leaving behind a wife and two small boys.[2]

Only quite a bit later did we come to realize that what lay at the center of a Schwarzschild geometry might be a true singularity, mass truly condensed to a mathematical point. At the time that Schwarzschild worked it out and for years afterward, most physicists assumed that the Schwarzschild singularity is a convenient approximation, not to be taken literally. After the work of Robert Oppenheimer and his colleagues in 1939, we began to wonder whether mass could truly collapse to a point. But it wasn't until the 1960s that gravitational collapse—whether or not to a true point—became generally accepted. (Good experimental evidence for black holes had to wait another quarter century.)

The problem on my mind when I met Regge in 1955 was this: Suppose that true singularities can exist. Will they be stable? Something that is stable one might expect to find in nature. Something unstable one would be unlikely to find. To see this point, imagine one hundred pendulums free to turn through a full circle. Start with fifty of them hanging with their bobs straight

---

[2] Martin's younger brother, Alfred, born in 1914 after their father had left for the Russian front, wound up with nine names (ten counting Schwarzschild). Karl and his wife had agreed on the name Alfred in case it was a boy, but when some family members objected to that name, Martin's mother gave the new baby no name at all. Soon the police arrived at the door to enforce a German law that a baby must be named soon after its birth. Called before a judge, she provided another "Schwarzschild solution" by registering the baby with a string of nine names, one of them Alfred. When Karl returned on leave, he confirmed "Alfred" as his choice, but Alfred was never able to rid himself officially of the other eight names.

down, and fifty very carefully positioned with their bobs straight up. All one hundred are in equilibrium, meaning that no net force acts to move them from where they are. The downward-hanging bob will not spontaneously start to swing. Nor will the upward-pointing bob spontaneously move, provided it has been positioned with exquisite care and there are no breezes or other disturbances. But there is a big difference between the two bobs. The downward-hanging one is stable: Even if some disturbance does come along, the bob will at most swing through a small angle. It will remain close to where it was. The upward-pointing bob is unstable: The slightest disturbance will cause it to swing all the way around and eventually, after friction acts, to come to rest, or nearly to rest, pointing down. As to what we "find in nature," suppose that after you carefully arrange the hundred bobs, fifty up and fifty down, you go away for a week. When you come back, you will almost surely find all hundred bobs pointing down. Natural disturbances will have worked to eliminate the unstable, upward-pointing bobs. There are none left to observe.

Schwarzschild had shown that the spacetime geometry surrounding a spherically symmetric concentration of mass is in equilibrium, but had not shown that it is stable. For instance, if the geometry changed ever so slightly away from perfect spherical symmetry, would the geometry go on changing dramatically into a wholly different form (like the upward-pointing bob after it is disturbed)? Or would it move back toward its original form, perhaps with small oscillations (like the downward-pointing bob)?

My intuition told me that the Schwarzschild singularity should be stable, but I had not yet been able to prove it. It was important to find out. There is no use looking for such entities in nature if they are not stable. They wouldn't last long enough to be seen. Regge, with his mathematical power, seemed to be just what I needed. I outlined the problem and my thinking about it to him, and he agreed to work with me. Then we went off to our respective institutions—I to Princeton and he to Turin, Italy. Since I had a vision of how the whole problem should be tackled and how it would work itself out, I sat down and wrote a paper with spaces left for equations, and sent it to Regge. He rose to the occasion, and filled in the blanks. Then, early in 1956, I was able to round up travel funds from the Friends of Elementary Particle Research in Princeton (the fund created by my Du Pont friends) to bring Regge to Leiden, where we spent ten days together and made the pieces of our work fit smoothly together. Indeed, the Schwarzschild singularity is stable. This was one more piece of evidence to encourage continuing work that led finally to a rather full understanding of the black hole and its properties.

My paper with Regge was one of ten that I published in 1957, most of them coauthored with students and other younger colleagues on aspects of general relativity and gravitation physics. Teaching the relativity course in 1952–1953 had been my investment. These papers were the payoff—returning the investment many times over, thanks to my numerous students and to the time I found in Leiden for thinking deeply. General relativity has the characteristic, more than any other branch of physics, that its basic equations are of the most concentrated essence, requiring massive mental assault to yield up their implications.

A January 1957 conference at the University of North Carolina, Chapel Hill, organized by two leading researchers in general relativity, Bryce DeWitt and Cécile DeWitt-Morette, on "The Role of Gravitation in Physics," proved to be the stimulus for me and my students to get some of our work written up for publication. We were at the receiving end of good-natured finger-pointing by colleagues, who accused us of "taking over" *Reviews of Modern Physics*. We accounted for eight papers in its July issue that year. Papers with Edwin Power of University College, London, and with my students Richard Lindquist, John Klauder, and Dieter Brill were among these. So was my paper with Joe Weber on gravitational waves, those evanescent cosmic twitches that stirred Joe's passion.

Since my first paper on geons, I had been concerned with both classical (nonquantum) and quantum gravity. On the one hand, I knew that within Einstein's classical theory of relativity, there were still riches to mine. Geons, wormholes, gravitational radiation, and possible infinite concentrations of mass were all within the domain of the classical theory, in which space and time are continuous and can be infinitely subdivided, and the lumpiness, the granularity, the fluctuations of quantum theory are lacking.

On the other hand, I knew that any eventual theory of gravity must take quantum theory into account. The world of the very small is a quantum world, and that must be as true of space and time and gravity as of electrons and photons and quarks. With my astonishingly capable graduate student Charles Misner, I was pushing on both of these fronts when we were together in Leiden in 1956. He and I also "took over" another journal, the December 1957 issue of *Annals of Physics*. This was a new venture, whose editor, Phil Morse, was receptive to our relatively long, discursive paper (unlike the editors of *Physical Review*, where conciseness was the greatest virtue). Charlie and I together produced a seventy-nine-page paper, "Classical Physics as Geometry," in which we developed geometric aspects of electromagnetism. I like to call it "already unified theory" (unlike the theory to which Einstein devoted so much time in his later years) because we showed how to extract electric and magnetic fields from the curvature of spacetime itself. And in this paper, we further developed the ideas of wormholes into which electric field lines disappeared to reappear somewhere else.

Next to this joint paper in *Annals of Physics* appeared my (shorter) paper on quantum aspects of gravitation theory, which I titled "On the Nature of Quantum Geometrodynamics." *Geometrodynamics* is an awful mouthful, but I liked the term for its symbolic link to its sister theory, electrodynamics. The idea is that the geometry of three-dimensional space is something that can undergo change as time passes, and propagate from one place to another, just as electromagnetic fields can do. Applying quantum theory to this idea leads to wormholes that pop in and out of existence and to space roiled into a quantum foam, much as water in a breaking wave sprays into foam. As I mentioned in Chapter 11, these quantum fluctuations of space occur over dimensions incredibly small even when compared with the size of a single proton.

Misner and I learned after completing our "classical" paper (we don't lay claim to its being a "classic" paper) that some of the ideas in it that we thought were new had been developed back in 1925 by George Rainich. One of the great pioneers of relativity theory, Peter Bergmann, then at Syracuse University, with whom we kept in touch, pointed this out to us. So our seventy-nine-page paper turned out to be partly new and partly review. Misner had put an enormous effort into the paper, only to find that some of the key ideas he contributed had been preempted. To many a graduate student, this would have been a devastating blow. Not to Charlie. He and I agreed that for his Ph.D. thesis, he should pick another topic so that he could present wholly original work to his committee. He rose masterfully to the challenge. He turned from classical to quantum gravity, and in the space of very few months produced a wonderful paper (and thesis) entitled "Feynman Quantization of General Relativity." The paper was added to the collection of our group in the July 1957 issue of *Reviews of Modern Physics*.

One very deep paper by my student Hugh Everett in this *Reviews* issue was so impenetrable that I was moved to publish next to it a short paper entitled "Assessment of Everett's 'Relative State' Formulation of Quantum Theory." Everett was an independent, intense, driven young man. When he brought me the draft of his thesis, I could sense its depth and see that he was grappling with some very basic problems, yet I found the draft barely comprehensible. I knew that if I had that much trouble with it, other faculty members on his committee would have even more trouble. They not only would find it incomprehensible; they might find it without merit. So Hugh and I worked long hours at night in my office to revise the draft. Even after that effort, I decided the thesis needed a companion piece, which I prepared for publication with his paper. My real intent was to make his thesis more digestible to his other committee members.

Everett's paper and my interpretation of it were concerned with the basics of quantum theory, only very loosely linked to relativity. The standard approach to quantum theory then (and now), assigns probabilities to the possible outcomes of quantum events. The actual outcome for a particular experiment is ascertained by a measurement that uses a "large," nonquantum detector—a piece of laboratory apparatus, for example, or the human eye. Only by replicating the same experiment many times can one confirm that the outcomes follow the predicted probabilities. And until the actual measurement is made, there is no way to know which among the possible outcomes will be realized.

A difficulty with this "Copenhagen interpretation," a difficulty that still deeply troubles me and many others, is that it splits the world in two: a quantum world, in which probabilities play themselves out, and a classical world, in which actual measurements are made. How can one clearly draw a line between the two? By how much must a quantum event be magnified to become a classical observation? When does probability give way to actuality?

Everett, in a tour de force, sought to get around these troubling questions by describing a totally quantum world in which there was no such thing as a classical observer, only quantum systems at all levels of size and complexity. Everett's "observer" is part of the quantum system, not standing apart from it. An oversimplified way to describe the outcome of his reasoning is to say that all of the things that might happen (with various probabilities) are in fact happening. Since there is no classical measuring apparatus in his formulation to determine which among the possible outcomes occur, one must assume that all the outcomes are occurring, but with no communication among them.

To see what this means, think of yourself driving down a road and coming to a fork. According to classical physics, you take one fork, and that's that. According to the conventional interpretation of quantum mechanics, you might take one fork or you might take the other, and which one you take will not be known until something happens to pin down your location, such as stopping at a gas station or restaurant, where some outside "observer" ascertains your location. There is something ghostly about even the conventional quantum interpretation, since it assumes that you travel "virtually" (as opposed to "really") down both roads at once, until it is established that you "really" traveled down a particular fork. According to the Everett interpretation, you go down both roads. If you later stop for gas on the left fork and someone observes you there and you are yourself aware of being there, that doesn't mean that there isn't another "you," uncoupled from the left-fork you, who stops to eat on the right fork, is observed by people there, and is aware of being there. Bryce DeWitt, my friend in Chapel Hill, chose to call the Everett interpretation the "many worlds" interpretation, and DeWitt's terminology is now common among physicists (although I don't like it). The idea has entered into the general public consciousness through the idea of "parallel universes." Although I have coined catchy phrases myself to try to make an idea memorable, in this case I opted for a cautious, conservative term. "Many worlds" and "parallel universes" were more than I could swallow. I chose to call it the "relative state" formulation.

To me the important thing was not what analogies or fanciful visions one might spin out of Everett's work, but two basic questions: Does it offer any new insights? Does it predict outcomes of experiments that differ from outcomes predicted in conventional quantum theory? The answer to the first question is emphatically yes. The answer to the second is emphatically no.

Should scientists care about new insights or different ways of looking at things if nothing new is predicted that can be measured? Yes, they should. We need to be always looking at what we already know in new ways. It's like an artist examining a piece of sculpture from every angle. The scientist, like the artist, might get a new idea or at least a deeper appreciation of what is already "known." There is no limit to depth of understanding. Different ways to describe the same set of equations can add insight. Maybe one way is clearly more economical, more "elegant" than another. Then we adopt that way. Maybe having two ways enriches our understanding by letting us examine the same domain of nature in two different ways. Maybe using one description will sometimes seem clearer at one time or for one purpose while using an alternate description will seem clearer at a different time or for a different purpose. In general relativity, for instance, it is sometimes easier to talk of the three-dimensional geometry of space evolving through time, and sometimes easier to talk of the four-dimensional geometry of spacetime that just "is." It is not a question of one description being right and the other wrong, or even of one being better than the other. They are simply two ways to describe the same physics. For some applications, it may prove easier to use one approach, and for other applications the other approach.

What we have in Everett's work is a mind-stretching new way to look at quantum theory, one that triggers some very provocative thinking about the nature of the world even if it predicts no new experimental results. It may one day help germinate a better quantum theory or a better merger of quantum theory and relativity. Quantum theory has been around for most of the twentieth century, and its successes are legion. But the last word has not been written on it. I think about it every day.

Everett went off not to a mathematics or physics department, but to the Pentagon, to devote his great talent to more applied problems than parallel universes. Some years later, I had the pleasure of being taken on a tour of the Pentagon by him. I learned that he had reprogrammed nearly all the computers there. A heavy smoker who avoided exercise, he died prematurely of a heart attack in 1982.

The Soviet Union's launch of *Sputnik* in October 1957 shocked Americans, just as Joe 1 (the first Soviet atomic bomb) had done eight years earlier. A good deal of the ferment that resulted from *Sputnik* has been good for the United States. Apart from the boost it gave to our own space program, it set in motion numerous educational reforms, such as those funded by the National Science Foundation, that have continued to this day. Regrettably, *Sputnik* also intensified the Cold War and resulted in increased spending for defense on both sides.

But the Cold War was more than a period of high international tension fueled by hundreds of billions of dollars in military expenditures. It proved to be also a guarantor of peace in Europe. The Cold War bridged the abyss of suspicion, fear, and mistrust that lay between two eras: Soviet-American cooperation in war in the 1940s and the still somewhat uneasy Russian-American cooperation in commerce and international affairs in the 1990s. For as long as there is a historical record, Europe has been torn by one war after another. Until the present period, the longest period of peace in Europe stretched from the end of the Franco-Prussian War in 1871 to the outbreak of World War I in 1914. But records are made to be broken. This forty-three-year record was surpassed in 1988, as peace in Europe continued to prevail. Future historians may well conclude that it was the Cold War that established a new era of long-term peace in Europe. I personally regard the hydrogen bomb, dreadful though it will be if ever used, as the policeman's stick that enforced the long peace of this era.

I was less surprised by *Sputnik* than many people were. The technological capabilities of the Soviet Union had worried me for many years. I was convinced that if the Soviet Union gained a clear enough military superiority, it would become adventurous, seeking expansion and conquest, possibly triggering World War III. Although my commitment to pure research remained always paramount, I felt it my duty to help my country marshal the best military capability it could, in our own defense. In World War II, my concern took the form of work on reactors to produce plutonium for atomic weapons. In the period 1950–1953, it took the form of work on thermonuclear weapons. After our success on the H-bomb project, it seemed to me that I should try to help the defense effort by working on a broader range of problems than just those concerned with nuclear weapons. Missiles, for example, were clearly going to become dominant militarily as well as in exploring and exploiting space. There were many other technical areas where the outlook of a physicist could be useful.

In the early 1950s, John von Neumann had justified his support for the development of thermonuclear weapons by citing the inevitable inaccuracy of missiles. Missiles, he argued, would never be able to hit a target thousands of miles away with enough accuracy to permit the effective use of ordinary atomic bombs. He thought that the much greater "yield" (destructive energy) potentially available from an H-bomb would be needed to assure destruction of a target that might be missed by a mile or more. This was one of the rare instances in which von Neumann's ability to foresee technical advance fell short. Steady developments in the accuracy of missiles have made multi-megaton yields irrelevant. The thermonuclear weapons that now sit atop missiles in the world's nuclear arsenals are smaller and lighter, with less explosive power, than the true superweapons that could be built. The missiles themselves can therefore be smaller and lighter, or they can carry multiple warheads.

My first foray into the aerospace industry came, I believe, in 1954. At the invitation of Marvin Stern, a brilliant young applied mathematician freshly transplanted from New York City to southern California, I accepted a consulting assignment at Convair Corporation in San Diego. (Most of the work was carried out in Princeton, with only occasional visits to San Diego.) Convair was, at the time, a major contractor for the Atlas missile—which is what gave Stern the funds to hire consultants. The range of problems that Stern outlined went beyond missiles; I was intrigued and decided to take part. One of the early challenges had to do with supplying water for the fast-growing population of southern California. Would it be practical to crack a giant floe from the Antarctic ice sheet and float it north to Los Angeles? Water from that ice, we knew, would be relatively fresh and pure. Our conclusion after a bit of calculation: not practical. Most of the ice would melt during the long tow northward.

The idea of a nuclear-powered airplane had captured the imagination of some people after World War II. It was one of the ideas presented to us for analysis by Convair. Our conclusion: a nuclear-powered submarine makes sense; a nuclear-powered airplane does not. The weight of shielding needed to protect the crew makes it completely impractical to carry a reactor aloft to power an airplane. A few people continued to toy with the idea for years, but nothing ever came of it. Nuclear-powered submarines, by contrast, are now common and have transformed the nature of undersea warfare and defense.

Indirectly, von Neumann had something to do with my work for Convair— and later for Lockheed Corporation. In the early 1950s, when fear of a Soviet lead in missile technology was beginning to take root in our intelligence services, he was appointed to a government committee charged to analyze the situation and recommend actions. One of the committee's final recommendations—strongly influenced, I believe, by von Neumann—was that every government contract for missile development should be augmented by a certain percentage to support broader company-directed research. In the course of this committee service, von Neumann visited various missile contractors in southern California. "They operate sheet-metal shops," he later told me. "The main part of their work is bending and welding metal."

So, as a result of the research recommendation inspired by von Neumann, Marvin Stern had enough money to hire some twenty consultants. Among those he rounded up in addition to me were my Princeton colleagues Eugene Wigner and Oskar Morgenstern, my old friend Edward Teller, and Stern's own recent Ph.D. supervisor, Richard Courant, from the Institute of Mathematical Sciences at New York University (NYU). Stern was lively, interested in ideas, fun to be with. He was one of those people, not unlike Carson Mark at Los Alamos, who combined a brilliant mind with a wonderful ability to work with technical people and get the best out of them.

In 1955, not long after my first affiliation with Convair, I accepted an invitation from Montgomery Johnson to contribute to Lockheed's comparable missile program in Van Nuys, California. My acquaintance with Johnson went back to 1933, when he and I were both working with Gregory Breit at NYU. Older than Stern and with a more phlegmatic disposition—he slightly resembled a teddy bear—Johnson was a solid researcher and manager who also knew how to bring in talent. Edward Teller was among his stable of consultants, as well as my younger Princeton colleague Sam Treiman and my former student Ken Ford. Johnson made my stay in Van Nuys especially agreeable by providing me a room in his house for a month.

Based on my observations of the U.S. missile program and on what I knew of the Soviet program, I was convinced by 1956 that a "missile gap" between the Soviet Union and the western democracies, if it did not already exist, could easily exist before long. *Missile gap* is really shorthand for "military technology gap." It was not only about missiles that I worried, but about many other aspects of warfare that were dependent on advanced technology and science.

To me, the answer to a military technology gap, real or potential, was simple: talent. First defining problems that needed to be addressed and then finding persons or teams to solve them was not the way to go. What was needed was to find and recruit the best scientific minds, bringing them in to help define the problems and to work on solving them. That is what we had done in World War II. It had worked brilliantly then. In the mid-1950s, it was not happening. The most creative scientists were back in their offices and labs and classrooms, doing their own work, and paying no attention to national security. The best of the young scientists, in particular, were disconnected from defense applications. This situation was hardly surprising. Most scientists felt that they had done their civic duty during the recent war. They saw no urgency in the international scene that would make them want to take up defense work, even part-time.

I was among a smaller group that perceived a possible international crisis. I wanted to give some of my own time to defense work, and I wanted to do what I could to bring in creative young minds to help in this effort. I don't mean to disparage the talented people who were working at Los Alamos and Livermore and other defense laboratories. A lot of good work was going on at those places, and a few of the really top people who got involved during the war stayed on. Yet overall, the United States effort fell short. One reason for this was that the brightest young scientists earning their doctoral degrees were not considering defense work, either full- or part-time. Another reason was that we had no labs comparable to Los Alamos and Livermore devoted to work on defense problems other than those associated with nuclear weapons.

I knew, realistically, that getting the sustained, full-time commitment of leading scientists to defense work in the 1950s was out of the question. I myself was prepared to work only part-time, and I could hardly expect more of others. The challenge, therefore, was to create structures that would attract more good scientists into defense work on a broad range of problems without asking them to sacrifice their commitment to pure research. Over a three-year span, 1957–1960, I worked doggedly to stimulate interest in creating the kinds of national defense labs that could succeed in the climate of those times. With a shifting base of allies in the academic community and in the government, and a shifting set of target agencies that needed to be influenced, I had to be light on my feet. Ideas about how to accomplish the goal mutated and evolved rapidly. What came out of all that effort in the end was not a new set of defense labs (which at one point we had advocated) or even one new defense lab, but an advisory group called Jason, which I shall discuss presently. Jason continues to function even today as a vehicle for tapping the talents of some of the nation's very best scientists in the interests of national defense.

During or shortly before that period, other useful agencies came into being, to whose inception I made little, if any, contribution. These included the Institute of Defense Analyses (IDA), the Advanced Research Projects Agency (ARPA) — both creatures of the Department of Defense — and the President's Science Advisory Committee (PSAC). PSAC and ARPA were in large part reactions to *Sputnik*.

My first chance to influence science policy came not at home but abroad. In January 1957, Scoop Jackson, who had just been appointed to chair NATO's Standing Committee on Scientific and Technical Personnel, invited me to serve on an advisory body to this NATO committee. Perhaps because

I pitched into this assignment with zeal, I was soon asked to chair the advisory body. Among my fellow members were Edward Teller (who kept showing up in my life); James Killian, president of MIT and chair of PSAC; nuclear physicist Maria Mayer, whom I had known since my student days at Johns Hopkins; Richard Courant, the applied mathematician from NYU; J. Kenneth Mansfield, an executive at Combustion Engineering; and Ruben Mettler, a technical supervisor of weapons systems at Ramo-Wooldridge Corporation.

It seemed to me that the need to bring the most talented scientists into defense work was no less urgent in the other NATO countries than in the United States. Even prior to that need, as I saw it, was the need to attract top students into science and technology in the NATO countries, to increase the pool available for applied as well as basic research. This thinking led me to propose to my fellow advisors—and, informally, to Senator Jackson—the idea of a NATO Center for Advanced Science and Technology. When I ran this idea up the flagpole, to use some slang of that time, not many people saluted. Maria Mayer was keen on creating a new organization to unite European science (she called it "an Academy"), but most of the others whom I consulted, on and off the advisory group, preferred more modest measures.

In our final report to Jackson's committee, we recommended graduate fellowships in science, prizes for promising students and their teachers in both high school and college, and a program of NATO summer institutes in science. These recommendations reached the committee in September 1957, shortly before *Sputnik*. I do not recall what, if anything, happened to our recommendation about prizes for students and teachers, but the other two recommendations were accepted and gave rise to programs that continued for many years. NATO fellowships have supported not only young European scientists but also Americans doing postdoctoral work in Europe. NATO summer institutes have provided some of the most stimulating, productive international conferences, drawing together the best talent from many countries in various specialties. The effect of these programs on military preparedness is at best indirect, but they have surely encouraged internationalism in science.

For me, *Sputnik* confirmed the reality of a missile gap, and also confirmed the threat that the Soviet Union posed to world peace. At the same time, it gave me and others who thought as I did a much more receptive audience in Washington. I began pushing on two fronts: to strengthen science education in America, and to create one or more laboratories devoted to working on a broad range of defense problems. Yet, being still a relative outsider in Washington, I wasn't sure where to push for best results. I had contact with Senator Jackson and two of his top aides, William Borden and Dorothy Fosdick. Jackson was at that time the most influential Senator on defense matters. I had come to know Kenneth Mansfield, Albert Hill, and James McCormack, all associated with IDA. Once PSAC was formed, I had a channel to the White House through PSAC members such as James Killian and I. I. Rabi. I adopted the principle that if one pushes in enough different places, some will turn out to be the right places.

As I turned my attention from Europe back to America in the fall of 1957, my first focus was on science education. Less than two weeks after *Sputnik*, I wrote to William Wright, head of the Nuclear Physics Division of the Office of Naval Research (ONR), "Don't you think the attached program for NATO countries ought to be only a sampler of a much larger science talent training program for the U.S.A.?" I picked ONR as a place to advocate this idea because, since World War II, it had been a principal supporter of basic research and education in science. In more colorful language to Senator Jackson a couple of days later, I suggested that his "atomic" level of science training in NATO countries should be supplemented by a "thermonuclear" level of science-technology-management training in America.

Ideas as ambitious as Oswald Veblen's November 1957 call for a "national university for advanced study" found insufficient support to go anywhere. Veblen, a professor at the Institute for Advanced Study, may have been influenced by the European model of universities supported by central governments. (George Washington had had no better luck in 1794–1796, even though he was willing to endow the national university himself.)[3] Yet numerous government programs to strengthen advanced education in science did in fact come into being—notably doctoral and postdoctoral fellowships and increased support for basic research. These needs were "in the air," perceived by many. My influence, if any, in bringing them about was slight.

By December 1957, I began to change the focus of my thinking on science in the national interest. The increased support for education and basic research that I had been advocating was important—indeed critical to the health of the nation in the long term. But in the short term, I became convinced, more dramatic steps were needed to overcome the terrifying Soviet lead in missiles and other military technology that I perceived. On December 11, I sent a long telegram to Killian, Rabi, and other members of PSAC endorsing the proposed but not-yet-approved ARPA (Advanced Research Projects Agency) and urging them to go beyond ARPA to create a national advanced research projects laboratory. I visualized such a laboratory—actually an interconnected set of laboratories—being linked to the Department of Defense in the same way that the Los Alamos and Livermore Laboratories

[3] Washington had the support of John Adams and Thomas Jefferson, and the idea was later championed by James Monroe and John Quincy Adams, all to no avail.

were linked to the Atomic Energy Commission. In the telegram, I referred to a "whole campus of laboratories," including a "program initiation laboratory," and added: "Without some centralizing of research like this, I don't see where we are going to get the scientific manpower to do what has to be done."

What made me think I could move giant bureaucracies? I simply felt I had to try. I was convinced that we faced a crisis in national security, and that something had to be done. I had read enough history to know that change does not occur without the strong push of a stubborn advocate, someone willing to think big. I had learned in my limited experience in the corridors of power (and in watching the performance of Edward Teller and other scientists in those same corridors) that scientific achievement carried more weight in Washington than one might suppose. It was not only my fellow scientists on PSAC who were willing to listen. So were politicians, government managers, and business executives. They seemed to regard me as a solid scientist (in truth, they would probably say brilliant scientist) but with a practical head that was screwed on right. I have often reflected that if I had chosen a career in public service with the goal of reaching and influencing people at the highest levels of government, I would very likely have had less impact than I did by pursuing a career in pure science with, at first, no thought of such a goal.

What made me think my ideas were right? Since boyhood, I had not been short on confidence. It is, after all, a necessary condition to succeed in science. And whatever successes I had had, in applied as well as basic work, no doubt reinforced that confidence. It helped that I had the support of numerous scientists whom I respected. Even though mine was a minority opinion among scientists, it was not a lone voice. I also turned to history for support. The progressive movers in history, the ones whose actions were applauded by later generations, were willing to step aside from the main stream and push for their minority points of view.

I followed up my first message to PSAC members with a more detailed proposal, which I sent to James McCormack, president of IDA. I also sent it for review to various physicists around the country. Marvin Stern at Convair and Kenneth Watson at the University of Wisconsin were among those who sent back supporting comments. Watson, thirty-six, had the wiry build of a basketball player and the sardonic wit common among mathematicians and theoretical physicists. He had already gained a reputation for his work in elementary-particle physics. Later he moved to the University of California, Berkeley, where he made important contributions to defense research, both as a Jason regular and in the aerospace industry. I never knew to what extent Watson was motivated by patriotism and to what extent he simply found defense problems challenging and "fun" as a counterpoint to his basic research.

Another young theorist who took seriously a responsibility to serve the nation was my Princeton colleague Marvin (Murph) Goldberger. We had not only our Princeton link in common. We had both been taught (at different times) by some of the same teachers at Rayen High School in Youngstown, Ohio. Goldberger was a contemporary of Ken Watson. Together, they later wrote a book on collision theory. Goldberger and his attractive, scholarly wife, Mildred, enjoyed good food, good wine, and good company in much the way that John von Neumann did. But I had much less resonance with Goldberger than with von Neumann—even though Goldberger supported my ideas on defense laboratories. Despite his brilliance as a physicist, Goldberger reminded me sometimes of a business executive impatient with scientists who seem to be groping indecisively as they search for answers. When he later chaired the Princeton Physics Department, he caused me a few sleepless nights because we differed in our thinking about the development of the department. As one always striving for the smoothest relationships with those around me, I suffer when those relationships get bumpy. Later Goldberger became president of the California Institute of Technology (Caltech) and, after his retirement from that position in 1987, returned to Princeton to fill a position once held by Oppenheimer, director of the Institute for Advanced Study.

In mid-January 1958, the month after my first defense-lab proposal, Senator Jackson scheduled an informal meeting in Princeton with a set of scientific advisors from academia, industry, and the government. I was invited, along with my Princeton colleagues Oskar Morgenstern, Eugene Wigner, and Lyman Spitzer. (I will say more about Morgenstern and Wigner shortly.) Among those attending from elsewhere were Marvin Stern (Convair, San Diego), Harvey Brooks (Division of Engineering and Applied Science, Harvard), Arthur Kantrowitz (AVCO, a defense contractor in Massachusetts), and Alvin Weinberg (director of the Oak Ridge National Laboratory). Oddly, Edward Teller was not among those in attendance. I don't know if he was invited.

At the last minute, illness prevented Senator Jackson from attending. He was represented by his aide Dorothy Fosdick, a knowledgeable, sharp-witted woman with whom I always enjoyed working. Her abilities were recognized outside of Washington. She had turned down four college presidencies to stick by Senator Jackson. The daughter of the famous minister Harry Emerson Fosdick, Dorothy had tried lobstering in Maine during the summers of her college years. The tight-knit fraternity of Maine lobstermen did not welcome her, and she frequently found the lines to her lobster pots cut—until she saved the life of a fellow lobsterman who was dragged overboard by a rope caught on his leg. Dorothy told me that she once took her father to hear the sermon of a minister in Washington whom she greatly admired. When the sermon started, she and her father looked at each other. They were listening to one of Harry Emerson Fosdick's own sermons.

Despite Senator Jackson's absence, the Princeton meeting was long and productive. The group had ample time to discuss my considerations about a defense lab or cluster of such labs. Following the meeting, half a dozen of the participants sent me their detailed thoughts on this subject. Most of them were supportive, but considered my grandest plan too grand.

Morgenstern and Wigner resonated most strongly with my thinking. Since they were in Princeton, it was convenient for us to join forces. For more than a year, we worked together trying to achieve a major advance in the way that defense research was carried out, an advance that would escalate the level of effort and also make it easier to bring the brightest young scientists into the work. I will leave the details out of this narrative, because our efforts involved so many people and so many agencies, and had so many twists and turns—not to mention many climate changes as the reactions to our ideas

Oskar Morgenstern on the grounds of the
Institute for Advanced Study, 1950.
(Photograph by Dorothy Morgenstern, courtesy of the
Archives of the Institute for Advanced Study.)

varied from warm to cool to warm again.[4] Having a sabbatical leave in the spring of 1958 enabled me to pursue this campaign more vigorously than I otherwise could have.

Morgenstern was a mathematical economist who had coauthored with John von Neumann the famous book *Theory of Games and Economic Behavior*. He was an Austrian who, when I knew him, still felt more comfortable speaking and writing in German than in English—although his English was nearly flawless. He loved being on boards advisory to the military, and took great pleasure in relating insider stories about his experiences in Washington. Ahead of his time as an academic, he formed his own consulting company in Princeton. Only when he lay dying of cancer in 1977 did I learn that he was the grandson of Kaiser Friedrich III of Germany. When I came down the stairs in his Princeton home after my last visit with him in his upstairs bedroom, I saw on the wall a picture of a much bemedaled figure. "Who is that?" I asked his wife, Dorothy. "That's Oscar's grandfather, Kaiser Friedrich III," she answered. "In his youth, the Kaiser had an affair. A respectable marriage to someone else was arranged for the young woman, whose daughter grew up to become Oscar's mother."

Eugene Wigner has entered this narrative earlier. He had a giant intellect, and he shared with his fellow Hungarian Edward Teller a fear and hatred of Communism. Wigner was polite to a fault but could harbor dark thoughts about people. Finding himself in a minority of scientists in his views on national defense and the Soviet threat, he was more sensitive to criticism and more hurt by it than I was. Both Morgenstern and Wigner had many high-level contacts in the government and in industry—as well as in the academic world—but I think they liked having me, as a native American, out front and visible as we worked together to secure an intensified commitment to defense research.

Morgenstern took it as a given that since I believed so strongly in the idea of a new defense laboratory and worked so hard to secure it, I should be willing to head it. When he came to realize that I really meant it when I said that I was not interested in that job, he became irritated with me. Wigner showed no such irritation, making clear his own disinclination for research management also. In the end, all three of us—Morgenstern, Wigner, and I—pulled back from any full-time commitment of more than a limited duration to a new lab. Various other qualified people—Watson and Goldberger, for instance—also turned away feelers about their interest in heading it. As the

[4] The historian of science Finn Aaserud has made a detailed study of this period. His article "Sputnik and the 'Princeton Three': The National Security Lab that Was Not to Be" was published in *Historical Studies in the Physical and Biological Sciences*, vol. 25, part 2, 1995, pp. 185–239. Aaserud is at the Niels Bohr Archive in Copenhagen.

new lab finally hung in the balance in 1959, my final decision to stay clear of the top job may be what tipped the lab into oblivion. Realizing that the final success of all our efforts might hinge on my willingness to accept the lab's directorship, I stewed in an agony of indecision. I remember saying to Janette one evening, "I just wish I would get hit by a truck and break some bones so the decision would be made for me." In the end, my commitment to pure research—and the knowledge that I was much better suited for research than management—prevailed.

But before that moment arrived (in the spring of 1959), we created a pilot operation in the summer of 1958 to see what good could come from having a group of leading scientists (who turned out to be mainly physicists) range freely without restriction over problems of potential military significance, after being briefed by insiders. The idea for such a summer study came from Al Hill of MIT and had the backing of the president's Science Advisor, James Killian. They accepted my proposed name—Project 137—and asked me to head the activity. (I'll explain the name shortly.)

Twenty-two participants in Project 137 convened in Washington in July and worked for three weeks. Wigner, at the last moment, was unable to take part. Morgenstern's participation seemed a bit half-hearted. However, it was a sterling group and included some of the best of the crop of physicists still in their mid-thirties: Goldberger, Watson, Stern, Val Fitch, and Sam Treiman. Fitch, then at Columbia, joined me at Princeton later. Treiman was already a Princeton faculty member in 1958 and is now my frequent luncheon companion. We share a conviction that much of the best research comes out of teaching, and much of the best teaching, out of research. Also part of the group was Fred Reines, talented in both theory and experiment, a bear of a man given to thinking big about nearly impossible problems as he paced up and down in his oversize shoes. Just two years before Project 137, he and Clyde Cowan had been the first to identify the neutrino experimentally—an achievement finally recognized with a Nobel Prize in 1995.

I had better describe the meaning of the summer study's name, since it was my suggestion. Some of the numbers that physicists deal with have magnitudes that depend on the units chosen. The speed of light, for example, is 300,000 kilometers per second and also 186,000 miles per second. The mass of a proton is 1.007 atomic mass units and also 938 million electron volts. But some numbers, called dimensionless numbers, have the same numerical value no matter what units of measurement are chosen. Probably the most famous of these is the "fine-structure constant," which is the square of the charge of the electron divided by the product of the quantum constant and the speed of light. In symbols, it is $\frac{e^2}{\hbar c}$, and numerically it is approximately the inverse of 137 (more exactly, of 137.036).

Physicists love this number not just because it is dimensionless, but also because it is a combination of three fundamental constants of nature. Why do these constants come together to make the particular number 1/137.036 and not some other number? No one knows. The fine-structure constant continues to allure physicists.[5]

You can see from the following list that the problems addressed by Project 137 were wide ranging (and did not include nuclear weapons): chemical sensing (à la butterflies and other insects), fuel supply for troops in limited warfare, information processing in the field, wireless transmission of power, generation of intense infrared radiation, undersea beacons, and worldwide communications to and from submerged submarines. Some of these were problems that teams in conventional labs might have tackled and solved; some required new thinking and might not have arisen in conventional settings.

Coming up with intelligent conclusions about all these topics in only three weeks was a tall order, but the marvelously talented participants rose to the challenge. When I reported orally on the project's work in August to Secretary of Defense Neil McElroy and other government officials, some of my colleagues were more nervous than I was. In fact, many interesting conclusions had been reached, and my report was well received. McElroy began to see how the free investigation of a group of scientists could strike some sparks and be useful to his department. He wanted to see Project 137 transformed into something more permanent. In essence, he became a supporter of the defense initiation lab that Morgenstern, Wigner, and I had been advocating. At the same time we gained another supporter, Herbert York. York, a protégé of Teller and former director of the Livermore laboratory, had recently been named director of ARPA and was soon to become director of Defense Research and Engineering. I recommended York for the directorship of the new defense laboratory. He was approached, but, like me, declined. For lack of a director—Goldberger, Wigner, York, and I were not the only ones approached—the lab idea finally shriveled.

Years later, York, by then chancellor of the University of California, San Diego, had second thoughts about his participation in defense work, and wrote several books critical of defense projects, including the H-bomb project. He was a man of principle, and his vision changed. I believe that distancing himself ideologically from Edward Teller, his close friend and patron, was difficult and painful for him.

From the ashes of the "national security lab that was not to be" (to quote

---

[5] Wolfgang Pauli was reportedly desolated to be put into Room 136 at the Vienna Hospital for his last illness. He arranged to be moved, a few days before his death, to Room 137.

Carving in a Princeton sidewalk, which I made (with the
construction superintendent's permission) c. 1993.
The photograph is from 1995.
(Photograph by Robert Matthews, courtesy of Princeton University.)

Aaserud—see footnote 4 on page 280) and from the embers of Project 137
arose, two years later, a structure that has proved to be quite robust and of continuing value to the Defense Department and other parts of the U.S. government. It is Jason, a group that conducts technical studies, mainly summer
studies. Jason might best be described as a club. On the one hand, Jason
membership operates much as does membership in an exclusive country
club. Once you are in, you are very likely to stay in. To get in, you must be
invited by those who are already members. As a mode of providing scientific
and technical advice to the government, this system has the drawbacks of
any "old boy's network." Some good people, especially women, can be over-

looked. On the other hand, Jason members are a highly talented group, and have produced analyses and recommendations that would not likely have emerged from larger, more democratically chosen teams. In other words, those who are in deserve to be in. Some of those who are out also deserve to be in.

The origin of the name *Jason* is not so clear as the origin of the name 137. According to Murph Goldberger, who would have reason to know, the name was suggested by his wife, Mildred, to symbolize the brave young men of this era searching for the modern equivalent of the Golden Fleece. Mildred was right that the number of brave young women engaged in the search was distressingly small.

In the interest of full disclosure, I must relate how I was personally reprimanded at the specific insistence of President Eisenhower. It happened in January 1953, and fortunately did not result in my suffering the fate that befell Robert Oppenheimer in this period. For all I know, the incident may have helped me. By making me a bit notorious in government circles, it perhaps increased my effectiveness as I pressed my views on official Washington in the following years. (As entertainers say, any publicity is good publicity.)

Following the successful Mike shot at Eniwetok in late 1952, it became clear that the isotope lithium 6 was going to be a critical part of the fuel for future thermonuclear weapons. Lithium 6 makes up only 7.5 percent of naturally occurring lithium. Separating it from the more abundant isotope, lithium 7, is nowhere near as difficult as separating uranium 235 from uranium 238, yet it requires an industrial-scale operation, which, at the time, Oak Ridge National Laboratory was prepared to undertake. Nothing happens without a pusher, however, and in this case the pusher was General James McCormack, director of the Division of Military Applications of the AEC. He had the backing of Senator Jackson, then chair of the Joint Congressional Committee on Atomic Energy. Jackson's aide William Borden was also actively interested. I kept in touch with McCormack, Jackson, and Borden, and strongly supported the need to get the production of massive amounts of lithium 6 on track.

I believe it was in 1952 that Bill Borden, having heard about Project Matterhorn, came up to Princeton from Washington to visit me. He asked me what we needed in order to move still faster in thermonuclear weapons development. I answered "lithium 6." That triggered Borden's focus on this problem, and also led to my getting acquainted with Senator Jackson and Dorothy Fosdick.

Borden was like a new dog on the block who barked louder and bit harder than the old dogs. Wherever he looked, he saw conspiracies to slow down or derail weapons development in the United States. When it seemed that planning for lithium 6 production was going too slowly, he imagined a traitor somewhere slowing it down. Oppenheimer was one of the people he least trusted, and the next year (on November 7, 1953), after he had left Washington to join Westinghouse Electric Corporation in Pittsburgh, he wrote a letter to FBI director J. Edgar Hoover that was eventually to block his ambition to return to Washington. In that letter he declared that "more probably than not J. Robert Oppenheimer is an agent of the Soviet Union." At the time of my security violation, however, Borden was still in Washington and the Oppenheimer hearing lay in the future. (That hearing occurred in April and May of 1954, and resulted in Oppenheimer's loss of security clearance.)

One evening in early January 1953, I boarded an overnight train in Trenton, New Jersey (a stop near Princeton), en route to Washington. (The trip takes much less time than is needed for a good night's sleep, but the sleeping car sits in Washington's Union Station until its passengers are ready to disembark around 8:00 A.M.) I had with me, in violation of standard regulations, at least one secret document that I wanted to review on the way to Washington. It was part of a report prepared by John Walker, a Congressional staff member and associate of Bill Borden, covering the questions of why lithium 6 was important, how much might be needed, and how it might be produced. According to what I have read elsewhere (but can't personally recall now), Walker's full ninety-two-page report contained H-bomb design information and addressed the question of what thermonuclear information Klaus Fuchs might have been able to give to the Soviets. I studied the excerpt of the Walker report in my berth on the train, and arrived in Washington refreshed and primed to discuss issues related to hydrogen weapons development.

When I pulled my things together in the morning and prepared to get off the train, I could not find the report on lithium 6. I looked everywhere—in my berth, under my berth, on the floor, in the lavatory. I enlisted the aid of the porter and spoke to several passengers. The report was nowhere to be found. The idea that lithium 6 was to be a key ingredient of future H-bombs was highly classified at the time. We learned later that Soviet scientists had already arrived at the same conclusions as we and that they used lithium 6 in their first thermonuclear test in 1953. But at the time of my train trip to Washington, earlier in 1953, we didn't know where the Soviets stood in their H-bomb development. I was more than a little distressed.

I went into Union Station and called Bill Borden at home. He immediately got the security people to work. The car in which I had ridden was disconnected from the rest of the train and was rolled to a siding a few blocks away, where agents scoured it looking for the document. They didn't find it, and it never turned up later either. I am told that agents even scanned the train's roadbed from Trenton to Washington, to no avail.

Word of this serious security violation reached President Eisenhower soon after his inauguration, and he blew up. He summoned the five Atomic Energy Commissioners to the Oval Office and kept them standing while he lectured them. "You can't allow these security violations to take place!" he thundered (according to my friend Harry Smyth, who was there). He insisted that I be given an official reprimand, and I was.

It is interesting, even now, to wonder whether my document was purloined by a Soviet agent. It could hardly have vanished into thin air. If Soviet scientists saw it, they would have learned only that their side and our side were thinking along very much the same lines (and perhaps that, on our side as well as theirs, it takes more than a recommendation by scientists to make something happen).

About half of my life was behind me when I first ventured into sociology. At a 1958 conference, "Neighborhood Goals in a Rapidly Changing World," I spoke on "National Survival and Human Development." My remarks were not far removed from my concern at that time about the Soviet threat and the prospect of another war. I emphasized the importance of developing the full capacities of all citizens, both in their own interest as individuals and in the national interest. I spoke also of the mutual benefit that comes from links to less developed parts of the world. Fortunately, pontificating on the state of humankind did not become habit-forming. Although I later did considerable speaking and writing outside of pure research, I stayed generally close to science and scientists.

My consulting work for Convair culminated in 1959 with the preparation of a long report entitled "A Doctrine for Limited War." In preparing the report, Marvin Stern, Chalmers Sherwin, Henry Kissinger, Oskar Morgenstern, and I got together in southern California for a day and a half every month and a half for a year and a half. Kissinger wrote the first draft of this report, and I did most of the writing to pull it into final form. During the course of our work, I came to admire Kissinger greatly for his sense of history and the incisive clarity of his thought. My own thinking, although based on far less study and practical experience in international affairs, meshed closely with Kissinger's. I still remember the gusto with which he quoted what the young Otto von Bismarck reportedly told his fiancée, "We must mount the wild horses of life and go wherever they take us."

To write on limited war, we had to define the very concept of "limited war." It is a conflict in which neither party considers its national interest to be at stake. By calling its troops in the Korean War "volunteers," for example, the Chinese signaled that they did not consider the existence of China to be at stake.

In our report, we took the position that it is foolish to rule out the use of nuclear weapons in all circumstances. To do so rendered them useless as instruments of diplomacy. We advocated the design and building of small, low-yield nuclear weapons for battlefield use, and a policy that would permit their use in special circumstances. It seemed to us that the two companion goals — to avoid war wherever possible, and to minimize death and destruction when war did occur — could best be achieved if the United States and its allies were prepared to employ nuclear as well as conventional weapons in limited war. It seemed likely to us that limited wars would be common in the second half of the twentieth century, and indeed they have been. Whether their number or destructiveness might have been less with a policy that permitted the use of small nuclear weapons remains a matter of debate.

Convair, the sponsor or our report on a doctrine for limited war, developed a case of butterflies in its stomach once the report was finished in the fall of 1959. Its executives had no qualms about actually making instruments of war. That was a major part of the company's business. But they feared the "merchant of death" label that might follow the release of a report on such a volatile subject as the use of tactical nuclear weapons in limited war. The report was never issued. Most of its content reached the public, however, through Kissinger's later writings — no doubt a more effective avenue than a Convair report would have provided.

Now, in my older years, as the time ahead shortens and my brain chugs along more slowly, I try to focus on the deepest questions. Every morning, whether I am surveying the ocean from the edge of the cliff at our summer retreat in High Island, Maine, or riding the bus from our home in Hightstown, New Jersey, to my Princeton office, I ask myself: How come the quantum? How come the universe? How come existence? With luck, perhaps I will contribute a little to the answers to these questions in what remains of my life. Not all of my colleagues consider these questions quite respectable. But if they are not respectable in the twentieth century, they will be in the twenty-first. They are questions that should be engaging the next generations of physicists. They are questions that should fall — will fall — within physics and will be answered as matters of science, not of philosophy or theology — or speculation.

In the years just following my Leiden professorship, I was not so focused. Then, as now, I was driven to understand the universe, driven to understand how quantum theory and spacetime geometry touch one another. But I had practical obligations to my students; their careers could not await answers to the deepest questions. I had obligations of citizenship to my country, which, in my view, was threatened. And I had obligations of duty to the public that ultimately provided the resources making possible my way of life as a scientist and professor. As I look back, I wonder at the range of activities I undertook as I tried to meet all these obligations. Perhaps I was too fragmented. Perhaps I was running too hard. But it was not in my nature to slow down. Examples— my father's and his father's before him—of selfless service are embedded somewhere in my soul.

13

·

THE BLACK HOLE

IN HIS classic 1913 paper on the quantum theory of the hydrogen atom, Niels Bohr introduced a fundamental principle that has come to be known as the "correspondence principle" (not to be confused with his later principle of complementarity). Most of the principles of physics are set forth in equations. But not the correspondence principle. The best we can do is describe it in words. And even the words have to be a little vague. The principle is nevertheless of the greatest significance in joining together the classical Newtonian world and the modern quantum world.

Every quantum system has some disjointedness, or granularity (what the mathematician calls discreteness). Its energy, or its momentum, or its angular momentum (a measure of the strength of rotation), or its electric charge can take on only certain distinct values, not the continuous range of possible values that would be true of a classical system. The correspondence principle says that a quantum system will follow approximately the laws of classical physics when the granularity of the quantum system is relatively small—that is, when only a small percentage change is required to go from one distinct quantum value of some quantity such as energy to the neighboring distinct quantum value. Moreover, the transition from quantum to classical behavior is gradual; there is no sharp dividing line. As the granularity becomes ever more insignificant by relative measure, the classical description of the quantum system becomes ever more accurate.

For example, when an electron in a hydrogen atom moves in a highly excited state of motion, the neighboring quantum states that are available to it differ from one another by only a small percentage in energy and a small percentage in distance from the nucleus. Here the electron's motion is well approximated by that of a solid speck of matter moving in a planetary orbit about a "sun" (the nucleus). By contrast, when the electron occupies one of the states of motion closest to the nucleus, where relatively large changes in energy and distance from the nucleus are required to reach the next available quantum state, the classical planetary description has practically no validity at all. The description of the electron then involves clouds of probability with no precise definition of the electron's speed or location.

The gradual transition from quantum to classical behavior is not like the change of color of mixed pigments as one color is added slowly until it dominates another. Rather, it is like the change of behavior of sugar as more crystals are added. For a few crystals, any change in the number of crystals represents a large percentage change (from four crystals to five, for instance, which is a 25 percent change). The sugar is then "quantized," or granular. It behaves as a collection of distinct pieces of matter. When the number of crystals is measured in the millions, adding or subtracting a crystal has practically no measurable effect. The sugar then behaves like a fluid. It can be poured. It flows. It takes on the dimensions of its container. Its granularity becomes unimportant. Yet we must remember that a barrel of sugar, for all its "continuous" behavior, is ultimately granular. And every physical system is ultimately a quantum system. Its classical behavior under certain conditions is only an approximation.

Going back as far as my student days, I had an interest in the correspondence principle and the fascinating transition from quantum to classical behavior. In 1958, I came back to this topic and invited my former student Ken Ford to work with me on the semiclassical analysis of scattering. Semiclassical analysis could equally well be called semiquantum analysis. It involves that ill-defined boundary between the quantum and classical worlds.

When a classical particle scatters, it approaches its target (such as an atom or a nucleus) along a well-defined track, changes direction, and flies away from the target along a different, but equally sharply defined track. It's as if a little girl shoots a dried pea through a pea shooter at the back of her brother's head and the pea bounces off. A quantum particle is a fuzzy ball of probability waves that may be larger than the target. It moves toward its target like a pillow thrown at the brother's head. Unlike the pillow, however, the quantum particle rebounds and flies away after interacting with the target. As the wavelengths get smaller or the target gets larger, the quantum fuzzball gets smaller relative to the target. This is the intermediate, or semiclassical zone, where the particle flying toward its target can be compared to a pin cushion hurled at the poor brother's head. The challenge for the physicist is to fashion a theory that takes advantage of both the quantum and classical aspects of scattering in this intermediate zone so that the results of the scattering can be used to figure out facts about the target.

Ken and I turned out three papers on semiclassical scattering. For one of them, we brought in two other colleagues to help. One was David Hill, who had completed his Ph.D. work with me a decade earlier on the mechanics of nuclear fission. The other was Masami Wakano, then my graduate student, later my colleague and coauthor on numerous papers. When I submitted our three papers to *Physical Review* for publication, I got a cool reception—not for the first time—from that journal's editor, Sam Goudsmit. He and George Uhlenbeck, codiscoverers of electron spin, arrived in America in the 1930s as part of the intellectual exodus from Europe. These two Dutchmen could hardly have been more different. Uhlenbeck, who settled at the University of Michigan, was tall, slender, friendly but stern, and conservative in his dress and manner. His lectures were polished masterpieces. Goudsmit, who joined Brookhaven National Laboratory on Long Island, was shorter and rounder, a bon vivant whose shirttail had trouble staying tucked in. He loved women and wine. His lectures, likely to be disorganized, were masterful in their own way. One of the great surprises for most physicists is to learn that Goudsmit and Uhlenbeck did *not* win the Nobel Prize. Most of us think they should have.

Goudsmit had a simple goal for *Physical Review*: to make it the best physics journal in the world. He succeeded. *Phys Rev*, as it is affectionately known throughout the world, is generally regarded as number one. But Goudsmit didn't like wordiness and he didn't like pedagogy. I probably published less in *Physical Review* than most American physicists did, because I liked to be discursive and I liked to teach. Many of my papers appeared in another journal of the American Physical Society, *Reviews of Modern Physics*. The papers on semiclassical scattering Goudsmit found to be both too long and too pedagogical. Instead of abbreviating them, as he suggested, I submitted them to *Annals of Physics*, whose editor, Phil Morse, had been receptive in the past. There they appeared, unedited, in 1959. These papers turned out to strike a chord of utility among atomic, molecular, and nuclear physicists. Surely not my most significant papers, they nevertheless elicited more reprint requests and gained more citations in other people's work than almost any other work I have done.

The correspondence principle, which we harnessed in the semiclassical scattering papers, is striking because it is both utilitarian and fundamental. Why utilitarian? Because at the fuzzy boundary between the quantum and classical worlds, one must develop techniques to deal with systems that lack the simplicity of a purely quantum system and the clean predictability of a purely classical system. Our work, and other work like it, was directed at finding practical methods to analyze problems in this twilight zone. It turns out that many phenomena of interest, such as one molecule striking another or one large nucleus striking another (or even a massless wave striking a black hole) lie in this part-quantum, part-classical domain.

And why fundamental? Because the quantum-classical boundary separates two worlds that seem to have nothing to do with one another. The quantum world is counterintuitive. It substitutes granularity for continuity, probability for certainty. It deals with waves that are particles and particles that are waves. It replaces fixed locations and fixed speeds with miasmas of uncertainty. The classical world, by contrast , is perfectly intuitive. It is the world we see around us. It is consistent with common sense because it is the world in which our common sense was nurtured. Yet these two worlds are linked. They must be linked. Looked at closely enough, the classical world around us is nothing but a collection of quantum systems. How, then, can a boundary be defined? Where is it? That question still lacks a deep and satisfying answer. We know only that in certain limits, quantum behavior gives way to classical behavior.

It is hardly possible for someone interested in nuclear physics and relativity, as I was and am, not to get interested in stars. My Princeton friend Martin Schwarzschild drew me into stellar atmospheres. From there it was natural to fall, so to speak, into the center of stars. Martin, his fellow astrophysicist Lyman Spitzer, the Princeton statistician John Tukey, and I met for lunch regularly in the 1950s. The "Chowder and Marching Society," Tukey called our luncheon meetings, at which we shared ideas on science and the state of the world.

John Tukey, like John von Neumann, was a bouncy and beefy extrovert, with interests and skills in physics and astronomy as well as mathematics. He was also an organizer of the Princeton folk-dance group. As a child in New Bedford, Massachusetts, he had been educated at home by his mother. Brown University was the first school he attended. Tukey had contributed to Project Matterhorn, as had Schwarzschild. Spitzer was, of course, a principal leader in Matterhorn. In the course of his consulting work for Bell Labs, Tukey had coined the term "bit" for the smallest unit of information—a ticket to immortality. (Credit for byte, a mouthful of bits, usually eight, goes to Werner Buchholz, who reportedly coined the term in 1956.)

Back in 1946 or 1947, when Schwarzschild was still at Columbia University, he and I had done some work together on the escape of radiation from relatively cool stars. In the outer layers of such stars, some electrons find rest from their torment by attaching themselves to hydrogen atoms, creating thereby negatively charged atoms (ions), each consisting of a single proton and *two* electrons (instead of the usual one). Such atoms absorb radiation much more strongly than ordinary hydrogen does, and so inhibit the star from radiating away its energy. Just how they do so was a question I was prepared to address, because, in my Johns Hopkins thesis work, I had studied the absorption of radiation by helium, also an atom with two electrons. In the end, Schwarzschild and I were stopped from publishing our thinking about the outer reaches of cool stars by a bugaboo of that time, insufficient computing power. We laid the work aside. But at one of our Chowder and Marching Society luncheons, we came back to a discussion of the subject, and I found myself thinking about stars again.

In for a penny, in for a pound. Now I wanted to go to the star's center. What interested me was not the center of an ordinary star like our Sun, cooking away and generating thermonuclear energy. I was interested in the center of a cold, dead star. What is the final fate of a star after it has consumed all of its nuclear fuel and can burn no longer? Does it blow itself apart? Does it condense to a nugget of dense matter, and if so, how small and how dense? Does it collapse to a black hole, and is there a singularity at the center of the black hole?

Others had thought about these questions before me. Back in the 1930s, Fritz Zwicky of Caltech had postulated neutron stars; Subrahmanyan Chandrasekhar, then at Cambridge University, had calculated the largest mass of a white dwarf that could withstand the crushing imperative of gravity; and Robert Oppenheimer with George Volkoff and Hartland Snyder at Berkeley had discussed the collapse of heavy stars. But much more remained to be learned about the ultimate fate of stars of all sizes. What they do depends on whether they are small, medium, large, or extra large. I set some of my good students to work on questions of stellar death, and they came up with answers. By 1958, Kent Harrison, Masami Wakano, and I had learned something about three conceivable ultimate fates of burned-out stars. A star like our own Sun, we reckoned, would shrink to become a white dwarf. There is nothing extreme about a white dwarf. It is small, but not terribly small; dense, but not terribly dense. Spacetime is "flat" within it (that is, relativity plays no special role), and the atoms in its core remain atoms.

A second fate awaits more massive stars. Stirling Colgate, an energetic weapons scientist turned astrophysicist, used the increasingly powerful computers becoming available in the late 1950s at Livermore Laboratory to trace the final stages of a star appreciably more massive than the Sun. His calculations showed that such a star becomes a supernova. Its calculated properties matched observed properties of supernovas that astronomers had seen from time to time in other galaxies. (In July 1054, Chinese astronomers had seen an outburst that was, in fact, a supernova in our own galaxy. Its residue, the Crab Nebula, remains an object of great interest.)

A supernova is a star that both collapses and explodes. The energy released by its collapse powers the explosion of its outer layers. Left behind, according to theory, is an object far smaller and denser than a white dwarf. It is called a neutron star because it is literally one giant nucleus made up of neutrons. Its atoms have been crushed out of existence, with their electrons and protons forced by the extreme gravity to unite, becoming neutrons. What Wakano's calculations showed is that the neutron star, within a certain range of size and density, is a stable object. Like the much larger white dwarf, it is a possible end state of a burned-out star. The white dwarf and the neutron star last indefinitely, unable to release more energy and not inclined to shrink further. A stable neutron star, we calculated, would have a diameter of only about 15 miles and a density a million billion times that of our Sun. Its mass would be comparable to that of the Sun, at most about 2 solar masses.

When we reported this work in 1958, the neutron star was still a theoretical object. It would be another decade before Antony Hewish and Jocelyn Bell in England would report the discovery of pulsars—quickly identified as rotating neutron stars. Then we could bolster our belief in these astronomical oddities with real observational evidence. Today we know that at the center of the Crab Nebula, exactly where the sky brightened so dramatically in July 1054, sits a neutron star, blinking at us 30 times per second as it spins on its axis.

The third possible fate of a dead star is the most dramatic. We are convinced now that if a star is massive enough, it will neither contract gently to a final state as a white dwarf nor transform itself into a neutron-star nugget after exploding as a supernova. It will collapse to a point of infinite or near-infinite density from which neither light nor anything else can escape. Only its gravitational aura will remain.

For some years this idea of collapse to what we now call a black hole went against my grain. I just didn't like it. I tried my hardest to find a way out, to avoid compulsory implosion of great masses. At first I thought that a collapsing star might radiate so much light and throw off so much matter in the early stages of its contraction that its mass would shrink below the value needed to keep the collapse going. It would stop contracting as it became a white dwarf or neutron star. Before long we learned that this mechanism won't save the day for a star of great mass. So I looked to elementary-particle interactions for salvation. Is it possible, I asked, that repulsive forces between particles can prevent the particles from being squeezed to densities much greater than the density of an atomic nucleus? In due course we learned that this mechanism won't save the day either. The enormous pressure that builds within collapsing matter actually adds to the mass of the material and contributes more to the gravity that is driving the collapse inward than to the outward forces that would stop the collapse.

At this point in the evolution of my thinking (the early 1960s), I realized that nothing could prevent a large-enough chunk of cold matter from collapsing to a dimension smaller than the "Schwarzschild radius." Within a sphere of this radius, everything, including light, is trapped. The surface of this sphere defines what we call the "horizon," the boundary of a black hole. Yet I could still entertain the idea that within the black hole, "new physics" stops the collapse short of a mathematical point of infinite density. In 1964, Roger Penrose offered a powerful theorem showing that such new physics would have to be new indeed, for his theorem establishes that, for just about any description of matter that anyone has imagined, a singularity must sit at the center of a black hole.

Where I stand now is to imagine that the yet-to-be-discovered true blending of the quantum with general relativity will indeed provide that new physics, operating at the "Planck scale," the incredibly small dimension where quantum foam makes its appearance, and that the core of a black hole will prove to have some structure, albeit tiny beyond all imagining.

Early in my study of relativity, I reached an incorrect conclusion, that if light shines exactly straight up, perpendicular to the surface of a collapsing star, it can always escape, carrying away energy (and mass), no matter how strong the gravity. Some work of David Finkelstein and Charlie Misner in 1959 implied that strong-enough gravity can trap all light, but several years passed before I had convinced myself in my own way that, indeed, sufficiently strong gravity closes every escape hatch. Stimulus to my thinking was provided by a pair of 1962 senior theses by two Princeton undergraduates, Allen Mills—the same Allen Mills who later detected the tri-electron (an entity made of a positron and two electrons)—and David Beckedorff. Beckedorff, working under Charlie Misner's guidance, brought new clarity to the old idea of Oppenheimer and Snyder (1939) that sufficiently heavy stars can collapse totally. Mills, working with me, studied the paths of light rays moving near immense concentrations of mass, paths bent this way and that like the tracks of meteorites approaching the earth. As so often happens, beginners slice through the fog, letting one see clearly. All radiation of all kinds, I soon realized, can be fully trapped around a sufficiently concentrated mass.

Beckedorff, in his work, drew on a 1960 paper by Martin Kruskal that bore my fingerprints. One day in 1956 or 1957, just when I was getting deeply into relativity research, Kruskal, a mathematical physicist then working on plasma physics at Princeton's Project Matterhorn, told me about an idea he had come up with for overcoming a mathematical difficulty in the treatment of spacetime around a "Schwarzschild singularity." He demonstrated that nothing out of the ordinary occurs at what we now call the horizon of a black hole. Spacetime inside and outside the horizon mesh smoothly. I found Kruskal's simple but elegant discovery quite interesting, although I didn't fully comprehend its significance right away. It turned out to relate to wormholes tunneling between remote parts of spacetime. In 1959, since Kruskal had not published this work, I decided to see to its publication myself. I wrote a paper "by M. D. Kruskal" and submitted it to *Physical Review.* Only I neglected to tell Martin what I was up to. The following spring or summer, while he was on leave in Germany, he received—"out of the blue," as he puts it—galley proofs of the paper to review. He was mystified only briefly. Perhaps the large drawings that illustrated the text and the long captions that went with them gave away the source. Martin regained his composure quickly enough to suggest that we coauthor the paper, but since the important ideas were his, I demurred.

The work of Kruskal, Beckedorff, and Mills came together for my graduate student Robert Fuller and me when we showed in a 1962 paper that, as we put it, "there exist points in space-time from which light signals can never be received, no matter how long one waits." In other words, a collapsing star can trap light as well as matter. We also demonstrated in this paper that a signal moving more slowly than light can never make an "end run" through a wormhole to reach a destination sooner than light can get there. To see what this means, think of a band of outlaws who can take a short-cut through an arroyo to get from one point on a railroad track to another point, heading off a train even though they move more slowly than the train. If the train is a ray of light, it can't be done.

In the fall of 1967, Vittorio Canuto, administrative head of NASA's Goddard Institute for Space Studies at 2880 Broadway in New York City, invited me to a conference to consider possible interpretations of the exciting new evidence just arriving from England on pulsars. What were these pulsars? Vibrating white dwarfs? Rotating neutron stars? What?[1] In my talk, I argued that we should consider the possibility that at the center of a pulsar is a gravitationally completely collapsed object. I remarked that one couldn't keep saying "gravitationally completely collapsed object" over and over. One needed a shorter descriptive phrase. "How about black hole?" asked someone in the audience. I had been searching for just the right term for months, mulling it over in bed, in the bathtub, in my car, wherever I had quiet moments. Suddenly this name seemed exactly right. When I gave a more formal Sigma

---

[1] Jocelyn Bell, the British student who found the first evidence for pulsars in 1967, began to refer jokingly to the source of the pulses as LGMs, or little green men.

Xi–Phi Beta Kappa lecture in the West Ballroom of the New York Hilton a few weeks later, on December 29, 1967, I used the term, and then included it in the written version of the lecture published in the spring of 1968. (As it turned out, a pulsar is powered by "merely" a neutron star, not a black hole.)

I decided to be casual about the term "black hole," dropping it into the lecture and the written version as if it were an old familiar friend. Would it catch on? Indeed it did. By now every schoolchild has heard the term. Richard Feynman, when he saw the term, chided me. In his mind, it was suggestive. He accused me of being naughty. In fact, the name *black hole* has a lineage. That's why it caught my fancy. Since at least the 1890s, the term "black body" has been used in physics to describe an idealized body that absorbs all radiation that falls upon it, and emits radiation at the maximum rate possible for a given temperature. The black body is a perfect absorber and as perfect an emitter as it is possible to be. A black hole has one of these characteristics, but not the other. It absorbs everything that falls upon it. It emits nothing.[2] Thus *black hole* seems the ideal name for this entity. The geometry of spacetime near a black hole, funneling into ever greater curvature, adds to the appropriateness of the name.

Several years later, Feynman called my language unfit for polite company (well, actually, he called it obscene) when I tried to summarize the remarkable simplicity of a black hole by saying, "A black hole has no hair." I guess Dick Feynman and I had different images in mind. I was thinking of a room full of bald-pated people who were hard to identify individually because they showed no differences in hair length, style, or color. The black hole, it has turned out, shows only three characteristics to the outside world: Its mass, its electric charge (if any), and its angular momentum, or spin (if any). It lacks the "hair" that more conventional objects possess that give them their individuality. There is no way to tell from outside whether a black hole was created using neutrinos, or electrons and protons, or old grand pianos—or indeed whether any matter at all went into it. It could be nothing but a wrinkle in spacetime. No hair stylist can arrange for a black hole to have a certain color or shape. It is bald.

Since I first embarked on my study of general relativity, gravitational collapse has been for me the most compelling implication of the theory—indeed

---

[2] Stephen Hawking, the notable British astrophysicist who holds the chair at Cambridge University once held by Isaac Newton, demonstrated in 1974 that through subtle quantum effects, a black hole can actually emit some energy. Given enough time, Hawking's mechanism could allow a black hole to evaporate. Back in 1964, my solid, sensible students Kip Thorne and David Sharp talked me out of making this crazy suggestion myself in the book on gravitational collapse that Thorne and I were writing.

the most compelling idea in all of physics. At first, as described above, I resisted the idea, struggling against it in every way I knew how. When I finally
concluded that there *is* no way out—that gravitational collapse is an
inescapable implication of the theory, an inescapable fact of nature—I shifted my focus. No longer did I look for ways to avoid the fate that awaits any
great mass of cold matter. Rather, I looked for ways to exploit gravitational collapse. What could it teach us about the machinery that drives the universe?

It teaches us that space can be crumpled like a piece of paper into an infinitesimal dot, that time can be extinguished like a blown-out flame, and that
the laws of physics that we regard as "sacred," as immutable, are anything
but. I first surmised in the early 1960s that the content of a black hole—the
matter falling into it—could lose the characteristics that it had outside the
black hole. In the world of our laboratories, for instance, we have a law of
baryon conservation—conservation of the total number of protons plus neutrons plus certain other particles in that family. Their number remains always
the same, never shrinking, never growing. I surmised that this law would cease
to be valid for a black hole, that if a chair containing a certain number of
baryons fell into a black hole, the black hole would not gain that number of
baryons—indeed "number of baryons" would have no meaning for a black
hole. When the chair falls into the black hole, the number of baryons that
we can count in the universe decreases. Some cease to be. The law of baryon conservation fails. Other similar particle-number conservation laws also
cease to be valid. I prefer to say that these laws are transcended rather than
that they are violated.

One day in 1967, when I was on the brink of christening the black hole, a
friend at the Naval Research Laboratory near Washington, D.C., wrote to
me to tell me about a bright Greek boy named Demetrios Christodoulou,
whom Princeton ought to consider. Knowing that our department was flexible
in its admissions policies and would look favorably on anyone with talent, I
asked for and gained approval to check out this young man myself. Since I was
going to be in Paris for the spring semester 1968, I wrote to Christodoulou to
ask if he might be able to come to Paris for an interview. He agreed, showing
up in Paris in January soon after my arrival there. We worked out an arrangement for him to take a succession of oral exams over several days. To serve as
a fellow examiner I roped in a Paris colleague, Achille Papapetrou, himself a
Greek.

Demetrios was sixteen. He had not yet finished high school, but had studied so much physics and mathematics on his own that he was far beyond
anything his high school could offer, even beyond the courses that he had
tried at the University of Athens. We wanted to test him on more than book
learning, but could not give him a laboratory task. So we gave him a recent

copy of the journal *Physical Review* and asked him to report back in two days on an experiment detailed there, in which William Fairbank of Stanford measured whether matter and antimatter are attracted equally by gravity.

The results of the exams were more than favorable. I reported to my Princeton colleagues and recommended admission on a trial basis. A couple of days later Christodoulou and I were attending a seminar together when I was handed a cable affirming his admission. With apologies to the speaker—my friend Isaak Khalatnikov from Moscow—I interrupted the meeting to announce that Christodoulou would be attending Princeton. Demetrios says that he was too elated to pay much attention to the rest of Khalatnikov's talk. He went at once to Princeton and did so well in courses that spring that he was admitted as a regular student in the fall. His dissertation, completed when he was nineteen, opened an unexpected chapter in the story of the black hole. Now he is a leading contributor to black-hole physics.

The story of Christodoulou's dissertation begins with the British physicist Roger Penrose.[3] One Friday in 1968, during a long Underground (subway) ride from his home outside London and a walk from the Underground station in London to his office at Birkbeck College for his regular end-of-the-week get-together with his research group, Penrose, as he tells it, was mulling over what topic he might bring to the group to stimulate some good discussion— something "new and true." Reflecting on a rotating black hole, he asked himself if there were not some way to tap its rotational energy. Yes, he mused to himself, if I toss a planet (or other massive object) toward a black hole and the object breaks apart, with one part falling into the black hole and the other part zooming away, the departing piece may carry away more energy than the original object brought in. Robbed of some of its spin, the black hole loses some of its mass.

Christodoulou and I were fascinated by this "Penrose process." How efficient could it be made? How much mass could be extracted from a spinning black hole? Demetrios soon learned that hurling planets at a black hole would more likely increase the black hole's mass than decrease it, but that the black hole's mass *could* be decreased and, as suggested by Penrose, at the expense of the black hole's spin. In the best tradition of ancient Greece, Demetrios discovered a wonderfully simple relationship that could be laid out along the three sides of a right triangle. For a black hole with no electric charge, the hypotenuse of the triangle measures the mass of the black hole, the altitude

[3] When Penrose worked in Princeton as a NATO postdoctoral fellow in 1959 and 1960, I saw firsthand a direct payoff of the science program that my NATO committee had recommended and that the Third Annual Conference of NATO Parliamentarians had adopted in 1957.

Demetrios Christodoulou, 1971. The writing on the blackboard is mine
(see photograph in center insert), except that Demetrios has replaced
"THE MAGIC MOUNTAIN THOMAS MANN" with "PURGATORIO DANTE."
(Photograph by Robert Matthews, courtesy of Princeton University.)

measures the spin, and the base measures the "irreducible mass," which is the
smallest mass the black hole can have, the mass it ends up with after processes such as those suggested by Penrose have extracted all the spin.

For a black hole that has not only spin but also electric charge,
Christodoulou and Remo Ruffini, then a postdoc at Princeton, found that a
simple triangle is still meaningful, with the base of the triangle now measuring a combination of irreducible mass and charge. The irreducible mass
remains the rock-bottom mass of the black hole (not counting quantum
effects), the mass it possesses after all its charge and all its spin have been
reduced to zero. As long as the black hole has any charge or any spin, its
mass is greater than its irreducible mass. It is marvelous that such simple
Pythagorean geometry can display rules even of black-hole dynamics.

That's not the end of geometry that would have pleased the ancient Greeks.
If two balls of putty collide and stick together, the mass of the new, larger
ball is the sum of the masses of the balls that collided. Not so for black holes.
If two spinless, uncharged black holes collide and coalesce—and if they get
rid of as much energy as they possibly can in the form of gravitational waves as
they combine—the *square* of the mass of the new, heavier black hole is the
sum of the *squares* of the combining masses. That means that a right triangle

Roger Penrose with his son Christopher
in Austin, Texas, 1963.
(Courtesy of Roger Penrose.)

with sides scaled to measure the masses of two black holes has a hypotenuse
that measures the mass of the single black hole they form when they join.
Try to picture the incredible tumult of two black holes locked in each other's
embrace, each swallowing the other, both churning space and time with gravitational radiation. Then marvel that the simple rule of Pythagoras imposes its
order on this ultimate cosmic maelstrom.

When Glenn Seaborg, chairman of the Atomic Energy Commission, reached
me by phone in a Vienna hotel one evening in the summer of 1968 to tell me
that I had been selected to receive that year's Enrico Fermi Award, Janette,
hearing only my side of the conversation, thought I was being asked to serve on
yet another committee. "Say no, say no," she whispered. I joked about this later

Four generations of Wheelers with President Johnson, December 3, 1968.
My father, Joseph Wheeler, is at the left. My son, Jamie Wheeler, and his son
John McGehee Wheeler are at the right.
(*Courtesy of the White House.*)

to Seaborg, who told the president. On December 2, the anniversary date of
Fermi's self-sustaining chain reaction in Chicago, I found myself, with Janette
at my side, facing Lyndon Johnson in the White House. After some very kind
remarks about my achievements, he said, "Well, with all due respect to Mrs.
Wheeler, we're glad that you decided to say yes." Then, in obvious high spirits, he started to hand me the check but held onto his end of it. "If you don't
want this," he said, "I have two daughters who would be glad to have it."

I was allowed to invite members of my family and two friends to the ceremony. My mother, Mabel Archibald Wheeler, was unfortunately no longer
living. She had died in 1960, shortly after celebrating her fiftieth wedding
anniversary. My father, Joseph Wheeler, was still going strong (he died in
1970). He was present as a proud parent, along with Janette, our three children—Letitia, Jamie, and Alison—and two grandchildren. From among our
academic and professional friends, Janette and I found it impossible to select
only the two allowed by White House rules, so we turned to a different friend,
our longtime gardener, Harry Davis, inviting him and his wife, Sarah, to the
ceremony.

Among some two hundred guests present at the reception following the
award ceremony, I saw only two black faces, those of the Davises. As instruct-

ed by a book of etiquette, Mrs. Davis said simply, when she reached Lyndon Johnson in the receiving line, "How do you do, Mr. President." Nothing in the etiquette book prepared her for what followed, a long conversation with the president. Johnson turned on the charm, and Sarah Davis enjoyed the experience. When the Davises returned to their home in Pennington, New Jersey, neighbors asked them where they had been.

"In Washington," they answered.

"What were you doing there?"

"We were visiting the White House and we met the President." Not until pictures arrived from the White House two weeks later were the Davises able to convince their neighbors that they weren't spoofing.

Four years later, the Davises wrote to Lyndon Johnson at his ranch in Texas to thank him, not only for the invitation to the White House, but for all that he had done to promote civil rights. Johnson replied courteously, saying he remembered them well. A week later he was dead.

I was the twelfth recipient of the Fermi Award — not counting Fermi himself, who had been presented the first such award by the AEC in 1955, before his name was attached to it. Winners before me included those astonishing Hungarians von Neumann, Wigner, and Teller, who gave so much to their adopted country; the American-born titans of Berkeley, Lawrence and Seaborg; the German-born Bethe; as well as Hahn (posthumously), Meitner, and Strassmann, the discoverers of fission. Oppenheimer, brought low in the riptide of the McCarthy era, must have derived special satisfaction in his Fermi award of 1963, the visible symbol of his rehabilitation.

I, too, was smarting a little, both from the derision with which a few of my colleagues had greeted my decision to work on the hydrogen bomb in the early 1950s, and from the fact that my old-fashioned brand of patriotism was in short supply in the 1960s and scarcely respected on college campuses. I was pleased when President Johnson, in his remarks, chose to comment on my contributions to national defense as well as to pure science. Having the award presented by the president of the United States made it special to me, as did the fact that it recognized my efforts, both during and after World War II, to keep America strong. And I must confess that I felt forgiven, at last, for losing the secret report on the train.

14

•

TEXAS AND THE UNIVERSE

I HAVE always been deadline-driven: a talk to give, a chapter to complete, a class to prepare, page proof to review, a promised thesis topic for a student. Looking over my published work in the latter part of my career, I see that most of it resulted from the pressure to deliver. Promises, promises. Without them, I would have accomplished less.

Scheduled talks provide the most inflexible deadlines. One must show up at a certain place at a certain time. Too often I venture onto the podium with ideas still only half-formed. Yet I am amazed at how often, standing before an audience, some new idea or connection occurs to me. Maybe that is why, over the years, I have accepted so many invitations to speak. I keep expecting, each time, to learn something new. Afterward, facing still another deadline for an article based on the talk, I try to write what I wish I had said.

Of course, I don't meet every deadline. After the 1964 Dallas Conference on Relativistic Astrophysics, I set to work with my young colleagues Kent Harrison, Kip Thorne, and Masami Wakano preparing an expanded written version of what I had presented on behalf of us all at the conference. As time went on, we kept finding more and more that we wanted to put into the report, which was to appear as part of the conference proceedings. The publisher, the University of Chicago Press, dunned us from time to time as we worked furiously to complete the paper. When I pleaded for mercy, the press granted us an extension of time. Finally, after a flurry of late-night work and consultation with Kip and the others, I got the manuscript into the mail. Then came back the response from the editor at the press. We had a spacetime problem. The manuscript took up too much space and the time had expired.

But there is a happy ending. "We would be happy to publish your report as a separate book," said Bette Sikes. So was born *Gravitation Theory and Gravitational Collapse*, published in 1965. This book summarized all that we knew at the time about collapsing stars, including the final graveyards where they end their days: as white dwarfs, neutron stars, or the fully collapsed objects that we did not yet dare call black holes. It was a source of great personal pleasure to me when, two years later, my friend the distinguished Soviet astrophysicist Yakov Zel'dovich organized the translation of this book into Russian. Zel'dovich and I had remarkably parallel interests, from hydrogen bombs to neutron stars to black holes.

In the summer of 1968, I suggested to my recent former students Charlie Misner (by then heading a relativity research group at the University of Maryland) and Kip Thorne (then a professor at Caltech) that they write a book. We were strolling along the bank of the river Seine at the time. (Conferences bring people together in pleasant places.) "The classic texts on relativity theory by Christian Møller and Peter Bergmann," I pointed out, "excellent and authoritative though they may be, are decades old and badly out of date. It's time for a book that incorporates all of the recent developments, one that emphasizes the physics and not just the mathematics. You are the ideal people to write such a book." They were not easily persuaded. (The young don't always know what a gigantic task it is to complete a book, but Misner and Thorne, who were ahead of their peers in so many ways, sensed right away that it was not something to be undertaken lightly.) Yes, they said, a new book on relativity and gravitation is badly needed. But we can't do it without you. They put me on the spot. I knew that I hardly needed another project at that moment, but I knew also that the best way to focus one's thinking and to come up with new questions and perhaps new answers is to try to summarize all that one knows about a given topic. I couldn't say no. So we stopped at a corner and shook hands to seal our union.

During a break at a NASA-sponsored meeting in New York in 1969 that brought the three of us together again, we went off to a Chinese restaurant to talk about the project. There we fashioned what we called afterward the "Chinese Restaurant Treaty." We talked about the length and coverage of the book, and the division of labor. A concise book of 200 pages or so, with six to eight chapters, ought to do it, we agreed. The most important part of our treaty—insisted upon by my astute young colleagues—was how we would define the completion of the book. So we agreed: When any two of us said it was done, it would be done. (They had seen too often my penchant for tinkering with a piece of work, adding a little more, then adding a bit more, then polishing and reworking a little more—a legacy, perhaps, of my apprenticeship with Niels Bohr.)

William Kaufmann, then president of the publishing house W. H. Freeman, had been influential in bringing out my earlier book with Edwin Taylor, *Spacetime Physics*, and helped to get this one started down the tracks. When the patient, long-suffering editors and designers at Freeman—notably Earl Tondreau, Beth Eddy, Aidan Kelly, and Bob Ishikawa—brought out the book in 1973, it had 1,279 pages divided into nine sections and forty-four chapters—and it offered two tracks through the forest. Naturally, we are happy that more than twenty years and more than 60,000 English-language copies later, *Gravitation* is still selling. Freeman's Adam Kudlacik deserves much of the credit for this. It was his daring idea to turn our tome into a reasonably priced paperback.

In the Soviet Union, our friends and colleagues Vladimir (Volodya) Braginsky and Igor Novikov edited the Russian edition of *Gravitation*; Mikhail Basco did the monumental translation job. (Novikov had himself been a translator of the earlier University of Chicago Press book on gravitational collapse.) The 16,000 copies printed by Mir Publishers sold out in four days. Never reprinted, Mir's edition of *Gravitatsia* has become a collector's item. (Recently, Braginsky tells me, a pirated edition has made its appearance in Moscow.)

Like most textbook contracts, ours called for the publisher to decide when a second edition was needed, and, if we were unable to prepare it, to select someone else to do it. Some five years after the book's publication, Peter Renz of Freeman asked if we were prepared to update it. Well, we said, we are awfully busy with many other things. We are not eager. "I have checked with our production people," Peter said. "They tell me that if you bring it up to date and cut the length in half, they will only have to double the price." That made a decision sound easy. But Peter made the decision a little less easy by proposing instead that we write a separate slender supplement. On January 27, 1980, Kip, Charlie, and I converged on the Philadelphia airport and spent several hours discussing what to do. In the end, with Peter's agreement, we decided that we would let *Gravitation* fade away gradually as an old classic, unrevised.

The most recent translation of the book is into Chinese, for an edition published in Taiwan. The translator is Li Shuxian, a physicist who, with her husband Fang Lizhi, came to the United States in 1991 after escaping from China. At a refreshment break at the Stanford conference on astrophysics in 1994, I asked her about the translation: "Will it ever reach mainland China?" She reached for a napkin from the refreshment table and wrote on it with her pen, "the Book will be sold in Mainland by smuggle."

Kip Thorne, Charlie Misner, and me in 1993, the twentieth anniversary
of the publication of *Gravitation*.
(Photograph by Susanne Kemp Misner.)

Charlie Misner has described our three years of serious work on the book as
"fun." Indeed it was, not least because he and Kip were such extraordinarily
capable scientists—as well as conscientious workers and good writers. Charlie
came to Princeton for a semester as a visiting professor. I sat in on his lectures and took notes. He was the strongest of the three of us in the mathematics of relativity. All three of us got together several times at the Wheeler
compound on High Island, South Bristol, Maine, and at various other places
around the world. Whenever two of us were scheduled to be in the same
place at the same time, the third would try to come too, and often we stayed
a few extra days after some conference just to work on the book. The three of
us met at our respective home bases of Princeton, College Park, and Pasadena, as well as in Austin (Texas), Copenhagen, Moscow, Leningrad, and Kiev.
I sat down with Charlie in Dublin and Kyoto and with Kip on the beach near
San Felipe, Mexico. Between meetings, we circulated drafts of chapters until
they "converged"—until the number of changes by the next reader
approached zero.

I met Fang Lizhi and Li Shuxian first in China in October 1981. Fang was
designated by the Chinese government to translate my lectures and to serve as
interpreter, guide, and source of wisdom for Janette and me as we traveled

Fang Lizhi and Li Shuxian, Tucson, Arizona, 1993.
(Courtesy of Fang Lizhi and Li Shuxian.)

about the country. He was, at the time, vice president of the Chinese National University of Science and Technology at Anhui, some 600 miles south of Beijing, where Li taught physics. Fang has an infectious laugh, and he laughs often. Li is much quieter.

In the years before I knew him, Fang had already become widely known for his outspoken advocacy of freedom and human rights. During the great Mao-inspired Cultural Revolution of 1966–1976, when some 20 million Chinese died, Fang had labored for two years in the rice fields. Later, as a university vice president, he brought about his own form of revolution, introducing democratic ways into the governance of the university. He encouraged faculty members to speak up, disagree, make motions, vote for and against them. It may well be that the seed of the larger democracy movement in China was planted at the Chinese National University of Science and Technology. Fang's ideas and ways of doing things were catching. Eventually, in 1989, came the Tiananman Square demonstration, so cruelly put down by tanks and rifle fire that killed hundreds of unarmed demonstrators.

Immediately after the Tiananman Square demonstration and massacre,

Fang and Li were put on the most-wanted list and placards were distributed all over China to help apprehend them. Friends smuggled them into a Beijing hotel under false names, and from there to the U.S. embassy, where they remained for a year and a half. Ambassador Winston Lord welcomed them and gave them a work room as well as living space. They kept their window curtains drawn to prevent observation from outside. Concerned colleagues from other parts of the world sent them current copies of leading journals of physics and astrophysics. Soon the embassy sheltered a budding research institute.

Ambassador Lord negotiated with the Chinese government for a face-saving way to free Fang and Li. On June 25, 1990, they stepped from the embassy's American soil into the American territory of the ambassador's car and from that onto the American territory of an American airplane at Beijing airport. They flew first to Alaska, then came for a time to the Institute for Advanced Study in Princeton, where I got reacquainted with them and heard their story. Now they are in Tucson, at the University of Arizona—a long way from Anhui and Beijing.

When I visited Fang and Li in their Princeton apartment, I noticed the separate computers on which they worked on opposite sides of the room. "They are useful," Fang explained. "When one of us doesn't want to interrupt the other, we send an e-mail message."

Once, when Charlie, Kip, and I were working on our book in Princeton, we escaped from the university to a more peaceful office across town at the Institute. When we were about to take a twenty-minute break, I said, "Why don't we go and call on Gödel?" So I took them to Kurt Gödel's office and introduced them to him. Gödel was a mathematician and logician, most famous for his proof of the undecidability of mathematical propositions.

It was a beautiful spring day. The sliding glass door in Gödel's office offered a view of the Institute's cooling pond and the trees beyond. The door was shut and Gödel sat there in his overcoat with an electric heater going. Kip and Charlie were more startled than I was. I knew that Gödel, always fearful about his health, studied medical treatises before taking medicine prescribed by his doctors. I explained that we were doing a book on gravitation, but we wanted most of all to know about the connection, if any, between his famous proof of undecidability in mathematics—the greatest advance in mathematical logic since Aristotle—and Heisenberg's principle of indeterminacy.

Gödel was totally against discussing this question. Instead, he asked us what we were going to say in our book about his theory of a "rotating universe." (The term doesn't mean literally that the universe as a whole is rotating—for what would it rotate relative to? In Gödel's theory, individual galaxies rotate more in one direction than another—just as the hands of clocks on a wall rotate more in one direction than another.) "Nothing," we had to confess. This response distressed him. He was so passionately interested in the subject and so desperate for facts and figures, it turned out, that he had taken down the great Hubble photographic atlas of the galaxies, lined up a ruler on each galactic image to estimate the galaxy's axis of rotation, and compiled statistics of the orientations. He found no preferred sense of rotation.

About a year after Kip and Charlie and I visited Gödel, I was in the office of my Princeton colleague James Peebles discussing cosmology, when his door burst open and his student Dan Hawley walked in, announcing, "Here it is, Professor Peebles," as he plopped the bound copy of his dissertation on the low table between us.

"What's it about?" I asked him.

"Whether there's any preferred sense of rotation among the galaxies," he answered.

"Oh, how wonderful!" I exclaimed, "Gödel will be so pleased."

"Gödel? Who's Gödel?"

"The greatest logician since Aristotle," I answered.

"Are you kidding?"

"No."

"What country does he live in?"

"He was born in Czechoslovakia and studied in Vienna, but now he lives right here in Princeton. Let me call him up." I rang Gödel's house and found him at home. Gödel asked so many questions about the work that I turned the phone over to Dan Hawley. Soon Hawley got out of his depth and turned the phone over to Jim Peebles. When Peebles finally hung up, he said, "My, I wish we had talked to him before we started this work."

Why was Gödel so passionately interested in this subject? I first heard him speak of it in 1949 at a seventieth birthday celebration for Einstein. In a universe with an overall rotation, he concluded, there could exist world lines (spacetime histories) that closed up in loops. In such a universe, one could, in principle, live one's life over and over again. I had to conclude that Gödel's passionate concern for his own health, so openly visible, was matched by an equally passionate, if less visible, wish to defy death and live again.

About a year after the encounter with Peebles' student, I was able to talk to Gödel a little more intimately at a cocktail party given by Oskar and Dorothy Morgenstern. There, at last, Gödel confessed to me why he had been unwilling to talk with Kip Thorne, Charlie Misner, and me about any possible connection between the undecidability he had discovered in the world of logic and the indeterminism that is a central feature of modern quantum mechanics. Because, he revealed, he did not believe in quantum mechanics. Gödel was a friend of Einstein and apparently the two had walked and talked

Celebrating Einstein's seventieth birthday, 1949, *left to right:* Eugene Wigner,
Hermann Weyl, Kurt Gödel, I.I.Rabi, Albert Einstein, Rudolf Ladenburg,
Robert Oppenheimer.
(Photograph by Howard Schrader, courtesy of Princeton University.)

so much that Einstein had convinced him to abandon the teachings of Bohr
and Heisenberg.

Next to Einstein, Morgenstern may have been the closest friend of the
reclusive Gödel. Morgenstern told me of the time that he went to visit Gödel
at his home. After knocking on the door and getting no response, Morgenstern eventually opened the door and looked in. He could see through to the
kitchen, where a cup of still steaming coffee sat on the table. But there was no
Gödel to be seen. Morgenstern went in to explore and found Gödel hiding
in the cellar behind the furnace.

Gödel died in 1978 at the age of seventy-one. (His famous proof had been
advanced in 1931, when he was twenty-five.) Some time after his death, I

found myself chatting across a lunch table at the Institute with a young man visiting from New York. "What brought you to Princeton?" I asked.

"I am editing the papers of Kurt Gödel," he answered.

So I couldn't help asking, "Have you come across any work related to the rotation of galaxies?"

"Interesting question," he responded. "Recently I found a great pile of sheets filled with numbers, totally unlike the rest of his papers. It took me some time to figure out what they were all about. They were the foundation of his statistical analysis of galactic rotation."

Of all the entities I have encountered in my life in physics, none approaches the black hole in fascination. And none, I think, is a more important constituent of this universe we call home. The black hole epitomizes the revolution wrought by general relativity. It pushes to an extreme—and therefore tests to the limit—the features of general relativity (the dynamics of curved spacetime) that set it apart from special relativity (the physics of static, "flat" spacetime) and the earlier mechanics of Newton. Spacetime curvature. Geometry as part of physics. Gravitational radiation. All of these things become, with black holes, not tiny corrections to older physics, but the essence of newer physics.

Starlight passing the Sun is deflected through a barely perceptible angle. Starlight near a black hole, by contrast, can be reined into a closed orbit, circling the black hole forever. Mercury, in its orbit around the Sun, deviates ever so slightly from its Newtonian path. Near a black hole, Mercury's altered path would be the least of its problems if the black hole's tidal gravity tore it to bits. Light, making its way from the Sun to Earth, is "stretched" so that a single vibration lasts a tiny bit longer than if the light were radiated from one point to another on Earth. As an atom disappears into a black hole, on the other hand, the light it emits is infinitely stretched until it appears not to vibrate at all.

But what of the black hole itself, apart from its effect on things around it? In 1970, Remo Ruffini and I postulated that it is an extraordinarily simple object, seen from the outside. It can influence the world around it only through its mass, charge, and spin (angular momentum)—nothing else. (It has no hair.) It actually took years of work by others—including the Canadian Werner Israel, the British scientists Stephen Hawking and Brandon Carter (both students of Dennis Sciama), and David Robinson—to prove the truth of this conjecture. The black hole, we now know, is stripped of superfluous frills. It has as few "options" as a Model T Ford. Inside may be a different story. The "secret life" within black holes became a serious subject of study in 1972, when my student Jacob Bekenstein studied their entropy.

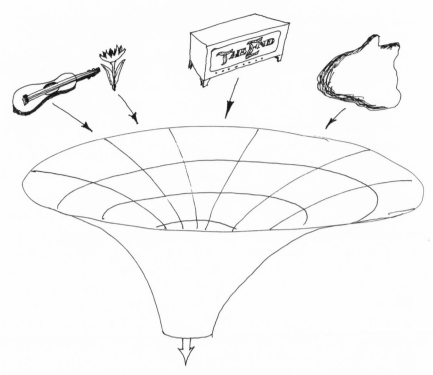

Most of the identifying characteristics of things that fall into a black hole are lost.
(Drawing by John Wheeler.)

Bekenstein's parents were Jews who had emigrated from Poland to Mexico. Jacob told me how, when they gained permission to continue on to the United States, they had traveled by bus from Mexico to New York. He and I liked to joke, but Jacob had his serious side. He was orthodox in his religious practices, yet ready to question any orthodoxy in physics. He wasn't swayed by glib reasoning or by authority. He was just the right person to tackle the question of the entropy of a black hole.

Entropy measures disorder, or complexity. Whatever is composed of the fewest number of units arranged in the most orderly way (a single, cold molecule, for instance) has the least entropy. Something large, complex, and disorderly (a child's bedroom, perhaps) has a large entropy. The second law of thermodynamics states that the entropy of a system left to itself will never decrease. Except in quite special circumstances, it will increase. In other words, the natural trend of events in nature is toward greater disorder.

In 1970, Stephen Hawking had discovered a "law of areas" for black holes that bears a certain resemblance to the second law of thermodynamics. A black hole's defining characteristic is its horizon, a spherical shell that cloaks whatever is within it. Outside the horizon, light can escape. Inside, it cannot. (For a black hole with the mass of our Sun, the horizon has a circumference of about 12 miles.) What Hawking discovered is that the area of the horizon of a black hole never decreases. Since the horizon is a measure of how much mass lies within, it is hardly surprising that it never shrinks, for a black hole can suck up mass (causing its horizon to expand) but it cannot expel mass (which would cause its horizon to contract). Hawking and everyone else concerned with such matters saw no connection between the nonshrinking horizon and entropy. The black hole was, after all, a nearly "hairless" entity of great simplicity. Some even argued that its entropy was zero.

The idea that a black hole has no entropy troubled me, but I didn't see any escape from this conclusion. In a joking mood one day in my office, I remarked to Jacob Bekenstein that I always feel like a criminal when I put a cup of hot tea next to a glass of iced tea and then let the two come to a common temperature, conserving the world's energy but increasing the world's entropy. My crime, I said to Jacob, echoes down to the end of time, for there is no way to erase or undo it. But let a black hole swim by and let me drop the hot tea and the cold tea into it. Then is not all evidence of my crime erased forever? This remark was all that Jacob needed. He took it seriously and went away to think about it.

Several months later, Jacob appeared at my door with his response. The area of a black hole's horizon, he said, not only resembles the entropy of the black hole—it *is* the black hole's entropy (to within a constant of proportionality). I had learned often enough in my career that nature has a way of being a little stranger than we think it ought to be. So I said to Jacob, "Your idea is so crazy that it might just be right. Go ahead and publish it." Well, it has turned out to be right. This has given me the greatest pleasure, especially because Bekenstein's rightness proves to be founded on *quantum* features of nature. Except for a numerical factor, the entropy of a black hole is the number of "Planck areas" it takes to cover the area of the horizon. The Planck area is the area of a square whose sides equal the Planck length, the fundamental quantum length of gravitation physics (see footnote 1 on page 247). Since this length is incredibly small compared even with the diameter of a single proton, it is clear that it takes a lot of Planck areas to cover a horizon of even a few square miles. Indeed, an astronomical number of Planck areas would be needed even to blanket a single proton!

Even in the forefront field of black-hole physics, dominated by the young, there is an establishment, and this establishment at first could not accept Bekenstein's idea. Stephen Hawking was visiting another British physicist, Brandon Carter, then working at the Paris Observatory, when they came across Bekenstein's paper. His conclusion is so preposterous (or so I understand they said to each other) that we must write a short report to show why it is wrong. But Hawking must have said, "Wait a minute. This deserves a closer look," for they did not write that report. Two years later Hawking discovered what we now call "Hawking radiation" coming from a black hole and realized that a black hole has a temperature. Then he saw clearly the compelling logic of Bekenstein's conclusion. Even then, some others were not converted.

Entropy is a subtle concept, but the message of Bekenstein's insight is easy to grasp. It tells us that a black hole is, after all, a richly complex, disordered entity. Simple though it may appear from outside, its interior need not be simple at all.

One reason it took time to make true believers of the black-hole establishment was that there exists an inescapable link between a large entropy and the ability of a black hole to radiate. Turning the argument around, nearly everyone argued that since a black hole, almost by definition, does *not* radiate, it cannot have a large entropy. Actually, a year earlier, in 1971, Yakov Zel'dovich, in an informal meeting with Kip Thorne in Moscow, had put forth the argument that quantum fluctuations in spacetime should make it possible for a black hole to radiate energy—perhaps only an infinitesimal amount, but not a zero amount. The black hole, he said, is not completely cut off from communicating with the world outside. In a startling and brilliant piece of work in 1974, Hawking proved that every black hole radiates and that the radiation, though tiny, is just enough to bring about consistency with Bekenstein's entropy postulate.

Stars of a certain size, perhaps greater than about three suns in mass, collapse eventually to black holes, and we now have good evidence for the existence of some of these "stellar" black holes—that is, black holes resulting from the collapse of old, dead stars. There is every reason to believe that they exist by the billion. (Since, in the early stages of collapse, a star may throw off a substantial fraction of its mass, the eventual black hole may be quite a bit less massive than the original star.)

But there is no reason to think that black holes are all of stellar origin. They can come in all sizes. Within the dimensions of a single elementary particle, there may exist countless "mini-micro black holes," collapsed from wormholes of quantum origin. These black holes, like the wormholes themselves—all part of the quantum foam that stirs up spacetime at its smallest dimensions—are expected to be disappearing as fast as they are being created.

Remote from the mini-micro black holes are their mega-gigantic cousins. We now have good evidence for black holes at the centers of galaxies, including our own galaxy. A galactic black hole may have the mass of millions or even billions of suns. What circumstances lead to its formation we don't yet know.

As I approached my sixty-fourth birthday in 1975, I felt that I was just getting warmed up for all the work I still wanted to do and all the students from whom I still wanted to learn. Retirement, even at Princeton's then mandatory age of seventy, half a dozen years away, was not something I was ready to think about. So, when George Sudarshan approached me that summer to ask if I might be receptive to an invitation to move to the University of Texas in Austin, I was willing to talk. George, who is from India, is an immaculate dresser, with a mind as well organized as his wardrobe. He had done notable work in elementary-particle physics and mathematical physics, including some work in general relativity. Before moving to the University of Texas in 1969, he had worked at Harvard, Rochester, and Syracuse Universities.

At our invitation, George stayed a day or two with us in Princeton on his way to Europe that summer, and gave me a good sales talk about conditions for working in Austin, assuring me that as far as he knew, there was no mandatory retirement age in Texas. Janette was in on the conversations, which set us both thinking. She, despite our long and happy experience in Princeton, was more than willing to consider fashioning a new life on the frontier (even though we didn't yet have an offer!). George hinted at the possibility of hiring other people to build a stronger relativity group than I had been able to assemble in Princeton. To tell the truth, I had been disappointed in the assignment of new faculty positions as they came along in the Princeton Physics Department.

In every department, every researcher would like to see more of his or her own kind. It's a natural instinct, the urge to assemble a group of people with overlapping interests who can stimulate one another. Princeton was no exception to this rule. As colleagues discuss the directions they would like to see research go in their department, they have to be prepared for give-and-take and compromise as they decide on the recruitment of new faculty members. For more than twenty years, I had made my low-key arguments for adding strength in relativity and some of the other areas of theoretical physics that interested me. But always it seemed that some other priority was higher. So I remained a group of one (except, of course, for the wonderful set of graduate and undergraduate students and postdoctoral visitors who enriched and amplified my research).

My receptivity to Sudarshan's overture may have been enhanced as well by the minor scars I bore from the recently ended Vietnam era, if I may call it that, when I found myself out of step with some of my faculty colleagues, including my friend Murph Goldberger, the department chair. At one meeting of physics faculty and students during the height of the campus unrest, he used such strong language to characterize the villainy of the federal government and the Department of Defense that I felt compelled to rise and point out that no country operates more democratically than ours. Our military is controlled by elected and appointed civilians, I said. It is not someone else's Defense Department; it is ours. Of course no one cheered, but to the credit of the students and faculty who were there, no one jeered.

Just before leaving for a meeting at Florida State University in Tallahassee in January 1976, I received a letter from Tom Griffy, chair of the Department of Physics and Astronomy in Austin. He and his colleagues wanted me to come to Texas. I could expect to build a group in relativity and gravitation physics. He suggested a visit to Austin in the near future. Janette was accompanying me to Tallahassee. We took the letter along and agreed to discuss it in Florida. Perhaps a bit to our own surprise, we both felt a keen interest in the prospect. While still at the Tallahassee meeting, I drafted a letter to Griffy containing our positive reaction, and had it typed and mailed there. Then, even before we left Florida, came a follow-up call from Griffy. I agreed to make an immediate trip—Florida to Austin and back—to discuss prospects. I liked what I saw on that fast trip. By the time Janette and I were back in Princeton, we had decided to go, pending only coming to agreement on the terms of the offer.

I have never made an important decision without consulting Janette or without her concurrence. We don't discuss physics, but we discuss people and opportunities and obligations. I couldn't function without her good judgment. (And I couldn't be much of a social creature without her leadership.) She endorsed our first move, from Chapel Hill to Princeton, even though it wasn't easy for her. And she endorsed the move from Paris to Los Alamos, even though it disrupted a year that she was thoroughly enjoying. Often, on her advice, I have said no to invitations. The move from Princeton to Austin was different. The children were grown and gone, and she was as open to change as I. We were both quickly enthusiastic about the idea.

A key word in our relationship is commitment. I am committed to Janette and she to me. That mutual commitment has gotten us through some difficult times, especially in our early life together, before and during World War II, when, with small children to raise, my work seemed all-consuming and I was often away. Yet, because of my commitment, I always found time to write, to call, to talk, and to read aloud with her. She maintained an interest in all that I was doing and with everyone with whom I was working. She was free with her advice, not just her automatic support. And she helped, then and now, to humanize me. In our later years, we have reached an equilibrium with fewer sources of stress, and with rewards of mutual support as great as they ever were.

I assumed I would have to resign from Princeton, but my friend Aaron Lemonick, a physicist-turned-dean there, had a better idea. Aaron was a "lifer" at Princeton. He had received his Ph.D. there under Robert Dicke and had never left (despite invitations to consider the presidencies of two other prestigious institutions). He was one of those delightfully rare administrators who look for ways to solve problems, removing roadblocks instead of erecting them. "You shouldn't resign, John," he said. "You should retire." So that's what I did. With Aaron's help, I became a professor emeritus at Princeton as I was starting a new life in Texas. The summer of 1976, the summer of my move to Texas, was the summer I turned sixty-five.

Both of the houses we had built in Princeton were on land acquired from the Institute for Advanced Study. The first one, at 95 Battle Road, we built during my first year in Princeton. The second, at 30 Maxwell Lane, we built in 1962 after the children were grown and mostly gone. The Institute had the first right of purchase on these houses at a price to be set by an appraiser selected by mutual agreement. As our Maxwell Lane house was nearing completion, the Institute was eager to avail itself of its option to buy the Battle Road house, which it wanted to offer to the eminent theorist Tsung Dao Lee as an inducement to bring him from Columbia University to the Institute. For several months we lived in temporary Institute housing on Einstein Drive, and T. D. (as he is always called) did indeed make the move to Princeton. When Janette and I visited to welcome the Lees, I was able to show their children the secret closet opened by a hidden button in the library.

We decided that we should sell the Maxwell Lane house when we moved to Texas. If our future financial welfare had been the only consideration, we would probably have kept it. But psychological considerations were stronger. Moving to Texas, we wanted to move fully, not remaining tethered to Princeton. Karl Kaysen, retiring as the director of the Institute for Advanced Study, was interested in the house but in the end did not purchase it. He and his wife returned to Massachusetts, where he gained an MIT affiliation. Kaysen was not uniformly admired by the Institute's scholars. Perhaps that contributed to his change of mind about our house. But during the time that he was the likely buyer, air-conditioning was installed at the Institute's expense, to the benefit of the eventual purchaser, Marshall Clagett, a historian of science at the Institute. Clagett and his wife, Sue, had taken a fancy to our house and liked its convenient location. They were glad to get it. We had filled the house with bookcases. Its office contained dozens of built-in small drawers, which provide my favorite method of filing—one drawer per project. The house also had a bomb-shelter basement; I didn't find out whether the Clagetts valued that.

After a summer in Maine, our move to Austin came in the fall of 1976. I went ahead by car. Janette followed by air. I met her when she landed in Houston, and we checked into a hotel for the night. "Is it a special event?"

asked the clerk. "It certainly is," I answered. So we were given a Texas-sized room and a bowl of fruit to welcome us to Texas—the honeymoon special, according to the clerk.

I resonated with the can-do spirit in Texas, tangible at the university and in the state as a whole. Tom Griffy, a warm extrovert never at a loss for words, was a good physicist and a good leader for the department. I also admired his work on behalf of national security for the Defense Department and the Central Intelligence Agency. In the city of Austin, both Janette and I quickly made friends and came to appreciate the warmth of the people, who were so pleasingly free of snobbery. While house hunting, we lived in an apartment in Cambridge Towers. Then we saw, fell in love with, and bought without haggling a large house at 1410 Wildcat Hollow that had been designed by an associate of Frank Lloyd Wright. It was set into a hillside and stretched ninety feet from front to back. Janette called it "interesting." Although the name of the road hardly suggests it, our new house, with floor-to-ceiling windows in the front rooms, offered a panoramic view of Lake Austin and the city beyond. We added a swimming pool, and we shared both the house and the pool with occasional snakes.

It took about four years to work up a full head of steam on the ninth floor of the Robert Lee Moore (RLM) building in Austin, where our Center for Theoretical Physics was located. By that time, I was in the middle of a group—of faculty, postdocs, and graduate students—that made working in Texas so rewarding. I couldn't walk down the hall or stick my head into an office without encountering someone with a stimulating thought or question. With my energetic student Warner Miller, I didn't even wait until I reached the building. I picked up Warner on many a morning at his house and we talked about what was new as we drove to the campus. Only later did I learn that Warner, after a late night working, often rolled out of bed, splashed water in his face, and took a deep breath in order to sound wide awake, even as the phone was ringing with my call to let him know that I was on my way. He confessed later that he sometimes went back home for a nap later in the morning.

Building a research group, even in theoretical physics, takes money. Upon the sage advice of Alfred Schild, who already headed an established Center for Relativity in Austin, I proposed to create a separate Center for Theoretical Physics. When I first told Loreen Rogers, the redheaded dynamo who presided over the Austin campus, what resources the new center would need, she didn't fall over in a faint. She and Tom Griffy—and perhaps others unknown to me—arranged for generous support. Once we got going, the National Science Foundation helped out, too. One of our early tasks was to recruit an assistant professor interested in the fundamentals of quantum theory. Reviewing candidates with my colleagues Al Schild, Richard Matzner, and Bryce DeWitt, I related to them the question Eugene Wigner liked to ask about faculty candidates, "Has he [or she] discovered the gunpowder?" Which of our young candidates might, in the future, "discover the gunpowder"? To me, that meant answering the question How come the quantum? What is the deeper principle that lies behind the strange quantum behavior that rules our world? The person we selected was Philip Candelas, an Englishman then completing his doctoral work at Oxford. (During his first year in Texas, until the Oxford degree was awarded, he was classified as a research assistant.) Well, so far neither Candelas nor anyone else has discovered the quantum gunpowder, but he has had a distinguished career in Texas, working on mathematical physics and the intriguing "string theory," which envisions elementary particles as vibrating bits of string.

To fill another faculty position, I invited Claudio Teitelboim, a Chilean who had been my graduate student and then an assistant professor at Princeton before he moved across town to the Institute for Advanced Study. Claudio bubbled over with original and useful ideas. (To "soften him up" for my invitation, I gave him anonymously a subscription to *Texas Monthly*, a handsome magazine of my adopted state. It must have helped.) After several years in Texas, he wanted to leave to fulfill his dream of setting up a theoretical-physics center in Santiago, but we worked things out so that he could do that without resigning his Texas position. For half a dozen years, before settling in Chile, he divided his time between Santiago and Austin. Claudio's father, Volodia, a novelist, journalist, and activist, had been a senator and Communist Party leader in Chile until 1973, when Pinochet's coup caught him unawares during a visit to Moscow, where he stayed to save his life. When the elder Teitelboim returned to Chile in 1989, Claudio encouraged his career choice of writer, not revolutionary. I had the pleasure of meeting Volodia in 1993 when I attended a conference in Chile commemorating twenty-five years of black-hole physics. Although Claudio and I did not publish any joint papers, he was a great stimulus to my thinking through our innumerable conversations on both gravitation physics and quantum physics—conversations that continue even now when he returns for regular visits to the Institute for Advanced Study.

Another young Oxfordian, David Deutsch, I interviewed one summer's day at the Portland, Maine, airport. As we walked round and round outdoors, discussing the mystery of quantum theory and possible ways to attack it, I couldn't help seeing the airport as a metaphor for physics. Its spacious, well-tended flatness represents what we solidly know; the vastness of air and space above, what we have yet to learn. Only a worm could imagine that the earth is all there is. Deutsch did join us for a time in Austin, where he illuminated the

Some members of my research group in Texas, 1985. *Seated:* Ignazio Ciufolini.
Standing, left to right: Arkady Kheyfets, William Wootters, John Wheeler,
Wolfgang Schleich, Philip Candelas, Roberto Bruno, Warner Miller.

strange ways in which quantum theory affects the behavior of superconducting rings.

Realizing that recruiting graduate students is as important as recruiting faculty and research staff, I proposed a strategy and the department embraced
it: to invite top undergraduate physics students from the United States, Canada, and Mexico each year, all expenses paid, to a seminar in Austin. We hoped
that they would like what they saw and want to enroll. We hoped, too, for
"recruitment by resonance," that two or three of these visitors might take to
each other and decide they would like to be together at the same place for
their graduate work. The immediate results were modest. The pull of places
like Caltech, Chicago, MIT, and the Ivy League institutions was just too powerful. But over time, we saw a gradual increase in the number and quality of
applicants. Texas is now getting some of the best.

Adding great strength to the department a few years after I arrived was
Steven Weinberg, who had received the Nobel Prize in Physics in 1979 for his
work uniting the weak and electromagnetic interactions of elementary particles. He was surely one of the brightest theoretical physicists in the world.
Although still young, Steve had mellowed and was no longer as prickly as he
had been in his twenties. According to rumor, during negotiations for his

move from Harvard to Texas, he asked that his salary be no less than that of the football coach. I don't know whether the university authorities agreed to this outrageous demand. Steve had done some notable work in general relativity and had published an advanced text on the subject. In Austin, however, his interests in particle physics did not directly overlap mine in gravitation and quantum theory. So we didn't do any joint research, although he was a great resource for lunchtime and corridor conversation.

Later Weinberg campaigned eloquently for the Superconducting Super Collider (SSC), scheduled to be built in Texas. I shared his disappointment when it was killed by the U.S. Congress. Because of this action, the leading edge of particle physics may move to Europe—where, however, American scientists will no doubt continue to play a major part. Physics is no less international now than it was when I first went to Copenhagen in 1934.

The spirit of Texas seemed to bring out the imp in me. One day, I got so excited by some results that my student Roberto Bruno had obtained relating topology (a branch of mathematics) to Mach's principle (to be explored in the next chapter) that I decided a celebration was called for. "Here's five dollars," I said to Warner Miller. "Go and get some firecrackers. We must celebrate a breakthrough." Warner couldn't find any firecrackers, but he returned before long with some Texas bottle rockets that he had stashed at home. They were a bit more than I had bargained for, but I was game. I went down the hall, telling secretaries to stay in their offices, and I sent Warner and some of his fellow students to the end of the hall to stop anyone who might be about to venture onto the ninth floor. Then I lit a bottle rocket, and we all watched as it flashed noisily and satisfyingly the length of the hall to lodge itself under a door at the other end. The amazed secretaries ventured warily into the smoke-filled corridor. The students were enchanted. Both university officials and the fire marshal spared me the scolding that I deserved.

1 5
·
IT FROM BIT

ERNST MACH was a German physicist and philosopher who belonged to
the generation before Einstein (as generations are measured in the world of
science). His influential book, *The Science of Mechanics*, was first published
in 1883. Its seventh edition, in 1912, may be the one that influenced Einstein as he was working on his theory of general relativity. (Yet another translation of Mach's book appeared in 1933. Not many monographs last so long!)
Mach didn't merely lay out Newtonian mechanics for students of the subject. He thought deeply about the meaning of physics and raised provocative
questions. He is now best remembered for "Mach's principle," the postulate
that the source of an object's inertia is all the other matter in the universe.[1]

Inertia is another name for mass. It is the property of a body that makes it
resist acceleration. A shot-putter cannot get the shot moving fast without great
effort. A spacecraft accelerates into orbit only with the help of powerful rocket engines. If you stub your toe on a rock, the rock "resists" acceleration—it
doesn't immediately get out of the way. The shot, the spacecraft, and the
rock have mass, or inertia, which makes them not "want" to change from
rest to motion or from one speed to another. Before Mach, no one had seri-

[1] Because of his pioneering studies in high-speed motion in fluids, Mach has gained an
even stronger hold on immortality through our use of Mach numbers to specify the speed
of aircraft through the air. Mach 1 is the speed of sound, Mach 2 is twice the speed of
sound, and so on. The Concorde gained fame as the first commercial aircraft to surpass
Mach 1. Its cruising speed over uninhabited areas is Mach 2.04.

ously asked, Why inertia? It was like asking Why space? or Why time? These are givens. They are just there.

Mass is related to acceleration. For a given force, more mass means less acceleration. Less mass means more acceleration. (Earth, with a huge mass, accelerates hardly at all in response to the upward pull of a falling apple. A photon, with zero mass, has no resistance whatever to acceleration. When it is created, it leaps instantly — "accelerates" — to the maximum possible speed permitted in nature, the speed of light.) Newton, back in the seventeenth century, first pointed out that acceleration is apparently "absolute," whereas velocity is "relative." Consider a pail of water (an example used by Newton). If the pail is carried along at a smooth, steady rate, then the water in it, which is not moving relative to the pail, has a flat surface and is otherwise indistinguishable from water in a motionless pail. This illustrates the relativity of velocity. Nothing about the behavior of the water can tell you whether it is at rest or moving at a uniform velocity.

Suppose now that you tie the handle of the pail to a long rope suspended from a hook overhead. Wind up the rope by turning the pail through many turns, and then release it. The pail starts to spin as the rope unwinds, and, after a time, the water in the pail also spins. The surface of the water becomes curved. It is low in the middle, high at the edge. When this happens, the water in the pail is not moving relative to the pail, yet it behaves differently than if it were at rest. The rotation—a form of acceleration—has affected the water differently than it was affected by uniform velocity in a straight line. Its acceleration appears to be absolute.

Newton, who believed in a tangible ether filling all space, had no trouble interpreting the behavior of the water in the rotating pail. It is accelerating relative to the ether, he said. The ether, if it exists, makes it possible to define "absolute rest" and "absolute motion." The water in the stationary pail and the water in the rotating pail behave differently because one is at rest with respect to the ether and the other is accelerating with respect to the ether. (Newton's laws make it clear why a *steady* motion through the ether has no measurable effects, but we won't go into that here.)

By the time Mach published the seventh edition of *The Science of Mechanics* in 1912, ether had fallen on hard times. Experiments to search for it had all failed. Einstein's special theory of relativity dispensed with it. Mach, too, rejected it. Yet here was the fact that a rotating pail of water behaved differently than a stationary pail of water. Why? The postulate advanced by Mach—what we now call Mach's principle—is that the "fixed stars" (that is, all of the other matter in the universe) determine the inertia of the water in the pail and are responsible for its reaction to acceleration. Mass-energy there (as I have put it) determines inertia here. The surface of the water in the pail, said Mach, is curved when the water is rotating relative to the fixed stars and is flat when it is not.

Philosophical twaddle is what some scientist-critics have called Mach's principle. Others, more kindly disposed, say it is a provocative thought but doesn't lend itself to being incorporated into the body of science. Einstein took it seriously, realizing that the apparent absoluteness of acceleration requires an explanation. On June 25, 1913, as he was wrestling with the formulation of general relativity, Einstein wrote a note of appreciation to Mach,[2] referring to his "happy investigations of the foundations of mechanics." He added: "For it necessarily turns out that inertia originates in a kind of interaction between bodies, quite in the sense of your considerations on Newton's pail experiment."

Incorporating Mach's principle into relativity in a completely consistent way, Einstein discovered, is possible only if the universe is closed. This is one of the reasons that Einstein believed in a closed universe, and it is a reason that I, too, believe in a closed universe. "Closed" means that the universe has so much mass that its different parts, pulling gravitationally on each other, will prevent limitless expansion. It also means that a beam of light sent into space will travel a closed orbit, its path bent by gravity until the beam returns to its starting point. When I first argued, many years ago, that the universe is likely to be closed, there was an enormous amount of "missing mass" to be accounted for. All the stars in all the galaxies provide only a very small fraction of the total mass needed to close the universe. If these visible stars and galaxies alone provided gravity, the universe would be open; its expansion would never stop. As time has passed, more and more invisible matter has been detected. The quantity of mass "absent without official leave" is getting smaller and smaller. By now, many astrophysicists believe that sufficient mass exists to close the universe and that before long this missing mass will be detected.

One popular view holds that the universe contains just exactly enough mass to stop its expansion without causing it to contract. According to this view, the universe will coast gradually toward a stop over infinite time. This is a razor-edge condition, too perfect to be true, in my opinion. Any amount of mass less than this precise value and the universe must be open, expanding without limit and without stopping. Any amount of mass more than this precise value and the universe must be closed, reaching a maximum expansion, then contracting to a Big Crunch. The Big Crunch is the long-term fate of the universe that I find most persuasive.

Like Einstein, I took Mach's principle seriously. It had just the sweep that

---

[2] The full letter, in Einstein's original handwritten German, is reproduced in Misner, Thorne, and Wheeler, *Gravitation* (New York: W. H. Freeman, 1973), pp. 544–545.

appealed to me, and I was sure that it could be put on as firm a foundation as any part of general relativity. It becomes reasonable when one thinks of space as a "thing," an entity every part of which touches and is influenced by every other part. Like a giant circus tent in which a single tiny wrinkle can be caused by a tent pole two hundred feet away, space "here" reflects the influence of mass "there." This view of space as a dynamic entity—every part of which is influenced by, and influences, every other part—is quite different from Newton's conception of the ether. His ether was a passive background, more like the canvas of a painting hung in a museum than the canvas of a circus tent rippling in the wind.

I wrote first about Mach's principle in 1963, showing that the principle could be used to select from among all the mathematically possible solutions to Einstein's equations those that described nature as we know it. (It is not unusual for the fundamental equations of physics to have mathematical solutions that do not correspond to anything physical. The limitation that restricts the solution to an actual physical situation is called a "boundary condition.")

With my student James Isenberg in Austin, I returned to Mach's principle in 1979. Inspired by Plato's dialogues and Galileo's dialogues, we presented our paper in the form of a dialogue between a doubter, a quite competent but skeptical physicist called F, and a believer called W, who in the end naturally got the better of the argument, just as he surely would have in the hands of Plato or Galileo. In real life, the skeptics don't always bend so meekly to the force of one's arguments. Poking fun at those skeptics, I wrote in that paper, "Mystic and murky is the measure many make of the meaning of Mach."

It is easy to be incredulous at the idea that something as tangible as the mass of a baseball held in your hand can be attributed to the content of the universe millions and billions of light years away. Yet that is what Mach and Einstein tell us. I returned once again in 1995 to the fascination of inertia "here" resulting from mass-energy "there" in a book, *Inertia and Gravitation*, written with my old friend and former student Ignazio Ciufolini. On the advice of his older colleague Remo Ruffini in Rome, Ciufolini had arranged to enroll in Montana State University to work with Kenneth Nordtvedt on problems in general relativity. When he arrived in Bozeman, he found that Nordtvedt, like the Athenian statesman Pericles, had decided to put politics first. Nordtvedt had won a place in the Montana legislature—good for Montana but not for Ignazio. Having heard of our work in Austin, Ignazio switched schools and started working with me.

Ciufolini's dissertation concerned the hypothetical question How could a fleet of satellites circling a black hole determine all the properties of that black hole by laser ranging or other distance-measuring techniques? At the same time, he got interested in a very practical application of laser ranging. A group of aerospace engineers on campus was keeping track of the distance to the Moon, accurate to within inches, by bouncing a laser beam from it and timing the round-trip with precision. Analyzing their measurements required that every small effect that alters the Earth-Moon distance—the tides, for example—be taken into account. Ignazio decided to see if any effect of general relativity might play a role. He looked, in particular, at an effect of Earth's rotation called "gravitomagnetism" (a gravitational analog of magnetism in which circulating mass replaces circulating electric charge). Gravitomagnetism, he concluded, is not big enough to measurably alter the Earth-Moon distance, but his calculations got him interested in Mach's principle and the source of inertia, and led eventually to our book.

It turns out that if mass "there" is the source of inertia "here," then mass "here" also makes a small contribution to inertia "here." (It is as if a wrinkle in the fabric of a circus tent, although caused mainly by a tent pole two hundred feet away, is also slightly affected by another wrinkle two inches away.) A planned experiment to demonstrate the reality of Mach's principle[3] rests on this circumstance—that the inertia of a body, although largely determined by distant mass, is slightly affected by nearby mass. In this delicate and difficult experiment, Newton's spinning pail of water is replaced by four spinning quartz spheres, each about the size of a Ping-Pong ball. Circling Earth in a satellite about 400 miles up, each small sphere, acting as a gyroscope,[4] has its axis of spin carefully freed from every minuscule disturbance that could alter its tilt. In a perfect Newtonian world, the gyroscope's axis of spin would point rigidly and relentlessly in one direction relative to the "fixed stars." (This is what makes gyroscopes useful instruments for controlling airplanes and spacecraft.) But in an Einsteinian world, both the curvature of space and the "dragging" of space near Earth cause the axis of the gyroscope to swing slowly around. The smaller of these two effects, the frame-dragging effect produced by Earth's rotation, nudges the axis of the gyroscope about 32 billionths of a degree per day. If all goes well, this already-delayed, extraordinarily sophisticated (and quite expensive) experiment (called Gravity Probe B) will be launched in October 2000. Francis Everitt, a leader of the Gravity Probe B experiment, likes to point out that the angle through which the gyroscope is "dragged" each day is the same as the angular width of a human hair 90 miles away.

[3] The experiment is really testing an implication of Einstein's general relativity called "frame-dragging," which is ultimately related to Mach's principle.

[4] *Gyroscope* means, literally, an instrument for seeing rotation. Jean-Bernard-Léon Foucault invented the gyroscope in 1852 as a means of seeing Earth's rotation.

The significance of this experiment can be appreciated by considering an idealized variant of it. Imagine a gyroscope suspended above the North Pole, with its axis of spin parallel to the ground. At a given moment, we arrange for the axis to point along the zero-degree meridian, toward Greenwich, England. As Earth rotates, the gyroscope's axis will remain pointing toward some spot in the heavens. Relative to Earth, it will point along successively increasing degrees of longitude until, after 24 hours, it points again toward Greenwich. The remarkable prediction of relativity is that, in a hypothetical universe in which everything has been swept away except Earth, the gyroscope's axis would turn with Earth (insofar as turning could be defined). Once set spinning with its axis pointing toward Greenwich, it would continue to point toward Greenwich. The gyroscope would not "know" that it should remain pointing toward any other place in the universe. There *is* no other place.

Now, in the real universe that we inhabit, the great part of the gyroscope's inertia is contributed by distant matter, but a tiny part is contributed by Earth. So, as Earth rotates beneath the gyroscope, the axis of the gyroscope "tries" to follow Earth. It rotates through a very slight angle each day. If it points toward a mark on the Astronomer Royal's desk in Greenwich at noon today, it will point toward a spot about half an inch away from that mark at noon tomorrow—a shift of ⅙ of 1 millionth of a degree. The gyroscope in orbit swings through an angle five times less than that!

Once settled in Texas, I began to think more and more about quantum theory. Relativity, despite all its drama and challenge, does not stretch human understanding—or human credulity—in the way that quantum theory does. It's clear that if we are to understand our world more deeply in the twenty-first century, those two great theories of twentieth-century physics, relativity and quantum mechanics, must be harmoniously joined. Right now there exists at best an uneasy peace between them.

Planck's quantum principle—that nature is granular—dates from 1900. Five years later, Einstein solidified the principle by applying it to corpuscles of radiation—photons. These events occurred before I was born, before my parents had met. In 1913 (when I was a toddler), Niels Bohr extended quantum theory to the structure of an atom, and introduced more strange ideas—the "quantum jump" (which has made its way into common speech), the unpredictability of a particular quantum event, the existence of a "ground state" in which particles move at high speed but can lose no more energy, and the emission of radiation with a frequency of vibration different from the frequency of revolution of the electron that caused the radiation.

In the mid-1920s (when I was in high school), these "strangenesses" of quantum theory were welded into the theoretical structure we call quantum mechanics by Werner Heisenberg, Erwin Schrödinger, Niels Bohr, and others. Nothing that came out of that synthesis was more startling than Heisenberg's uncertainty principle, which denied the possibility of simultaneously measuring certain properties of motion. The uncertainty principle introduced us to quantum fluctuations, revealing empty space to be in fact a cauldron of activity, bubbling ever more vigorously as one looks at ever smaller bits of space and intervals of time. From quantum fluctuations come measurable properties of particles and atoms (as I discussed in Chapter 10), and not-yet-measurable properties of spacetime itself—wormholes, quantum foam, and uncertain geometry.

Since that grand synthesis in the 1920s, quantum mechanics has been largely a finished theory, applied in a myriad of ways but not altered at its core. It has fascinated me throughout my professional career. When I made the Center for Theoretical Physics on the ninth floor of the RLM Building in Austin my new home, I couldn't resist coming back to the quantum. Einstein, by inventing the photon and by laying the theoretical base for lasers, had contributed as much as anyone to quantum theory. Yet he died not believing in it. Bohr died its champion, but recognized that it was unfinished business. "Whoever talks about Planck's constant and does not feel at least a little giddy," said Bohr, "obviously doesn't appreciate what he's talking about."[5] Well, it made me giddy in 1933 and it still does.

Relativity has done some amazing things, showing the way to the expanding universe, black holes, gravitational radiation, dynamic geometry, and the origin of inertia. But without quantum mechanics, relativity, too, is unfinished business. If I wanted to tackle the largest questions, I told myself in Texas, I had better think harder about the quantum as well as about relativity.

Quantum mechanics is often described as the theory of the very small. A true statement, as far as it goes. Quantum mechanics is an absolute necessity, and an everyday tool, in explaining how molecules, atoms, photons, electrons, and other particles behave. It is of no consequence in explaining the motion of spacecraft, planets, comets, and whole galaxies. So what does the quantum have to do with the universe? Perhaps everything, because in any fundamental theory of existence, the large and the small cannot be separated. A Big Bang, with everything squeezed to infinite density, gave rise to our universe. The Big Bang was the original high-energy particle laboratory. A Big Crunch, with everything again infinitely squeezed, may end the universe. In between that beginning and that potential end, there seems to be no lack of

[5] Bohr's remark is as recalled by Edward Teller in A. P. French and P. J. Kennedy, Eds., *Niels Bohr: A Centenary Volume* (Cambridge, Mass.: Harvard University Press, 1985), p. 184.

infinitely squeezed matter and energy, in black holes of all sizes. A gravitational wave rolling across the cosmos can have its origin in a piece of matter as heavy as many suns and as small as a single particle. Whatever quantum fluctuations contort the tiniest dimensions can influence the fabric of space and time in the large. In short, there is no hope of comprehending the "big picture" unless one takes account of both relativity and quantum mechanics.

I have spoken already of quantum fluctuations that can stir up space and time with wormholes and quantum foam in the realm of the incredibly small. The very geometry of spacetime fluctuates, too. Saying that geometry fluctuates is the same as saying that gravity fluctuates. Over small-enough distances and short-enough times, the uncertainty principle rules. The smooth, predictable behavior of classical physics is replaced by random fluctuations. As Yakov Zel'dovich guessed and Stephen Hawking proved, such fluctuations even permit a black hole to evaporate (albeit slowly), thus evading the rule that nothing escapes from a black hole. The uncertainty principle with its implication of limitless fluctuations is one of the great messages of quantum mechanics.

The theory delivers its other great message at a different level, the level of the human observer and the measurement laboratory. No matter what the uncertainties of the small-scale world, no matter how chaotic the fluctuations, our knowledge of nature rests ultimately on perfectly definite, unambiguous observations—what we see directly or what our measuring apparatus tells us. How can this be? If the world "out there" is writhing like a barrel of eels, why do we detect a barrel of concrete when we look? To put the question differently, where is the boundary between the random uncertainty of the quantum world, where particles spring into and out of existence, and the orderly certainty of the classical world, where we live, see, and measure? This question, related to the correspondence principle discussed in Chapter 13, is as deep as any in modern physics. It drove the years-long debate between Bohr and Einstein. It has motivated international conferences. Books could be written about the quantum theory of measurement—and have been (Wojciech Zurek and I assembled such a book in 1983).

The common way of dealing with the question of measurement in quantum theory is to say that the act of measurement "collapses" uncertainty into certainty. The idea can be illustrated with a famous experiment—originally a thought experiment, but now a real one. A weak source of light sends photons, one at a time, toward an opaque plate that contains a pair of closely spaced slits. Beyond the opaque plate is an array of small detectors that can record the arrival of photons. The detectors produce signals that inform a human observer where each photon arrived.

Classically, there are no puzzles. Each photon passes through one slit or the other, and strikes a detector that tells which slit it passed through. The results of an experiment with only the first slit open added to the results of an experiment with only the second slit open gives the same results as an experiment with both slits open.

Quantum mechanically, the situation is much more interesting. Each photon is governed by laws of probability and behaves like a cloud *until it is detected*. It passes through *both* slits, not one or the other, and arrives *everywhere* at the detector array, with a large probability of arriving at certain detectors, a small probability of arriving at others, and zero probability of arriving at still others. Yet finally, if the detectors are sensitive enough to be triggered by single photons, each photon will be detected *somewhere*—somewhere quite specific. Where a particular photon will be detected is completely unpredictable. After many photons are detected, the calculated distribution of probability over the many detectors can be verified. The act of measurement is the transforming act that collapses uncertainty into certainty.

Watching this actual experiment in progress makes vivid the quantum behavior. One sees a flash at one point, then another, then another, then another. They seem random. At first no pattern is evident. Then, gradually, as the flashes accumulate—each one signaling a detection event—one sees places where many photons are detected, other places where none are detected, and a regular oscillation of intensity from strong to weak to strong to weak again. This oscillation precisely mirrors the original probability, a probability calculated on the assumption that each photon passes through both slits and

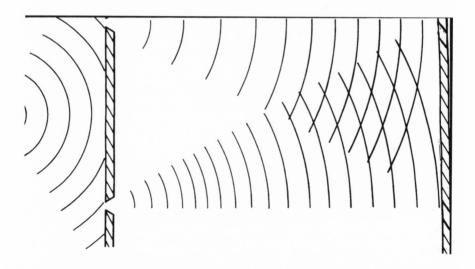

Interference of waves from two slits. Bohr's colleague Harald Høffding: "Where can the photon be said to *be*?" Niels Bohr: "To be? To be? What does it *mean* to be?"

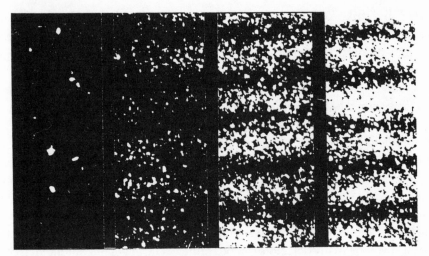

Four images showing the detection of electrons that have passed through a pair of closely spaced slits. (The electron, like the photon, is a particle with wave properties.) In the short exposure of less than one second at the left, the pattern appears random. As the exposure time is increased, until the two-minute exposure at the right is reached, the pattern resembles more and more the interference pattern expected of waves. From the spacing of the "fringes" in the two-minute pattern, one can measure the wavelength of the electrons. Because the pattern results from electrons flying one at a time toward the detector, one must conclude that each electron passes through both slits. The sum of results for two one-slit experiments would be quite different, with just two sharp lines, and no particles reaching the midpoint of the pattern, which here displays a maximum.
(*Courtesy of Hannes Lichte, Universities of Dresden and Tübingen.*)

that different parts of each photon cloud "interfere" with other parts. Moreover, unlike the classical situation, if the experiment is carried out first with only the first slit open, then with only the second slit open, the combined results do *not* mimic the results of the experiment with both slits open. Because the photons pass through both slits at once in the two-slit experiment, the result of this experiment is totally different from the summed results of two one-slit experiments.

I spoke above of classical behavior and quantum behavior. The quantum behavior is the one actually observed in experiments with photons or electrons. The classical behavior is only what *would* be observed if particles followed classical laws. They don't. But what of baseballs? Why do they follow classical laws if photons and electrons do not? Imagine a large metal barrier mounted near home plate on a baseball diamond, a barrier containing two holes, each, say, a foot in diameter, with a few inches of separation between them. When a pitcher throws baseballs toward the barrier, some go through one hole, some through the other (and some through neither). A baseball can't go through both holes at once. It seemingly has no cloud of probability, no interference, and no "collapse" of uncertainty to certainty. Well, in fact, it *does* have a cloud of probability, and there *is* collapse to certainty. But the baseball's cloud is of less than microscopic dimension. It is as if the baseball had a tiny skin of uncertainty, vastly less than the thickness of its actual skin, less even than the diameter of a single atom. Quantum fluctuations are large for small domains of space and time, and small for large domains of space and time. For the "huge" baseball, the quantum fluctuations are entirely below the threshold of observation.

Now, as to collapse of uncertainty to certainty. A photon goes entirely unobserved until it triggers a detector. From the moment of its emission until that moment of detection, one has no knowledge of its location or direction of travel. It remains an ethereal cloud of probability precisely because it is unobserved. The baseball, by contrast, can be observed repeatedly all along its course from pitcher to barrier to catcher. By any number of means—high-speed camera, radar, sonic rangefinder, even the human eye—its progress can be monitored. So detection, the thing that collapses uncertainty to certainty, is occurring repeatedly, providing certain information about where the baseball is and how fast it is moving and in what direction all along its path, not just at some final observation point. The photon and the baseball differ only in scale, not in principle.

(*Drawing by Chas. Addams;* © *1940 The New Yorker Magazine, Inc.*)

I want to pursue the double-slit experiment only a little further. Simple though it is in concept, it strikingly brings out the mind-bending strangeness of quantum theory. (It has been a touchstone for discourse about the implications of quantum theory for three-quarters of a century.) Imagine some object larger than a photon and smaller than a baseball—a large molecule, for instance. If fired at a barrier with slits, does it behave more like a photon or more like a baseball, more like a "pure" quantum object or more like a classical object? It could behave like either; it all depends on how we choose to examine it. If we subject it to no perturbing observations, its quantum properties can dominate. It can go through both slits at once, land at an unpredictable point, and reveal interference between one part of its cloud and another. Such behavior has been demonstrated for objects as large as whole atoms. If, however, we get overly curious, and follow it with observations along its path, it will respond by behaving classically. Then we cause its repeated "collapse" to certainty. By tracking it, we evaporate its cloud of uncertainty.

For twenty-eight years, in Europe and in America, Bohr and Einstein debated the meaning of quantum mechanics. These two giants, full of admiration for each other, never came to agreement. Einstein refused to believe that quantum mechanics provides an acceptable view of reality, yet he could never find an inconsistency in the theory. Bohr defended the theory, yet he could never escape being troubled by its strangeness. Reportedly, once when Einstein remarked, as he liked to do, that he could not believe that God played dice, Bohr said, "Einstein, stop telling God what to do."

A thought experiment that I first discussed in 1978 gets at the core of what fueled the Bohr-Einstein debate. Beyond illuminating that famous debate, this experiment may have something to tell us about the very machinery of the universe. I call it the "delayed-choice experiment." Here is how it works.

First, we borrow a baseball diamond.[6] On home plate we install a half-silvered mirror. This is a piece of glass with a thin reflective layer on one side, a gossamer layer of metal that reflects half the light that strikes it and lets the other half through. We put a light source nearby and arrange the light source and the half-silvered mirror so that half the light from the source is sent toward third base and half is sent toward first base. In quantum language, any photon has a 50 percent chance of being sent toward third base and a 50 percent chance of being sent toward first base. On the average, after many photons have been emitted, half will have gone each way.

Next, we mount fully reflective mirrors on the first and third bases. The one on third base reflects the light that hits it toward second base and on out into

---

[6] My apologies to readers not familiar with baseball. You may need to consult a baseball fan to help you visualize the placement of parts in this experiment.

right field. The one on first base reflects the light that hits it toward second base and on out into left field. If we mount detectors in left and right fields, they will tell us how many photons took each route. When the detector in right field clicks, signaling the arrival of a photon, we can conclude that a photon reached the detector via third base. When the detector in left field clicks, we can conclude that a photon has reached that detector via first base. Quantum mechanics predicts that the photons will arrive at random times at both detectors, but at the same average rate. Nothing strange yet. We have just arranged for photons to be sent randomly over two different routes, with equal probability, and whenever we detect a photon, we can determine what route it followed.

But wait. Quantum mechanics does more than say that photons may follow either route according to some random sequence. It says that the cloud of probability that *is* the photon until it is detected can take both routes at once! This is just like the double-slit experiment, where interference between the two paths shows that single photons go through both slits. Uncertainty collapses to certainty only when the measurement is made.

Now, for our baseball-diamond experiment, we can demonstrate that every photon does take both the first-base route and the third-base route. On second base we install another half-silvered mirror. This one is arranged so that half of the light reaching it from third base is reflected into left field and half is transmitted straight ahead into right field; and half the light reaching it from first base is reflected into right field and half is transmitted straight ahead into left field. This half-silvered mirror works in such a way that the two beams headed for left field *de*structively interfere with each other; that is, the crest of one probability wave overlaps the trough of the other probability wave, canceling it out, so no light reaches left field. The two beams headed for right field *con*structively interfere (meaning the crests of the two probability waves overlap and reinforce each other); all the light reaches right field. (By this time, if your ability to visualize baseball diamonds is not well honed, you may feel "out in left field" yourself, where there is no light.)

If we watch our two detectors for some time, we will find that the one in left field never clicks and the one in right field clicks at about twice the rate it did in our first baseball-diamond experiment. All the photons are going to right field. Now we have learned something new, that every photon went by *both* routes simultaneously, for otherwise there is no explaining the action of the half-silvered mirror on second base that sends them all to right field.

All is going well with our experiment when a dog wanders onto the field and onto the baseline between second and third bases. He interrupts the light path from third to second base, and suddenly the detectors change their behavior. The detectors in left and right field start clicking at about the same rate, each one at a quarter the rate of the right-field detector before the dog appeared. Half of the light is hitting the dog. The other half, all going via first base, is being split at second base to go half to one detector and half to the other. There is no interference, constructive or destructive, for only one path is now open. The mirror on second base simply splits the beam. Now the dog wanders across the infield, no longer interrupting any photon beam. Once again, the left-field detector goes silent and the right-field detector picks back up to its previous rate of recording photons, showing that every photon is again following both paths at once, not just one or the other. You can guess what happens when the dog crosses the baseline between home plate and first base. Both detectors are active at the same rate, as when the dog interrupted the other beam. Finally, the dog wanders off the field and we see again the result of each photon interfering with itself as it makes its way from home plate to right field.

The great lesson of quantum mechanics is that if we choose to measure one thing, we thereby prevent the measurement of something else. We can decide what we want to measure, but we can't decide to measure all properties of a system at once. The most elementary example of this limitation is that the position and speed of an electron cannot be measured at the same time. In our baseball-diamond experiment, we can choose to measure which path a photon followed (by having no half-silvered mirror at second base), but then we lose information about interference between different parts of the electron's probability cloud. Or we can choose to reveal the interference (by placing the half-silvered mirror at second base), but then we lose information about the path followed by the photon. More exactly, we make the whole idea of following a single path meaningless.

This is already mind-stretching. To make it more so, we come now to delayed choice. We will turn on the light source near home plate for only 1 billionth of a second, during which time it emits, say, 1,000 photons. At the speed of light, a photon travels only about 1 foot in a billionth of a second. So we can wait a leisurely 10 or 20 billionths of a second after the light source is turned off before we decide which experiment we want to do—that is, what we want to measure. During the time we are thinking it over, the photons are long gone from home plate, but none can yet have reached second base. They are somewhere en route. If we want to find out which route each photon followed, we need only remove the half-silvered mirror from second base and wait for the clicks of the detectors in left and right field to reveal the paths of each one. If we want to demonstrate that every photon followed both paths at once, we need only place our half-silvered mirror at second base (*after* all the photons have left home plate) and wait for the silence of the left-field detector and the increased rate of clicking of the right-field detector to tell us that every photon went by both paths at once.

As with some other thought experiments, the march of technology has caught up with and made it a real experiment. At the University of Maryland, Carroll Alley, Oleg Jakubowicz, and William Wickes—on a laboratory bench, not a baseball diamond—demonstrated delayed choice in 1984. The strangeness of the quantum world, from which Einstein incessantly sought escape and from which Bohr saw no escape, is real.

If delayed choice is real in the laboratory, it is surely real on a baseball diamond and real in the universe at large. We need only expand the dimension of the baseball diamond to a billion light-years, putting a quasar at home plate, galaxies at first and third bases, and Earth, with its telescopes and counters, at second base. Galaxies 1 and 3, as we may choose to call them, are capable of bending the quasar light so that it will reach Earth by two different paths. (Such galaxies really exist. Gravity can bend light as surely as it bends moving material particles.) If we point a telescope at Galaxy 1, we see photons from the quasar that were deflected as they passed near Galaxy 1. If we point a telescope at Galaxy 3, we see photons from the quasar that were deflected as they passed near Galaxy 3. But if we put a half-silvered mirror (like the one we had at second base) at the place in our observatory where light from both Galaxies 1 and 3 is directed, we can—in principle—cause quasar light from these two directions to interfere so that all of it goes off in one direction, none in the other. Moreover, the rate of arrival of light from the quasar could be so low—again in principle—that only one photon at a time is detected, with a waiting time until the next one arrives. What interpretation is then possible but that each single photon on its billion-year trip from the quasar to Earth followed both paths via both galaxies in the form of ephemeral clouds of probability spreading through remote reaches of space until we pin down that photon with our measurement? Since we make our decision whether to measure the interference from the two paths or to determine which path was followed a billion or so years after the photon started its journey, we must conclude that our very act of measurement not only revealed the nature of the photon's history on its way to us, but in some sense *determined* that history. The past history of the universe has no more validity than is assigned by the measurements we make—now!

Reasoning like this has made me ask whether the universe is a "self-excited circuit"—a system whose existence and whose history[7] are determined by measurements. By "measurement" I do not mean an observation carried out

---

[7] And perhaps even whose laws. But I leave a discussion of that possibility to the next chapter.

by a human or a human-designed instrument—or by any extraterrestrial intelligence, or even by an ant or an amoeba. Life is not a necessary part of this equation. A measurement, in this context, is an irreversible act in which uncertainty collapses to certainty. It is the link between the quantum and classical worlds, the point where what *might* happen—multiple paths, interference patterns, spreading clouds of probability—is replaced by what *does* happen: some event in the classical world, whether the click of a counter, the activation of an optic nerve in someone's eye, or just the coalescence of a glob of matter triggered by a quantum event.

The event that I am calling a measurement is what Niels Bohr called "registration." Bohr liked aphorisms. No elementary phenomenon, he said, is a phenomenon until it is a registered phenomenon. For example, a high-energy particle emitted in a radioactive event spreads in all directions as a cloud of probability. If it leaves a trail of disrupted atoms in a piece of mica, a trail that could be observed a million years later, its uncertainty (of position and direction) has collapsed to certainty. The unknown and unknowable spread of the particle in all directions has been replaced by a "measurement" or "observation" or "registration"—a macroscopic trail in mica that cannot be reversed or erased or modified by the future history of the particle. What the particle might have done—or, in a quantum sense, all the things it is doing simultaneously—is replaced by what it did in fact do.

Not all potentiality is converted to actuality in any finite time. There are innumerable clouds of probability running around in the universe that have yet to trigger some registered event in the macroscopic world. We have every right to assume that the universe is filled with more uncertainty than certainty. What we know about the universe—indeed, what is knowable—is based on a few iron gateposts of observation plastered over by papier-mâché molded from our theories. Those iron gateposts tell us not only what is, but what was. My diagram of a big U (for universe) attempts to illustrate this idea. The upper right end of the U represents the Big Bang, when it all started. Moving down along the thin right leg and up along the thick left leg of the U symbolically traces the evolution of the universe, from small to large—time enough for life and mind to develop. At the upper left of the U sits, finally, the eye of the observer. By looking back, by observing what happened in the earliest days of the universe, we give reality to those days.

The eye could as well be a piece of mica. It need not be part of an intelligent being. The point is that the universe is a grand synthesis, putting itself together all the time as a whole. Its history is not a history as we usually conceive history. It is not one thing happening after another after another. It is a totality in which what happens "now" gives reality to what happened "then," perhaps even determines what happened then.

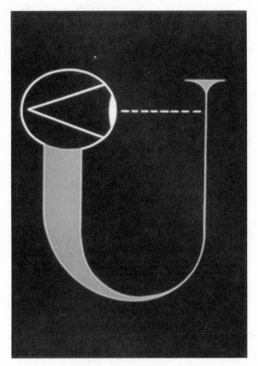

Does looking back "now" give reality to what
happened "then"?

Many students of chemistry and physics, entering upon their study of quantum mechanics, are told that quantum mechanics shows its essence in waves, or clouds, of probability. A system such as an atom is described by a wave function. This function satisfies the equation that Erwin Schrödinger published in 1926. The electron, in this description, is no longer a nugget of matter located at a point. It is pictured as a wave spread throughout the volume of the atom (or other region of space).

This picture is all right as far as it goes. It properly emphasizes the central role of probability in quantum mechanics. The wave function tells where the electron might be, not where it is. But, to my mind, the Schrödinger wave fails to capture the true essence of quantum mechanics. That essence, as the delayed-choice experiment shows, is *measurement*. A suitable experiment can, in fact, locate an electron at a particular place within the atom. A different experiment can tell how fast the electron is moving. The wave function is not central to what we actually know about an electron or an atom. It only tells us the likelihood that a particular experiment will yield a particular result. It is the experiment that provides actual information.

Measurement, the act of turning potentiality into actuality, is an act of

There once was a man who said "God
Must think it exceedingly odd
　　If He finds that this tree
　　Continues to be,
When there's no-one about in the Quad."

Ronald Knox

Dear Sir, Your astonishment's odd:
I am always about in the Quad.
　　And that's why the tree
　　Will continue to be,
Since observed by Yours faithfully, God.

Anon.

Does seeing make it so?
(*Limericks from E. O. Parrott, Ed.,* The Penguin Book of Limericks,
*New York: Viking Penguin, 1986, p. 55. Ronald Knox reprinted by per-
mission of A. P. Watt Ltd. on behalf of Earl Oxford & Asquith.*)

choice, choice among possible outcomes. After the measurement, there are roads not taken. Before the measurement, all roads are possible—one can even say that all roads are being taken at once.

Thinking about quantum mechanics in this way, I have been led to think of analogies between the way a computer works and the way the universe works. The computer is built on yes-no logic. So, perhaps, is the universe. Did an electron pass through slit A or did it not? Did it cause counter B to click or counter C to click? These are the iron posts of observation.

Yet one enormous difference separates the computer and the universe—chance. In principle, the output of a computer is precisely determined by the input (remember the programmer's famous admonition: Garbage in, garbage out). Chance plays no role. In the universe, by contrast, chance plays a dominant role. The laws of physics tell us only what *may* happen. Actual measurement tells us what *is* happening (or what *did* happen). Despite this difference, it is not unreasonable to imagine that information sits at the core of physics, just as it sits at the core of a computer.

Trying to wrap my brain around this idea of information theory as the basis of existence, I came up with the phrase "it from bit." The universe and all that it contains ("it") may arise from the myriad yes-no choices of measurement (the "bits"). Niels Bohr wrestled for most of his life with the question of how acts of measurement (or "registration") may affect reality. It is registration—whether by a person or a device or a piece of mica (anything that can preserve a record)—that changes potentiality into actuality. I build only a little on the structure of Bohr's thinking when I suggest that we may never understand this strange thing, the quantum, until we understand how information may underlie reality. Information may not be just what we *learn* about the world. It may be what *makes* the world.

An example of the idea of it from bit: When a photon is absorbed, and thereby "measured"—until its absorption, it had no true reality—an unsplittable bit of information is added to what we know about the world, *and*, at the same time, that bit of information determines the structure of one small part of the world. It *creates* the reality of the time and place of that photon's interaction.

Another example: The surface area of the spherical horizon surrounding a black hole measures the black hole's entropy, and entropy is nothing more than the grand totality of lost information. For a black hole whose horizon spans even a few kilometers, the number of bits of lost information is large beyond any normal meaning of large, even beyond anything we call "astronomical." Nevertheless, it is not unimaginable. We have an *it* (the area of the black hole's horizon) fixed by the number of *bits* of information shielded by that area.

Often quoted is the saying attributed to the architect Ludwig Mies van der Rohe, "Less is more." It is a good principle of design, even a good principle of physics research. In thinking about the world in the large, I have another phrase that I like, borrowed from my Princeton colleague Philip Anderson: "More is different." When you put enough elementary units together, you get something that is more than the sum of these units. A substance made of a great number of molecules, for instance, has properties such as pressure and temperature that no one molecule possesses. It may be a solid or a liquid or a gas, although no single molecule is solid or liquid or gas.

"More is different" may have something to do with "it from bit." The rich complexity of the universe as a whole does not in any way preclude an extremely simple element such as a bit of information from being what the universe is made of. When enough simple elements are stirred together, there is no limit to what can result.

The Vietnam era, extending roughly from the mid-1960s to the early 1970s, was a watershed in American society—and, to some extent, throughout the world. Many good things happened then: an increased awareness of the need for equity for minorities and women, an increased tolerance for diversity, a heightened concern for the environment, an intensified dialogue about values. We continue to reap the benefits of those changes. But many not-so-

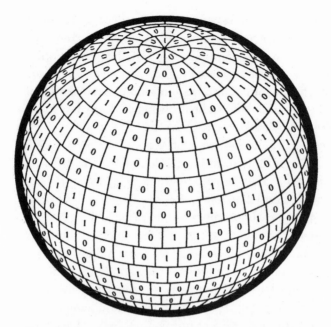

The number of Planck areas needed to tile the horizon of
a black hole determines how many bits of information
are shielded by the horizon.
(From A Journey into Gravity and Spacetime *by Wheeler*
© *1990 by John Archibald Wheeler. Used with permission of*
W. H. Freeman and Company.)

good things happened, too: cynicism that transformed some productive people into nonproductive ones, increased use of drugs, a decline in civility in
intellectual discourse, permissiveness that eroded ethical standards, antagonism toward science, and increased acceptance of the nonrational and the
nonscientific. The fact that I like to explore the very limits of science, that I
like to speculate on what directions future science might take, makes me
particularly aware of the boundary between science and nonscience and particularly intolerant of pseudoscience.

   To my discomfort and distress, as I got more and more involved in exploring the world of the quantum and as I wrote on subjects like the delayed-
choice experiment, I found myself being increasingly cited by the
pseudoscientists who were looking for scientific underpinnings for their
moonshine. There arose in the 1970s a move to "legitimize" parapsychology
by trying to find real science to explain it. Some competent scientists were
drawn into this effort, but they wholly missed the mark, mostly by misunderstanding the quantum theory and the work that had been done on measurement theory. I didn't mind that some of my respected colleagues in science

thought that I myself had gone a little bit around the bend. They were entitled to remain more conservative, as I tried to be daring. But it did bother me greatly when I found my work cited as supporting the paranormal.

In 1969, over my objection and over my minority vote on its board of directors, the American Association for the Advancement of Science (AAAS), under the presidency of the anthropologist Margaret Mead, admitted the Parapsychological Association as an affiliated organization. When I had the opportunity a decade later to speak at the annual meeting of AAAS in Houston (January 1979), I chose as my topic the quantum theory of measurement. I wanted to emphasize in this talk that the essential feature of an act of "measurement" is amplification from the quantum thing observed to the classical thing doing the observing, which need have nothing to do with human intervention or human consciousness. (My somewhat ponderous title was "Not Consciousness but the Distinction between the Probe and the Probed as Central to the Elemental Quantum Act of Observation.")

My friend and distinguished colleague Eugene Wigner shared the podium with me in Houston. We had learned, to our surprise and dismay, that we in turn were to share the podium with several parapsychologists. Too late to escape. I delivered my talk with its discussion of the essentials of quantum measurement and reference to some of the apparent paradoxes that so engaged Bohr and Einstein as they wrestled with the quantum. Wanting to escape guilt by association, I had prepared in advance two appendices intended for the press. After the talk, I wrote a letter to William Carey, then director of AAAS, suggesting that this most prestigious of American scientific organizations establish a panel to decide whether it was time to eliminate parapsychology from the organization. The blessing that AAAS had bestowed on pseudoscience was, I felt, a consequence of the permissiveness of the Vietnam era, and it was time to undo it.

My two appendices, as well as a copy of my letter to Bill Carey, reached a large audience when they appeared in the *New York Review of Books* in May 1979. Introducing them was an excellent essay by Martin Gardner, the popular writer on mathematics and science whose work included the book *Fads and Fallacies in the Name of Science*. My first appendix, "Drive the Pseudos Out of the Workshop of Science," was a defense of the rights of the parapsychologists coupled with a plea that they be banished from under the AAAS umbrella, whose shade was not necessary to protect their rights. The content of the second appendix is well summarized by its title, "Where There's Smoke, There's Smoke."

All of this had no effect on AAAS. The Parapsychological Association remains an affiliate to this day, and parapsychology nonscience continues to flourish.

16
·
THE END OF TIME

TIME IS anything but simple.

In reflecting on the course of one's own life, it's easy to think of time as Isaac Newton thought of it in the seventeenth century—as a river, something flowing relentlessly forward, incapable of being slowed down or speeded up or turned aside by anything that we do or that nature does. "Absolute, true, and mathematical time," Newton wrote in the opening section of his *Principia*, "of itself, and from its own nature, flows equably without relation to anything external." The flow of Newton's time is measured by repetitive motion: the swing of a pendulum, the oscillation of a quartz crystal, the circling of Earth around the Sun. It served admirably through the eighteenth and nineteenth centuries to provide temporal mileposts for anything that happened in classical physics—mechanical motion, heat, electricity and magnetism, light and radio waves. To some scientists, moving *through* time seems a more congenial image than time as a flowing river, but the result is the same—Earth and its inhabitants and all the world inexorably aging at the same rate. In common speech, we often talk of time passing (Newton's flow), but we talk also of moving from this point in time to that point in time (time sitting there as we move through it).

The most unsettling consequence of Einstein's 1905 special theory of relativity is that time is relative. This relativity of time is commonplace in the laboratory (for instance, we observe that high-speed radioactive particles live longer than low-speed ones), but it has so far not been directly experienced by humans. To see what it means, imagine a future in which high-speed space travel is an everyday matter. Suppose that on a certain future date, all of Earth's inhabitants agree that it is Saturday, January 1, 2050. If, on that date, tourists head off into space at various speeds (all near the speed of light) for galactic tours of various durations and then return to Earth, time will get rapidly out of joint. When you later call on your friends—some of whom made short trips, others long trips, some of whom traveled on tour buses that zipped at very close to the speed of light, others a bit less speedily, and still others who have stayed at home—you will see that some have aged a little, some a lot. On the walls of their homes, not only will the clocks show different times, the calendars will show different months and different years. In one family where the parents went on the deluxe tour and the children stayed home, the parents are now younger than the children. The question What time is it? has no single answer. Even the question What date is it? gets different answers—all correct—from different people. In that future world, every person's time depends on that person's travel history.

Only for the imagined speeds of human travel is this science fiction. Otherwise it is solid science. It is even "old-fashioned" science. If such speeds could be reached, there would be a myriad of dates and times on Earth, with no possibility of one date and one time.

The other thing that special relativity did for time is to join it with space into the four-dimensional entity spacetime. It was Hermann Minkowski, not Einstein himself, who in 1908 conceived this beautiful geometric merger. "Henceforth," said Minkowski, "space by itself, and time by itself, are doomed to fade away into mere shadows, and only a kind of union of the two will preserve an independent reality."[1] Minkowski's vision powerfully altered the way we look at the world around us. He showed that Einstein's theory unites not only space with time, but also the electric field with the magnetic field, and energy with momentum. "Event" acquired a new meaning in Minkowski's world. It is a point in spacetime—something occurring at a particular place and a particular time. Events are joined by "world lines"—tracks through spacetime. A consequence of this new spacetime view is that motion *through* time, or motion *of* time—whichever way you chose to look at classical time— is replaced by static time, time that is just "there" in the same way that space is there. A world line becomes a line on a map, a four-dimensional map on which the line traces a path through both space and time.

The spacetime of special relativity has been called a canvas, a canvas on which are painted the points and lines showing all that has happened, is hap-

---

[1] This excerpt from a talk by Minkowski appears in Alan L. Mackay, Ed., *Dictionary of Scientific Quotations* (Bristol, England: Institute of Physics Publishing, 1991), p. 174.

pening, and will happen. As I mentioned earlier, spacetime has also been called a stage, the stage on which all action takes place. Whether canvas or stage, it is the arena of action, not the action itself. *General* relativity, Einstein's monumental 1915 achievement, changed all this. Space and time join the action. Perhaps, as I have been driven to speculate, they are *all* the action. A world made only of spacetime is a conceivable world. Its hills and valleys and ripples and wormholes could mimic any matter and any field. As I've mentioned in earlier chapters, the formulation of general relativity that has appealed to me is the one I call geometrodynamics, which treats not just the paths of world lines through spacetime, but the motion of the entire blob of space—three-dimensional space—through time. The geometry of that space evolves as it moves, perhaps in marvelous, strange ways. So geometrodynamics is the dynamics of geometry.

Time as part of the action is only part of the story. There are still-stranger features of time that continue to intrigue me. One of these is the asymmetry of time. Another is the quantum fluctuation of time. Beyond those two is the strangest feature of all, the end of time.

We can go left and right, or forward and backward, or up and down, all with equal ease. Nothing in the laws of nature or in experience tells us that we are limited to moving through space in only one direction. Then why do we seem to move through time as if it were a one-way street? Why does time have an arrow? Why do we remember the past but not the future? These questions were sharpened by twentieth-century physics, but they were born earlier.

In Chapter 8, I introduced a feature of Newtonian mechanics called "time-reversal invariance." It means that for any motion that does occur or can occur, the motion in reverse is also possible. If a pendulum bob swings from left to right, it can swing from right to left, an exact reversal of the left-to-right motion in every detail. If you stroll from your house to the corner, you *could* (but not necessarily *do*) stroll backward from the corner to your house. A good way to think of time reversal is to imagine that some example of motion is recorded with a video recorder. When the scene is replayed, it is shown exactly as it occurred. But if the video is played backward, you are looking at something that did *not* occur. Newtonian mechanics says that the reversed display *could* have occurred and that by watching the video, you have no way to know whether you are seeing the actual motion or its reversal.

Imagine, for instance, an extraterrestrial who lives near the North Star trying out his new camcorder by focusing on our solar system. He sees, and records, a group of planets circling counterclockwise around the Sun. Then he invites some friends over and plays the recording backward for them. They see a group of planets circling *clockwise* around the Sun. Can they tell whether they are seeing the actual planetary motion or its reversed image?

No. All the planets of the solar system *could* move in the other direction, and if they did, their motion could be, in every particular, the exact reversal of what is their actual motion. That's what time-reversal invariance means.

But wait a minute. What if you recorded something as simple as a piece of paper being torn in two. If you looked at the video backward, you would see two pieces of paper, each with a ragged edge, miraculously being joined together, the ragged edges giving way to a smooth, seamless union. You would have no trouble at all knowing whether you are seeing a record of the actual occurrence or a time-reversed view of it. The forward-in-time view can happen (and did happen). The time-reversed view could never happen, you say. Well, Newton and those who followed him in the nineteenth century would say, you are not quite right. In fact, it *could* happen, albeit with an incredibly small probability. You won't live long enough to see it happen. The universe won't even last long enough for it to happen, yet in principle two torn sheets *could* weld themselves into a single smooth sheet. Two wrecked automobiles *could* back away from their point of impact, healing their damage as they do so. The molecules of perfume that have spread to fill a room *could* migrate back and regather on the starlet who entered the room.

So, for complex events, time does have an arrow. Most series of events occur in one direction, but not the other. For just about everything in the world around you, you can easily tell the difference between forward-in-time and backward-in-time. But the lesson of nineteenth-century physics is that this difference is a matter of probabilities, not absolutes. One direction might be quite likely, the other direction incredibly, ridiculously unlikely. Yet unlikely is not quite the same as impossible. So Newtonian time-reversal invariance is saved, even if it doesn't seem to have much to do with the world that we experience.

For simple systems, forward-in-time and backward-in-time motion might be equally likely. The solar system has such simplicity. Just one sun, nine planets, and no friction. The torn piece of paper, with its billions upon billions of atoms, is far more complex. For it, the discrepancy between the likely and the unlikely is so vast that, for all practical purposes, one is possible and one is not possible.

The second law of thermodynamics relates probabilities to time's arrow. It says that any system left to itself (free of outside influences) will tend toward greater disorder. There are more ways to be disorderly than to be orderly. Disorder is therefore more probable than order. If the system is quite complex, the probability of disorder exceeds the probability of order by an enormous factor—so enormous that, for all practical purposes, there is only one direction of change (for an isolated system): from order to disorder. Picture an adobe house in a desert community. Left untended, it deteriorates and erodes.

After enough time passes, it will be a mound of earth. Still more time—perhaps hundreds of years—and there will be no evidence that it was ever there. That is the second law of thermodynamics at work, order changing spontaneously to disorder.

The intriguing thought that follows from these considerations is that we are aware of a one-way flow of time only because we are ourselves complex systems interacting with other complex systems. We remember the past and not the future not because there is any fundamental asymmetry in time but because of the overwhelming disparity between the likely and the unlikely in everything that we are and do and see.

Back when Dick Feynman and I were talking about electrons moving with equal ease backward and forward in time, we realized that such a way of thinking made sense because of the extreme simplicity of the electron. You can tell by looking into a person's face something about what that person has been through. You cannot tell anything about an electron's history by looking at it. Every electron is exactly like every other electron, unscarred by its past, not blessed with a memory—of either the human or computer variety. The electron pays for its freedom to move forward and backward in time by remembering neither future nor past. We remember the past and are trapped in one-way motion through time.

In speaking of nineteenth-century physics, I am speaking really of the sciences of thermodynamics and statistical mechanics. Developments in those fields showed how the statistics of large numbers can convert the time-symmetric laws that govern simple events into the time-asymmetric laws that we see governing complex events.

For more than half of the twentieth century, it appeared that time symmetry in the small remained the rule, with time asymmetry being an artifact of complexity. As particles were discovered, as nuclei were explored, as quantum electrodynamics evolved, at first nothing marred this picture of perfect time symmetry for all basic laws. In the work that Feynman and I did on action at a distance, for instance, we found that the apparent one-way flow of radiation—forward in time, not backward—could be entirely accounted for by the large-scale distribution of absorbing mass in the universe, lots of it. It required no time asymmetry in the fundamental laws of electrodynamics.

Then in 1964, James Cronin (now at the University of Chicago) and Val Fitch, with their colleagues James Christenson and René Turlay, discovered that time-reversal invariance fails for the decay of the K meson, or kaon. Here, for the first time, was an example of time asymmetry at the elementary level of single particles, not the level of complex systems.

Since earliest times, science has been driven by a faith in simplicity. A faith that laws exist and that they are simple, unchanging laws. A faith that as we probe more deeply, to smaller and smaller units of matter, we find ever simpler systems and ever simpler laws governing them. Twentieth-century science has in part sustained this faith, and in part it has not. On the one hand, fundamental laws such as those of Einstein's general relativity and Dirac's electron theory are expressed by equations of breathtaking brevity and generality. Most physicists call these equations both simple and beautiful. They sustain the faith that nature at its core is simple. On the other hand, when we see time symmetry marred in an elementary process, when we contemplate the writhings of spacetime in wormholes and quantum foam, when we see tiny deviations from Dirac's predictions for the electron produced by quantum fluctuations, we realize that the "floor" of simplicity as we move to smaller and smaller domains is illusory. Beneath that floor, in still smaller domains, chaos and complexity reign again.

There is a story attributed to William James. Whether apocryphal or not I don't know, but I like it. James said, supposedly, in a lecture that the world is supported by an elephant. He was making the point, presumably, that beneath the superficial reality of the world around us is a deeper reality. A woman in the audience rose to ask, "What supports the elephant?"

"A turtle, madam," he replied.

"And, what," she persisted, "supports the turtle?"

"Another turtle," answered James.

"And what supports that turtle?" she asked.

"Madam, it's turtles all the way down."

No, it isn't turtles all the way down. To be sure, the observed mechanical properties of substances have been explained in terms of atoms, and atoms have been explained in terms of electrons and protons and neutrons, and protons and neutrons have been explained in terms of quarks. There are a few turtles standing on other turtles. But we have many reasons to believe that the layers of explanation do not continue indefinitely. For one reason, the bits of matter cease to be distinct as we go deeper. No two pencil leads are identical, but every atom of carbon 12 (the principal constituent of graphite) is identical to every other atom of carbon 12. If the number of layers (the number of turtles) were infinite, why would an atom of carbon not be as complex as a pencil lead? For another reason, quantum theory tells us that things get more chaotic, not more orderly, as we go deeper. The ultimate basis of reality is not likely to be found in the unpredictable fluctuations that characterize the smallest dimensions, the deepest layers. How can a turtle composed of seething quantum foam, in which even space and time cease to have meaning, be patiently holding up more solid turtles?

Time itself, I have come to believe, participates in the general complexity, in the fluctuations, in the uncertainty. The smooth flow of time—or our smooth passage through it—is an illusion that is shattered when we look at short-enough intervals of time, and when we ask about time at the moment of the Big Bang, at a moment of gravitational collapse, at the moment of the Big Crunch.

Students and others often ask what existed before the Big Bang. To say that we don't know is not to say enough. Even to say that we have no *way* of knowing is not enough. We really have to say that space and time came into existence, along with matter and energy and the laws of physics, at the moment of the Big Bang. If the universe expands to a maximum size, starts contracting, and eventually collapses to a fiery death—a fate that seems likely to me and to some other theorists, even though the evidence for it is still weak—then time, and space too, will end in this Big Crunch. I can reach no conclusion other than this: there was no "before" before the Big Bang, and there will be no "after" after the Big Crunch.

But we don't have to look back to the Big Bang to speak of the beginning of time or look forward to a Big Crunch to speak of the end of time. Black holes are more than likely coming into existence throughout the universe right now, and some may be evaporating. Every black hole brings an end to time and space and the laws of physics in itself as surely as the Big Crunch will bring an end to the universe as a whole. Even though we have yet to see a black hole in process of formation, we see plenty of evidence for the boiling ferment that characterizes the universe. The exploding stars we call supernovas pop up frequently in astronomers' telescopes. A recent near one was Supernova 1987A in the Magellanic Cloud, a close neighbor of our own galaxy. A supernova of July 1054 within our galaxy has left behind the beautiful Crab Nebula with a rotating neutron star (a pulsar) at its center. Astronomers have found numerous other pulsars. How can we doubt that black holes, too, are forming in great numbers. It would be a strange oddity if they were not.

Theory suggests also that black holes of incredibly small size, at the scale of the so-called Planck length, are forming and dissolving all the time by the trillion, within the dimensions of every elementary particle. At that scale, with spacetime churned into quantum foam, space and time in fact lose their meaning. When we blend the two greatest theories of the twentieth century, quantum theory and general relativity, we have to conclude that time is a secondary concept, a derived concept. It has meaning only at a scale large compared with the Planck length and only well away from black holes, the Big Bang, or the Big Crunch. It is not a river that rolls inexorably forward. It is not a lake across which we glide. It is more to be compared with temperature or with entropy, concepts that take their meaning only when large numbers of particles are involved. Time, we must conclude, is of statistical origin, valid only when dimensions are large enough and when conditions are not too extreme.

According to a graffito that I once saw in the men's room of the Pecan Street Cafe in Austin, Texas, "Time is nature's way to keep everything from happening all at once." Maybe that's as good a definition as any. Time is, in fact, an immensely complex idea that sits at the core of critical unanswered questions about the universe and existence, questions I can't stop pondering.

By 1970, I had become convinced not only that black holes are an inevitable consequence of general relativity theory and that they are likely to exist in profusion in the universe, but also that their existence implies the *mutability* of physical law. If time can end in a black hole, if space can be crumpled to nothingness in its center, if the number of particles within a black hole has no meaning, then why should we believe that there is anything special, anything unique, about the *laws* of physics that we discover and apply? These laws must have come into existence with the Big Bang as surely as space and time did.

Let me shift to a biological analogy. Life, we have every reason to believe, arose from nonlife. Why did it take the forms it has taken and evolve as it has? There is surely not one path only that it could follow. Chance played a major role. Chance that the life we are familiar with happened to get started on this particular planet near this particular sun. Chance that a moon provided tides, and a rotating Earth provided changes of wind and weather. We have no reason to believe that life on some other planet near some other star has much in common with life that we know here—and every reason to think that even on our own planet, life could very well have followed a quite different evolutionary track. Referring to one of his detractors, Charles Darwin said, shortly after publication of *The Origin of Species*, "[The astronomer Sir John] Herschel says my book is 'the law of higgledy-piggledy'."[2]

My conception of the origin of physical law in the Big Bang is similar. By whatever higgledy-piggledy chance, space arose, time arose, laws of physics arose. Perhaps, as with the origin of life on Earth, there were limits to what might arise. It wasn't "anything goes." But it also wasn't "only one thing goes." And, just as life arose from nonlife on Earth, something arose from nothing in the universe. That "nothing" from which something arose should not, however, be confused with the emptiness of a vacuum. It is nothing in a profounder sense. It is nothingness. Why is the universe what it is? What other way might it have been? These are questions that have not been answered, indeed scarcely addressed.

[2] Quotation taken from Charles Hershaw Ward, *Charles Darwin: The Man and His Warfare* (Indianapolis, Ind.: Bobbs-Merrill, 1927), p. 297.

High Island, South Bristol, Maine.
(Photograph by Jack Lane.)

For more than forty years, now, our summer home on High Island, Maine, has been the retreat that has best allowed me to puzzle over these and other questions. That is where much of the book *Gravitation* took shape, especially during visits by Charlie Misner and Kip Thorne, and where *Journey into Gravity and Spacetime* was finally completed under the persistently pleasant prodding of Susan Moran.

My very first summer get-away was in Benson, Vermont, where Janette and I honeymooned in 1935. Later, we went often to the summer home of Janette's parents in Salisbury Cove, Maine, on Mt. Desert Island. When the children had gotten older and we had put aside a bit of money, it was natural to look to New England as a place to find our own summer home. Our 1956 stay in crowded Holland provided a stimulus. Among our friends there, only the Oorts had found a way to escape to peaceful solitude. They spent Sundays on a houseboat in the middle of a lake.

Weekend after weekend in late 1956 and early 1957 we searched for a spacious, serene seaside property, starting in Connecticut and working our way up the coast. Finally, in April, only Maine remained to be examined. A brochure from the Maine chamber of commerce brought three possibilities to our attention. At my suggestion, Janette flew to Portland, rented a car, and drove some 70 miles to look at the properties. The first was run-down. The second was everything she had imagined: a beautiful big house with a wide lawn sloping down to the sea, all at a remarkably reasonable price. "Why are you selling?" Janette started to ask the owner, just as a squadron of Navy planes from the Brunswick Naval Air Station roared low overhead. The owner's reason for selling was our reason not to buy.

Snow still covered most of High Island when Janette phoned me from there that April, entranced with this third possibility. Hearing her enthusiasm, I said, "Why don't you go ahead and get it." It was a property of about 40 acres, half of High Island, with bluffs overlooking the sea, and including one house, a simple log cabin. My first look at it came in June. Janette was apprehensive about what my reaction might be. At the end of the school year, with the children filling up the back seat of our little car and a small rental trailer in tow, we headed for our new summer place. I plunked down a chair on the lawn and sat for most of an hour drinking in the view. Janette's apprehension vanished.

The log cabin, refurbished, served as our summer living place until the children got married and the family began to grow larger. Now there are seven living places on our part of the island, including the house designed for us by the Princeton architect Victor Olgyay (who had also designed our Maxwell Lane house). High Island is where many a family reunion has taken place and where Janette and I celebrated our fiftieth and sixtieth wedding anniversaries — each one complete with T-shirts designed for the occasion. After traveling the world over, I still find our High Island home the most refreshing, relaxing, inspiring place I know. We have never regretted for a moment the acquisition of the property or the design and placement of the new house. Whether indoors or out, I can look out over rocky headlands and blue sea toward Portugal.

Other places have been important to me: Copenhagen, where I joined the international family of physics and formed a lifelong attachment to Niels Bohr; Princeton, where I spent more time than anywhere else and had a long succession of inspiring students; and Austin, where, in a lively decade, aided again by notable students, I tried to blend relativity and the quantum. But High Island is special. If, in what time is left to me, I have any thoughts worthy of helping to guide physics into the twenty-first century, those thoughts are likely to arise on High Island.

At age seventy-five I marked ten years at the University of Texas in Austin. There I had let my mind run free over the nature of space and time, black holes and quanta, measurement and information. There I was not only tolerated as I pulled aside from the herd in theoretical physics to pursue my own

Janette and I celebrated our fiftieth wedding anniversary in 1985, surrounded by our three children, their spouses, eight grandchildren, and one grandchild's fiancé.

byways, I was encouraged and supported. But as my seventy-fifth birthday approached, I realized that it was time to think of retirement from the formal duties of a university professor. Janette agreed, and we began to plan a return to the Princeton area, close to our children and grandchildren and a few great grandchildren, close to my old institution, Princeton University. In Hightstown, 11 miles from Princeton, we found a congenial retirement community, Meadow Lakes, with services that can see us through any future contingencies. (By general agreement among Meadow Lakes' 300 residents, whose average age is eighty-two, those of us under ninety call ourselves "middle-aged.")

In March 1986, after completing the formidable task of winnowing down our books and other possessions to manageable mass and volume, Janette and I moved into a charming apartment at Meadow Lakes, where we now happily reside. From there I go almost daily to my office at Princeton, kindly provided by the university. Down the hall from my Jadwin Hall office is my stimulating colleague Val Fitch. Not far away, until Alzheimer's got the better of his mind around 1990—he died in 1994—was my old and good friend Eugene Wigner. Many other colleagues, too numerous to mention individually, keep Princeton a vital, attractive place to work.

In these later years, I have dared to think about and write about and ask about the physical world in terms that some of my colleagues consider outside the scope of science—science as it is now accepted, defined, and practiced.

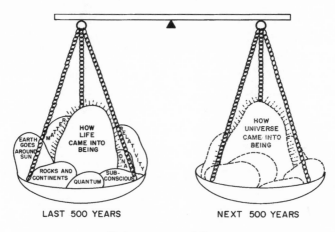

LAST 500 YEARS   NEXT 500 YEARS

Science past and future.

Is the universe a self-excited circuit, made real by observation? Does physics rest on yes-no bits of information? Do the laws of physics come into existence in the higgledy-piggledy of the Big Bang, and are they extinguished in the Big Crunch? Is time a secondary, derived concept? Can the laws of physics mutate in the way that living organisms mutate? What deeper principle will one day make quantum theory seem inevitable and natural? Will a comprehensive view of the physical world come not from the bottom up— from an endless tower of turtles standing one on the other—but from a grand pattern linking all of its parts? I do not apologize for questions like these and hope that they are not the result merely of a weakening brain. The student of physics—the future researcher—needs to be stimulated not only by learning of recent solid advances in the subject but also by hearing speculative thoughts on where the subject might be headed. Einstein liked to say that he had earned the right to make mistakes. I hope that I have earned the right to speculate on the future shape of the subject to which I have devoted my life.

I think back to Bernard de Spinoza in the seventeenth century; to Bernhard Riemann, William Kingdon Clifford, and Ernst Mach in the nineteenth; and to Niels Bohr in the twentieth. Spinoza's belief in the harmony, the beauty, and the ultimate comprehensibility of nature had a profound effect on Einstein. What else could have led Einstein to express his sympathy for the Good Lord if general relativity proved to be wrong? Through Einstein and his followers—Dirac, Murray Gell-Mann, Feynman, and others—it has become an article of faith in twentieth-century physics that if a theory is simple enough, comprehensive enough, "beautiful" enough, it must be right.

More than half a century before Einstein, the great German mathematician Bernhard Riemann dared to speculate that the geometry of physical space need not be God-given Euclidean space, but could well be some other geometry and should be determined by experiment, not by hypothesis. William Kingdon Clifford imagined that space could be a dynamic substance capable of deformation that could propagate like waves. Later, Mach postulated that inertia "here" results from mass "there." Einstein brought all of these speculations into the realm of "real" science. And it was not mere coincidence. He was influenced by the thinking, at least, of Riemann and Mach.

Niels Bohr, as creative a scientist as any in the twentieth century, dared to speculate, too, on the meaning of the quantum, on the role of the observer in determining reality. I was privileged to work with Bohr on a topic of startling immediacy, nuclear fission, and also to watch him wrestle with the paradoxes of the quantum as he looked toward what the physics of the future might look like. "Wrestle" was a term he liked. Physics was a battle. Nature was a worthy opponent. I have unquestionably been influenced by Bohr in the ways that I approach both the immediate, where, little by little, the frontier gets nudged back, and the hazy future, where questions far outnumber answers, where only intuition and vision guide progress.

In 1995 (in my eighty-fourth year), like all of my colleagues in physics at Princeton, I prepared a set of questions for distribution to students in the department. The idea was to acquaint students with what faculty members were puzzling over, to plant ideas for possible research topics in the minds of the students, and to help them decide with which faculty member they might like to work. Three of the six questions I wrote down were "conventional"— questions that could be addressed in the context of well-understood current theory and fit a standard mold of "suitability" for an academic department. One such question addressed an unsolved puzzle from the 1930s on the scattering of gamma rays by lead nuclei. Another was the question of black holes being created not by imploding stars or imploding galactic cores, but by intense concentrations of gravity waves. A third concerned the behavior of space geometry in certain cosmological models.

Then I diverged from the academically suitable to the battlefield and to outer space. How, I asked, could the energy from a small buried nuclear explosion be transformed into electricity to charge the batteries of electric-powered army vehicles? How could messages for all the world to read be displayed on the Moon free of the censorship of dictators?

Finally, I dared to put before the students two of my favorite questions: How come existence? and its subsidiary, How come the quantum? I hardly expected any student to say, "Aha! Those are the questions that I want to work on." As to my colleagues' reactions: Well, the questions are not for them. My goal was only to plant an idea deep in the minds of the students, an idea that might find some way to flower five or ten or fifty years later.

Throughout my long career of teaching and research and public service it has been interaction with young minds that has been my greatest stimulus and my greatest reward. That reward comes back again with compound interest as I hear now from so many former students who let me know what they are up to and how their early wrestling with deep questions in physics helped to shape their lives. Not just graduate students. Even more numerous have been the scores of undergraduates who brought their enthusiasm and fresh perspective to the questions I put to them and who helped me to see more clearly.

But I am still too busy, too busy searching, to spend much time looking back. As Niels Bohr's friend Piet Hein puts it in another of his grooks,

> I'd like to know
> what this whole show
> is all about
> before it's out.

APPRECIATION

TO RECEIVE and to respond to the love and help that came from parents, brothers, and sister in my early years hardly prepared my heart for the still deeper love and support that Janette has given me over the sixty-three years of our marriage. By now, our hopes and concerns are so woven together that life of one without the other is inconceivable.

ACKNOWLEDGMENTS

WE ARE indebted to Arthur Singer and the Alfred P. Sloan Foundation for the financial and moral support that helped to get us started, to Mary Cunnane for the early editorial support that kept us going, and to Drake McFeely for the final editing and thoughtful guidance that saw us through to the end. Susan Middleton superbly copyedited the manuscript. Others at W. W. Norton who provided notable help include Sarah Stewart, Timothy Hsu, and Jo Anne Metch. John Brockman, our marvelous agent, served not only as matchmaker but as valued critic and adviser.

Among those who generously agreed to be interviewed for the record are Bryce DeWitt, Cécile DeWitt-Morette, Tom Griffy, David Hill, Carson Mark (ill at the time but generous with his reminiscences; he died in 1997), Warner Miller, Charles Misner, David Sharp, Lawrence Shepley, Ted Taylor, and John Toll, as well as immediate family members Janette Wheeler, Alison Lahnston, Letitia Ufford, and James Wheeler.

Edwin Taylor was kind enough to read and criticize the entire manuscript in an early draft. Kip Thorne read large chunks of later drafts and offered invaluable suggestions and corrections. We turned also to Charles Misner to help us straighten out certain tricky problems of exposition. We are enormously grateful to these three for the time they gave to make this a better book. Our loyal family members also read large portions of the manuscript and offered good suggestions.

We owe a special debt to our incomparable fact checker, Caroline Eisenhood, who combed books, Web sites, and physicists' recollections by the hundred in her meticulous checking. (Remaining errors, we hasten to add, are ours, not hers.) Spencer Weart, director of the Center for History of Physics at the American Institute of Physics (AIP), assisted in many ways, not least through his encouragement. Others who helped with facts and photographs include Finn Aaserud and Felicity Pors at the Niels Bohr Archive in Copenhagen; Joe Anderson and Jack Scott at AIP; Beth Carroll-Horrocks, Tim Wilson, and Scott DeHaven at the American Philosophical Society; Roger Meade at Los Alamos National Laboratory; and Robert Matthews, staff photographer at Princeton University, who always dropped whatever he was doing to help us.

Scores of others kindly answered our questions, filling us in on physics or history (or, in one case, translating from Italian). They include David Cassidy, Margit Dementi, Francis Everitt, Val Fitch, Peter Galison, Marvin Goldberger, Lillian Hoddeson, Gerald Holton, William Kaufmann, Martin Klein, Willis Lamb, T. D. Lee, J. Kenneth Mansfield, Philip Morrison, Abraham Pais, Wolfgang Panofsky, Miriam Planck, Rudy Rummel, Steve Schwartz, Martin Schwarzschild (a friend whom we lost in 1997), Silvan Schweber, Roger Stuewer, Claudio Teitelboim, Edward Teller, Jayme Tiomno, Sam Treiman, Frank von Hippel, and Arthur Wightman. We wish we could name them all, for their help was important.

We thank Emily Langford Bennett, secretary to one of us (JAW), for her clerical support from beginning to end, and Michele Michael Reel for her faithful transcriptions of taped interviews. JAW is grateful to his longtime intellectual home, Princeton University, and to the University of Texas at Austin, for the splendid colleagues and students that both provided. KF thanks the University of Pennsylvania for a Visiting Scholar Appointment during work on this book.

Finally, our gratitude to our wives, Janette and Joanne, and our children, Letitia, James, Alison, Paul, Sarah, Nina, Caroline, Adam, Jason, and Ian for their loyal support and their frank advice.

John Archibald Wheeler
Kenneth Ford

And I must add my special gratitude to Kenneth Ford, whose persistence and judgment made possible the otherwise impossible undertaking of this book.

John Archibald Wheeler

INDEX

Page numbers in *italics* refer to illustrations.